Lecture Notes in Artificial Intelligence   2684

Edited by J. G. Carbonell and J. Siekmann

Subseries of Lecture Notes in Computer Science

T0223551

Springer
Berlin
Heidelberg
New York
Hong Kong
London
Milan
Paris
Tokyo

Martin V. Butz   Olivier Sigaud
Pierre Gérard (Eds.)

Anticipatory Behavior in Adaptive Learning Systems

Foundations, Theories, and Systems

 Springer

Series Editors

Jaime G. Carbonell, Carnegie Mellon University, Pittsburgh, PA, USA
Jörg Siekmann, University of Saarland, Saarbrücken, Germany

Volume Editors

Martin V. Butz
University of Illinois at Urbana/Champaign
Illinois Genetic Algorithms Laboratory (IlliGAL)
104 S. Mathews Av., Urbana, IL, 61801, USA
E-mail: butz@illigal.ge.uiuc.edu
and
University of Würzburg
Department of Cognitive Psychology
Röntgenring 11, 97070 Würzburg, Germany
E-mail: butz@psychologie.uni-wuerzburg.de

Olivier Sigaud
Pierre Gérard
Laboratoire d'Informatique de Paris 6 (Lip6)
AnimatLab
8, rue du Capitaine Scott, 75015 Paris, France
E-mail: {olivier.sigaud/pierre.gerard}@lip6.fr

Cataloging-in-Publication Data applied for

A catalog record for this book is available from the Library of Congress.

Bibliographic information published by Die Deutsche Bibliothek
Die Deutsche Bibliothek lists this publication in the Deutsche Nationalbibliografie;
detailed bibliographic data is available in the Internet at <http://dnb.ddb.de>.

CR Subject Classification (1998): I.2, F.1, F.2.2, J.4

ISSN 0302-9743
ISBN 3-540-40429-5 Springer-Verlag Berlin Heidelberg New York

Springer-Verlag Berlin Heidelberg New York
a member of BertelsmannSpringer Science+Business Media GmbH

http://www.springer.de

© Springer-Verlag Berlin Heidelberg 2003
Printed in Germany

Typesetting: Camera-ready by author, data conversion by Boller Mediendesign
Printed on acid-free paper   SPIN: 10927809   06/3142   5 4 3 2 1 0

Foreword

The matter of anticipation is, as the editors of this volume state in their preface, a rather new topic. Given the almost constant use we make of anticipation in our daily living, it seems odd that the bulk of psychologists have persistently ignored it. However, the reason for this disregard is not difficult to find. The dogma of the scientific revolution had from the outset laid down the principle that future conditions and events could not influence the present. The law of causation clearly demands that causes should precede their effects and, therefore, concepts such as purpose, anticipation, and even intention were taboo because they were thought to involve things and happenings that lay ahead in time.

An analysis of the three concepts – purpose, anticipation, and intention – shows that they are rooted in the past and transcend the present only insofar as they contain mental representations of things to be striven for or avoided. Purposive or goal-directed action could be circumscribed as action carried out to attain something desirable. In each case, the particular action is chosen because, in the past, it has more or less reliably led to the desired end. The only way the future is involved in this procedure is through the belief that the experiential world manifests some regularity and allows the living organism to anticipate that what has worked in the past will continue to work in the future. This belief does not have to be conscious. Skinner's rats continued to turn left in a maze where the left arm had been baited. They did so because the meat pellet they found the first time had "reinforced" them to repeat the turn to the left. But positive and negative reinforcement can work only with organisms that have evolved to act as though actions could be relied on to have constant results. The anticipation is implicit.

On the conceptual level, to anticipate means to project into what lies ahead a mental representation abstracted from past experience. In many cases we might not call such a projection an anticipation, although, in principle, it is. If, for instance, you are about to go for a walk, take a look at the sky, and pick up your umbrella, you do this because you have learned from experience that the kind of clouds you saw through the window forebode rain. You are not anticipating an event but merely its possibility.

Tools are another example. The material and shape of a hammer have been developed and refined over the course of many generations' experiences and you trust the tool you now hold in your hand to drive in future nails just as it drove in nails in the past. You may not actually anticipate its action, you simply believe that it will work.

If you have ever had the appalling driving experience of your foot going all the way to the floor board when you needed to brake, you will know just how unquestioningly you anticipated the brake pedal to do what it is supposed to do.

In one form or another anticipation pervades the fabric of our experience. As living organisms we constantly rely on a great deal of regularity in the world as we perceive it. It may not always work out, but apparently it works often enough for us to survive. To the examples I gave, many others could be added as illustration of the variety of the term's applications. The contributions to this volume spring from very different sources and are likely to provide a welcome starting ground for the classification and modeling of different kinds of anticipation.

Ernst von Glasersfeld
Scientific Reasoning Research Institute
University of Massachusetts
Amherst, MA 01003, USA

Preface

This book evolved out of the first Workshop on *Adaptive Behavior in Anticipatory Learning Systems*, held on the 11th of August in Edinburgh, UK, in conjunction with the *7th International Conference on Simulation of Adaptive Behavior: from Animals to Animats*. Although the matter of anticipation is a rather new and often misunderstood topic, the workshop yielded a lot of attention and interest among a large mixture of people including computer scientists, psychologists, philosophers, neuroscientists, and biologists. The workshop itself was a great success, starting from the very conceptual bottom and moving to first applications of anticipatory behavior systems at the end of the day. It became clear that there is more behind anticipation than the mere relation to prediction, expectation, or planning. Nor is there anything mysterious about the topic. Psychological as well as philosophical issues revealed the strong motivation of the topic and also initiated controversy and discussion. Fundamental distinctions between different forms of anticipations helped to structure thoughts and ideas. Conceptual reflections on representations and dynamical systems showed the different possible manifestations of anticipatory processes. Anticipations also seem to form the basis of attention, motivation, emotion, and personality. Solid approaches revealed that anticipations can, but not necessarily do, significantly improve the adaptive behavior of agents, including robots, stock traders, plot guidance agents, and animat-like adaptive agents.

Despite these exciting ideas and reflections many questions remain to be solved. Fundamental questions such as in what circumstances anticipatory behavior is actually beneficial or which environmental properties allow anticipatory behavior to be beneficial remain to be answered. To make it even more challenging, anticipatory behavior is not clearly defined at this point. Which types of anticipatory behaviors are most suitable for which environmental properties?

These questions form the major concern of this book. The articles are meant to stimulate thought as well as provide guidance for future, more detailed and revealing analysis, development, understanding, and creation. The book is structured similar to the workshop day. First, philosophical thoughts and concepts are meant to stimulate the reader's concerns about this topic. Fundamental cognitive psychology experiments then confirm the existence of anticipatory behavior in animals and humans and outline a first framework of anticipatory learning and behavior. Next, several distinctions and frameworks of anticipatory processes are discussed, including first implementations of those concepts. Finally, several anticipatory systems and studies on anticipatory behavior are presented.

March 2003

Martin V. Butz,
Olivier Sigaud,
and Pierre Gérard

Organization

We are thankful to the organizers of the 7th International Conference on Simulation of Adaptive Behavior (SAB VII) for giving us the possibility to hold the workshop on Adaptive Behavior in Anticipatory Learning Systems (ABiALS 2002) during the conference. The organizers were Bridget Hallam, Dario Floreano, John Hallam, Gillian Hayes, and Jean-Arcady Meyer. Special thanks go to Gillian Hayes for taking care of the workshop organization. The book emerged out of the workshop. It includes revised workshop contributions and further contributions in response to a second call for papers and a review process.

We are more than grateful to our program committee for providing us with careful reviews of the diverse contributions. Due to their hard work, we were able to provide at least three reviews for each contribution, significantly improving many. Due to the wide variety of contributions it was not always easy to judge the significance and impact of the studies in a review. However, we believe that all accepted contributions provide new insights into the realm of anticipatory behavior, and while some of the studies might be preliminary, all of them strongly suggest further research in the area of anticipatory behavior.

Organizing Committee

Martin V. Butz   University of Illinois at Urbana-Champaign, IL, USA
                      University of Würzburg, Germany
Olivier Sigaud   Université de Paris VI, Paris, France
Pierre Gérard   Université de Paris VI, Paris, France

Program Committee

Emmanuel Daucé   Université de la Méditerrannée, Marseille, France
Ralf Möller   Max Planck Institute for Psychological Research,
                      Munich, Germany
Wolfgang Stolzmann   University of Würzbug, Germany
Jun Tani   Brain Science Institute, RIKEN, Saitama, Japan
Stewart W. Wilson   Prediction Dynamics, MA, USA

Table of Contents

Systems, Evaluations, and Applications

Anticipatory Behavior: Exploiting Knowledge About the Future to Improve Current Behavior

Martin V. Butz[2,3], Olivier Sigaud[1], and Pierre Gérard[1]

[1] AnimatLab, Université de Paris VI, Paris, France
{olivier.sigaud,pierre.gerard}@lip6.fr
[2] Department of Cognitive Psychology, University of Würzburg, Germany
butz@psychologie.uni-wuerzburg.de
[3] Illinois Genetic Algorithms Laboratory (IlliGAL),
University of Illinois at Urbana-Champaign, IL, USA

Abstract. This chapter is meant to give a concise introduction to the topic of this book. The study of anticipatory behavior is referring to behavior that is dependent on predictions, expectations, or beliefs about future states. Hereby, behavior includes actual decision making, internal decision making, internal preparatory mechanisms, as well as learning. Despite several recent theoretical approaches on this topic, until now it remains unclear in which situations anticipatory behavior is useful or even mandatory to achieve competent behavior in adaptive learning systems. This book provides a collection of articles that investigate these questions. We provide an overview for all articles relating them to each other and highlighting their significance to anticipatory behavior research in general.

1 Introduction

Intuitively, anticipations are an important and interesting concept. Looking ahead and acting according to our predictions, expectations, and aims seems helpful in many circumstances. For example, we say that we are in anticipation, we are looking forward to events, we act goal-oriented, we prepare or get ready for expected events, etc.

Several recent theoretical approaches have been put forward in an attempt to understand and formalize anticipatory mechanisms. Despite these important approaches, though, it is still hardly understood why anticipatory mechanisms are necessary, beneficial, or even mandatory in our world. Therefore, this book addresses the following questions:

- When and in which circumstances are anticipations beneficial for behavior and life?
- Which types of anticipatory behavior are important to distinguish?
- Which environmental properties or rather which fundamental characteristics of our environment make which types of anticipatory processes useful?
- How can the different anticipatory processes be modeled and implemented in artificial adaptive systems?

M. Butz et al. (Eds.): Anticipatory Behavior ..., LNAI 2684, pp. 1–10, 2003.

Over the last few decades, experimental psychology research gradually started to accept the notion of anticipations beginning with Tolman's suggestion of "expectancies" [29, 30] due to his observation of *latent learning* in rats (learning of environmental structure despite the absence of reinforcement). More recently an outcome devaluation procedure [1, 9, 19] has been employed that provides definite evidence for anticipatory behavior in animals. Even more recently, cognitive psychology provides further evidence of distinct anticipatory mechanisms in, e.g., learning [14, 15], attentional processing [18], or object recognition tasks [22].

In theoretical biology [20, 21] and physics [21, 10] anticipations have been suggested to contribute to the essence of complexity and life itself as well as to the stabilization of chaotic control processes. Robert Rosen puts forward one of the first definitions of an anticipatory system:

> [...] a system containing a predictive model of itself and/or of its environment, which allows it to change state at an instant in accord with the model's predictions pertaining to a later instant.[20, p.339]

In Rosen's definition a *system* might be any entity in an environment, such as an animal, a human, or any other living being as well as inanimate physical entities such as machines, robots, or even weather systems. A predictive model is a model that provides information about the possible future state(s) of the environment and/or the system. The system becomes an anticipatory one when it has such a model and when it uses the model to change behavior according to the predictions in this model. For Rosen, anticipation is the fundamental ingredient to distinguish living from non-living systems.

Several recent attempts have been made in artificial intelligence to integrate anticipatory mechanisms into artificial learning systems in the framework of reinforcement learning [27, 16], learning classifier systems (as online generalizing reinforcement learners) and related systems [24, 4, 12, 31], as well as neural networks [8, 11, 28, 2]. So far, research in artificial intelligence has included anticipatory mechanisms wrapped in model learning systems such as the model-based reinforcement learning approach. Anticipatory processes were never analyzed on their own.

This book suggests the investigation of the characteristic properties and enhanced capabilities of anticipatory behavior in a distinct framework. We are interested in when anticipatory behavior is useful, which environmental properties enable effective anticipatory behavior, what types of anticipatory behavior can be distinguished, and what are the distinct behavioral impacts of anticipatory behavior processing. This introduction takes a general approach to these questions clarifying what we mean by anticipatory behavior and related questions. More concrete treatments of the questions, as well as first implementations and application studies of anticipatory behavioral adaptive learning systems, can be found in the successive articles. The provided overview to each article is meant to give guidance to the reader and relate the articles to the big picture of anticipatory behavior put forward herein.

2   What Is Anticipatory Behavior?

Without a conceptual understanding of what anticipatory behavior is referring to, scientific progress towards more elaborate and competent anticipatory behavior systems is impeded. The term "anticipation" is often understood as a synonym for prediction or expectation—the simple act of predicting the future or expecting a future event or imagining a future state or event. Merriam-Webster online provides the following definitions for anticipation [17]:

1. a) a prior action that takes into account or forestalls a later action
   b) the act of looking forward; especially : pleasurable expectation
2. the use of money before it is available
3. a) visualization of a future event or state
   b) an object or form that anticipates a later type
4. the early sounding of one or more tones of a succeeding chord to form a temporary dissonance

These definitions stress the look into the future rather than the actual effect of this look. The verb definition stresses the effect of the look into the future much more: [17]:

transitive senses
1. to give advance thought, discussion, or treatment to
2. to meet (an obligation) before a due date
3. to foresee and deal with in advance : FORESTALL
4. to use or expend in advance of actual possession
5. to act before (another) often so as to check or counter
6. to look forward to as certain : EXPECT

intransitive senses
− to speak or write in knowledge or expectation of later matter

In the understanding of this book, anticipation is really about the impact of a prediction or expectation on current behavior. Thus, anticipation means more than a simple lookahead into the future. The important characteristic of anticipation that is often overlooked or misunderstood is the impact of the look into the future on actual behavior. We do not only predict the future or expect a future event but we alter our behavior—or our behavioral biases and predispositions—according to this prediction or expectation. To make this fundamental characteristic of "anticipation" clear, we decided to call this book "*Anticipatory Behavior* in Adaptive Learning Systems" and not merely "Anticipations in Adaptive Learning Systems". To be even more concrete we define anticipatory behavior as follows:

Anticipatory Behavior: *A process, or behavior, that does not only depend on past and present but also on predictions, expectations, or beliefs about the future.*

The definition is kept fairly general not only to give the reader a feel of what this book is concerned with but also to immediately raise questions and point out the need for further distinctions.

In fact, any "intelligent" process can be understood as exhibiting some sort of anticipatory behavior in that the process, by its mere existence, predicts that it will work well in the future. This implicit anticipatory behavior can be distinguished from explicit anticipatory behavior in which current explicit future knowledge is incorporated in some behavioral process.

To give more intuitive understanding of the concept of anticipation we end this section with a great intuitive example, derived from Sjölander [23], that distinguishes different levels of anticipatory behavior and their resulting impacts. The example addresses the difference in the hunting habits of snakes and dogs. Essentially, a snake is not able to predict future movement of its prey. If the prey disappears, the snake's hunting behavior remains 'activated' meaning that it may actively start searching for the prey. However, it does not search for the prey where it should be by now but searches at the spot where the prey was sensed last. On the other hand, a dog hunting a hare (or rabbit) does not need to sense the hare continuously. If the hare, for example, disappears behind a bush the dog predicts the future location of the hare by anticipating where it is going to turn up next and continues its hunt in this direction. This behavior clearly indicates that the dog employs some kind of predictive model of the behavior of the hare predicting the movement of the hare and adapting its behavior accordingly. The snake, on the other hand, does not exhibit any predictive capabilities and consequently does not have a predictive model of the prey that it can employ. Thus, the best thing to do for the snake is to search for the prey where it was sensed last—in (implicit) anticipation to re-sense the prey and eventually catch it.

3   Overview of the Book

The book basically starts from the general and ends with the very concrete. First, philosophical considerations of and reflections on anticipations outline the complexity of the topic. Next, psychological observations of anticipatory behavior are provided which lead to early theories on anticipatory behavior characteristics and properties of anticipatory behavior systems. The following section regarding "Formulations, Distinctions, and Characteristics" develops several mathematical and computational frameworks of anticipatory behavior. Finally, "Systems, Evaluations, and Applications" investigates and develops concrete systems, elaborates on their behavior, and discusses their potentials. The following paragraphs introduce the contributions in somewhat further detail.

3.1   Philosophical Considerations

Seeing our intuitive belief that anticipatory behavior is present in many forms, it is important to investigate the impacts of anticipation on cognition and behavior. Why is anticipatory behavior useful in our world? Which are the cognitive consequences of anticipatory behavior?

This question reaches far back into history and is related to many interesting considerations. The philosophical part of this book provides important thoughts on the impact of anticipatory behavior. The interested reader is also referred to Ernst von Glasersfeld's thoughts on anticipations [13].

Riegler addresses the questions raised by the distinctions between implicit and explicit anticipations used in this book from a constructivist standpoint. He first stresses the importance of anticipations in our cognitive abilities and in our culture, before arguing that unconscious processes are playing a fundamental role in our anticipatory capabilities. Then, from a detailed exposition of the philosophical controversy raised by Libet's ideas on the so-called "readiness potential" and its implication on the possibility of free will, he concludes first that explicit anticipations cannot be equated with the kind of anticipations that a conscious subject actually feels and second that anticipations can only "canalize" our future cognitive processes at a level which is inaccessible to the subject.

Nadin takes a rather different stance on anticipation by regarding guessing, expectation, prediction, and planning as a counter-distinction to anticipation. He argues that anticipatory behavior can only be considered in conjunction with reactive behavior. Further, since anticipation is not reducible to deterministic sequences it is possible to improve predictions and forecasts but it is impossible to accurately predict the future according to past and present.

3.2   From Cognitive Psychology to Cognitive Systems

After comprehending what anticipatory behavior means and why anticipatory behavior can exist in our world, we want to know when and where anticipatory behavior is useful. That is, which environmental properties give rise to beneficial anticipatory behavior?

To approach these questions it is helpful to look at manifestations of anticipatory behavior in real life as well as experimental investigations that show the usefulness of anticipatory behavior in experimental and theoretical scenarios. The psychological section of this book provides many insights in how anticipatory behavior was (re-)discovered by psychology and how it is experimentally assessed in animal and human behavior. Moreover, two psychological-based anticipatory behavior models are derived.

Hoffmann stresses the impact of anticipations on behavioral execution and learning. Similar thoughts had been put forward already in the 19th century in the *ideomotor principle* but were then neglected by the behaviorist movement in the early 20th century. His theory of anticipatory behavioral control emphasizes the primacy of action-effect relations and the secondary conditioning on important contextual information. Behavior is triggered by a representation of its behavioral consequences, making it inherently anticipatory. The proposed mechanisms are supported by a large variety of experimental investigations. A first implementation of the anticipatory behavior control theory was realized in the anticipatory classifier system [24, 25, 26].

Witkowski distinguishes four psychological learning theories, integrating them into a dynamic expectancy model of behavior. The model distinguishes between four essential capabilities of anticipatory animats: (1) action independent future predictions; (2) action dependent future predictions; (3) reinforcement independent action ranking; and (4) guided structural learning by detecting unpredicted events (that is, biased learning of a predictive model). Five rules are put forward that guide the generation of predictions and the prediction-dependent action execution. The implementation of the framework in the SRS-E system shows many interesting behavioral properties.

3.3   Formulations, Distinctions, and Characteristics

Next, further frameworks of anticipatory behavior are put forward. In "Internal Models and Anticipations in Adaptive Learning Systems" we postulate further distinctions of anticipatory behavior suggesting (1) implicitly anticipatory behavior, in which predictions are only done implicitly in the control structure, (2) payoff anticipatory behavior, which compares expected payoff before action execution, (3) sensory anticipatory behavior, which alters sensory processing due to predictions, expectations, and/or intentions, and (4) state anticipatory behavior, in which the behavioral component is biased explicitly on future predictions, expectations, and/or intentions. Examples of all types are provided in a literature review on previous adaptive learning systems.

Dubois develops a mathematical theory of strong and weak anticipations. He defines a strong anticipatory system as a system whose predictive model is essentially represented by itself, whereas a weak anticipatory system is a system in which the model is an approximation of the system. Furthermore, he distinguishes between incursive and hyperincursive control. While hyperincursion allows the mathematical formulation of multiple possible outcome scenarios, incursive control results in system stabilization much like *model predictive control* [7] but in a more fundamental way.

Bozinovski sketches a framework of personality based on anticipatory behavior. He addresses the questions of what motivation and what emotion are in an anticipatory system. Motivations for anticipatory behavior are characterized by the anticipation of future emotional consequences. Thus, emotions are seen as the internal reinforcement mechanisms that shapes motivational driven behavior. This is a somewhat controversial but interesting view. In fact, other researches proposed rather the opposite in that emotions are designed to influence current activity selection by, for example, shaping current motivations in an implicitly anticipatory fashion [5, 6]. Further thoughts and elaborations on this matter are necessary.

Davidsson introduces the concept of preventive state anticipation. In this form of anticipation the agent continuously predicts future states. Behavior is altered only if a future state is undesirable. Experimental investigations show the efficiency of the simplest form of preventive state anticipation (i.e. linear anticipation) in a single agent world, cooperative multi-agent world, and a competitive multi-agent world.

Tani puts forward the dynamical systems perspective in terms of anticipatory behavior. He suggests that dynamical systems can prevent the curse of re-representation in the predictive model by learning an implicit dynamic representation of the world in the form of a recurrent neural network. Tani shows that the dynamic predictive model can be used efficiently to predict future states, even conquering the problem of non-Markov states. He explains that the dynamic representations form fractal attractors where each attractor represents a possible state, whereas the fractal structure of each attractor provides information about the past.

3.4 Systems, Evaluations, and Applications

With several concepts of the characteristics of anticipatory behavior and the most important distinctions in mind, this section looks at actual studies of anticipatory behavior in several frameworks including neural network systems, evolutionary computation models, as well as rule-based approaches. Useful characteristics of anticipatory behavior are identified. First applications are suggested.

Baldassarre introduces feed-forward neural net planners and reinforcement-learning based planners. The system can be regarded as a neural net extension of Sutton's Dyna-PI model [27] with additional goal representations (the "matcher") and goal dependent planning algorithms. Reactive and anticipatory behavior are integrated in one framework choosing either one according to current confidence measures. This confidence measure reflects the animat's belief in its own predictions and results in a controlled "thinking before acting". The paper highlights the noise-robust stability of the resulting predictive ANN. Forward and backward planning are applied.

Fleischer, Marsland, and Shapiro introduce a landmark detection mechanism that is based on anticipatory behavior, particularly sensory anticipatory behavior. The anticipatory landmark detection mechanism is shown to clearly outperform pure stimulus-based landmark detection. Moreover, it is shown that the established landmark categories improve behavior when used in a goal-oriented route-following task, pointing out the importance of efficient learning and representation of a predictive environmental model.

Hülse, Zahedi, and Pasemann base their investigation on an evolved minimal recurrent controller. They form macro-action maps to represent the encountered environment exploiting the structure of the recurrent controller. Although the discretization approach departs from Tani's dynamical system perspective, interesting behavior patterns are realized such as exploration, homing, and navigation behavior. Although no truly anticipatory behavior is shown, the discretization approach seems to have great anticipatory behavior potential.

Laaksolahti and Boman provide an interesting application scenario proposing the anticipatory guidance of plot. As anticipations can be seen as stabilization mechanisms as well as guidance down an inevitable path, it is only natural that this property may be extended to plot guidance in an interactive narrative scenario. In its wider sense, the idea of plot guidance is derived from Davidsson's idea of preventive state anticipation.

Edmonds investigates the usefulness of predictive information in an artificial stock market scenario. Both the predictive system as well as the trading system are learned by the means of genetic programming methods. The extensive experimental analysis provides an unclear picture of which scenarios actually benefit from predictive knowledge. As expected, though, predictive knowledge is not sufficient by itself to improve behavior. The study strongly points out the need for further structured investigations of when, where, and how anticipatory behavior is beneficial.

Butz and Goldberg enhance the anticipatory classifier system ACS2 with further state-anticipatory mechanisms. The paper addresses the online generalization of state values while learning a predictive model. State values reflect the utility of reaching a state given a current problem (in the form of a partially observable Markov decision process (POMDP)). For ungeneralized states, the values are identical to values that can be determined by the dynamic programming algorithm approximating the Bellman equation [3]. The resulting system, XACS, implements a predictive model learning module and a separate reinforcement learning module generalizing the representations of both modules online. Behavior is state-anticipatory in that future predictions and the values of those predicted states determine actual behavior. The interaction of multiple reinforcement modules is suggested, allowing for the design of a motivational or even emotional system .

4   Conclusions

Although the concept of anticipatory behavior has been appreciated over many decades, explicit research on anticipatory behavior began only recently. This book provides philosophical considerations, psychological manifestations, formal and conceptual foundations, and first investigations and applications of anticipatory mechanisms. Advantages, as well as possible drawbacks, of anticipatory behavior are revealed. Although none of the questions addressed in this book are answered completely at this point, the large variety of scenarios and examples presented herein are an important step towards a proper understanding and utilization of anticipatory mechanisms.

Anticipatory behavior appears useful in many situations allowing for previously impossible behavioral patterns. First, anticipatory processes can stabilize behavioral execution. Second, anticipations may guide, or canalize, behavioral flow. Third, anticipatory mechanisms can bias attentional processes enabling goal-directed focus and faster reactivity. Fourth, anticipatory behavior may result in advantages in hunting and other competitive scenarios. Fifth, anticipatory behavior may result in faster adaptivity in dynamic environments by the means of internal reflection and planning. Sixth, cooperative behavior may be improved and suboptimal behavior may be overcome by preventive state anticipatory behavior. Finally, anticipatory behavior appears to be an important prerequisite for social interaction.

In conclusion, anticipatory mechanisms can be beneficial in many different areas and in many different forms. Despite this strong diversity, the basic concept of anticipatory behavior is the same in all areas. Thus, the unified investigation of these systems in terms of their anticipatory properties and capabilities may enable prosperous interdisciplinary research advancements effectively sharing new ideas and insights. In particular, future research on anticipatory behavior may lead to (1) significant improvement of the behavior of adaptive learning systems; (2) further understanding of the function of anticipatory mechanisms in animals and humans; (3) the creation of social interactive systems with human-like anticipatory features; (4) the discovery of the processes underlying motivations and emotions; (5) the development of truly cognitive systems that do not only reactively move through the world but learn about important resemblances, contiguities, and causes of effects, and efficiently exploit this knowledge by anticipatory behavior mechanisms. We hope that this first survey on anticipatory behavior in adaptive learning systems will hold its promise and lead to an insightful and rewarding new research direction.

References

[1] Adams, C., Dickinson, A.: Instrumental responding following reinforcer devaluation. Quarterly Journal of Experimental Psychology **33** (1981) 109–121

[2] Baluja, S., Pomerleau, D.A.: Expectation-based selective attention for visual monitoring and control of a robot vehicle. Robotics and Autonomous Systems **22** (1997) 329–344

[3] Bellman, R.: Dynamic programming. Princeton University Press, Princeton, NY (1957)

[4] Butz, M.V.: Anticipatory learning classifier systems. Kluwer Academic Publishers, Boston, MA (2002)

[5] Cañamero, L.D.: Modeling motivations and emotions as a basis for intelligent behavior. In Johnson, W.L., ed.: Proceedings of the first international symposium on autonomous agents (Agents'97), New York, NY, The ACM Press (1997) 148–155

[6] Cañamero, L.D.: Designing emotions for activity selection in autonomous agents. In Trappl, R., Petta, P., Payr, S., eds.: Emotions in Humans and Artifacts. The MIT Press, Cambridge, MA (in press, 2003)

[7] Camacho, E.F., Bordons, C., eds.: Model predictive control. Springer-Verlag, Berlin Heidelberg (1999)

[8] Carpenter, G.A., Grossberg, S., Reynolds, J.H.: ARTMAP: Supervised real-time learning and classification of nonstationary data by a self-organizing neural network. Neural Networks **4** (1991) 565–588

[9] Colwill, R.M., Rescorla, R.A.: Postconditioning devaluation of a reinforcer affects instrumental learning. Journal of Experimental Psychology: Animal Behavior Processes **11** (1985) 120–132

[10] Dubois, D.M.: Computing anticipatory systems with incursion and hyperincursion. Proceedings of the First International Conference on Computing Anticipatory Systems, CASYS-1997 (1998) 3–30

[11] Gaudiano, P., Grossberg, S.: Vector associative maps: Unsupervised real-time error-based learning and control of movement trajectories. Neural Networks **4** (1991) 147–183

[12] Gérard, P., Sigaud, O.: YACS: Combining dynamic programming with generalization in classifier systems. In Lanzi, P.L., Stolzmann, W., Wilson, S.W., eds.: Advances in learning classifier systems: Third international workshop, IWLCS 2000. Springer-Verlag, Berlin Heidelberg (2001) 52–69

[13] Glasersfeld von, E.: Anticipations in the constructivist theory of cognition. Proceedings of the First International Conference on Computing Anticipatory Systems, CASYS-1997 (1998) 38–48

[14] Hoffmann, J.: Vorhersage und Erkenntnis: Die Funktion von Antizipationen in der menschlichen Verhaltenssteuerung und Wahrnehmung. [Anticipation and cognition: The function of anticipations in human behavioral control and perception.]. Hogrefe, Göttingen, Germany (1993)

[15] Hommel, B.: Perceiving ones own action - and what it leads to. In Jordan, J.S., ed.: Systems theory and apriori aspects of perception. North Holland, Amsterdam (1998) 143–179

[16] Kaelbling, L.P., Littman, M.L., Moore, A.W.: Reinforcement learning: A survey. Journal of Artificial Intelligence Research 4 (1996) 237–258

[17] Merriam-Webster: Merriam-webster online collegiate dictionary, tenth edition (2002) http://www.m-w.com/.

[18] Pashler, H., Johnston, J.C., Ruthruff, E.: Attention and performance. Annual Review of Psychology 52 (2001) 629–651

[19] Rescorla, R.A.: Associative relations in instrumental learning: The eighteenth Bartlett memorial lecture. Quarterly Journal of Experimental Psychology 43 (1991) 1–23

[20] Rosen, R.: Anticipatory systems. Pergamon Press, Oxford, UK (1985)

[21] Rosen, R.: Life itself. Columbia University Press, New York (1991)

[22] Schubotz, R.I., von Cramon, D.Y.: Functional organization of the lateral premotor cortex. fMRI reveals different regions activated by anticipation of object properties, location and speed. Cognitive Brain Research 11 (2001) 97–112

[23] Sjölander, S.: Some cognitive break-throughs in the evolution of cognition and consciousness, and their impact on the biology language. Evolution and Cognition 1 (1995) 3–11

[24] Stolzmann, W.: Antizipative Classifier Systems [Anticipatory classifier systems]. Shaker Verlag, Aachen, Germany (1997)

[25] Stolzmann, W.: Anticipatory classifier systems. Genetic Programming 1998: Proceedings of the Third Annual Conference (1998) 658–664

[26] Stolzmann, W.: An introduction to anticipatory classifier systems. In Lanzi, P.L., Stolzmann, W., Wilson, S.W., eds.: Learning classifier systems: From foundations to applications. Springer-Verlag, Berlin Heidelberg (2000) 175–194

[27] Sutton, R.S.: Reinforcement learning architectures for animats. From Animals to Animats: Proceedings of the First International Conference on Simulation of Adaptive Behavior (1991) 288–296

[28] Tani, J.: Model-based learning for mobile robot navigation from the dynamical systems perspective. IEEE Transactions. System, Man and Cybernetics (Part B), Special Issue on Learning Autonomous Systems 26 (1996) 421–436

[29] Tolman, E.C.: Purposive behavior in animals and men. Appleton, New York (1932)

[30] Tolman, E.C.: There is more than one kind of learning. Psychological Review 5b (1949) 144–155

[31] Witkowski, C.M.: Schemes for learning and behaviour: A new expectancy model. PhD thesis, Department of Computer Science, Queen Mary Westfield College, University of London (1997)

Whose Anticipations?

Alexander Riegler

VUB, CLEA
Krijgskundestraat 33, B-1160 Brussels, Belgium
ariegler@vub.ac.be

Abstract. The central question in this paper is: Who (or what) constructs anticipations? I challenge the (tacit) assumption of Rosen's standard definition of anticipatory systems according to which the cognitive system actively constructs a predictive model based on which it carries out anticipations. My arguments show that so-called implicit anticipatory systems are at the root of any other form of anticipatory systems as the nature of the "decision maker" in the latter cannot be a conscious one.

Introduction

The notion of anticipation is often linked with that of construction. Robert Rosen's (1985) standard definition of anticipatory systems, for example, characterizes an anticipatory system as one "containing a predictive model of itself and/or of its environment, which allows it to change state at an instant in accord with the model's predictions pertaining to a latter instant". Clearly, this ability requires the construction of the predictive model in the first place. In this paper, I want to go a step further and claim that constructing is all what a cognitive system is doing, making anticipation an integrative part of this continuing cognitive construction.[1] Whereas the standard definition tacitly assumes that the cognitive system is (consciously) constructing the predictive model, I want to challenge this presumption and try to reveal the agent and the processes behind the construction activity. The central question I seek to address is thus: Are we consciously creating anticipations on basis of which we plan and make decisions, or are anticipations and decision making made for us?

In the course of the paper I will first show that anticipation is the driving force in a wide range of cognitive behavior. From magic practices in so-called "primitive" cultures, to superstitious behavior in animal and human beings, to so-called "volitional" cognition. Based on the categorization of Martin Butz (2002) I will argue that what he called implicit anticipatory systems forms the foundation for all other forms of anticipations, whether strong or weak in the sense of Daniel Dubois (2000).

---

[1] Philosophers may argue that construction is an activity carried out by a conscious subject only and which can never be associated with passivity (Olivier Sigaud, personal communication). In this paper I consider construction a process by which a structure—physical or mental—is errected. Later on we will see that the (philosophical) distinction between conscious and unconscious as an a priori condition for cognitive construction may not hold.

M. Butz et al. (Eds.): Anticipatory Behavior ..., LNAI 2684, pp. 11-22, 2003.

Anticipation and the Unknown

Fishing and navigating in offshore waters is a game with the unknown. Weather conditions, sharks, and streams make it difficult if not impossible to anticipate the outcome of your trip, especially if your equipment is simple and your boat is small. For members of primitive cultures it has always been a challenge. Social anthropologist Bronislaw Malinowski set out in the early 20th century to live among islanders in the Pacific Ocean who fished both inshore and offshore. Staying there for several years, Malinowski noticed a sharp contrast in behavior. Offshore fishing beyond the coral reef was accompanied by many elaborate rituals and ceremonies to invoke magical powers for safety and protection. To his surprise, nothing like that he could observe among the inshore fishermen, who carried out their job with a high degree of rational expertise and craftsmanship. Based on his observations he drew the conclusion that "we do not find magic wherever the pursuit is certain, reliable and well under the control of rational methods and technological processes. Further, we find magic where the element of danger is conspicuous", and primitive man "clings to [magic], whenever he has to recognize the impotence of his knowledge and of his rational technique."(Malinowski 1948)

Malinowski considered magic as response to uncertainty. His claim was that magical rituals are carried out in unknown situations where the degree of freedom seems to transcend the degree of control.[2] They reduce the threat caused by the dangers and uncertainties of life. Phenomena for which the individual doesn't have an explanation can be made less threatening by anticipating that a known action pattern will eventually make the phenomena disappear. There is also an emotional aspect to it. Instead of getting overwhelmed by the details of a new situation, humans seek to replace them with familiar activity and behavioral patterns that show a high degree of predictability to putatively gain control again, to be able to anticipate the outcome. Thus, in order to fight the feeling of threat that emanates from the inexplicable humans try to find causes by which it can be made explicable. Often such causes are derived from a single (positive) experience that accidentally linked the cause with a result similar to the threatening phenomenon.

Such behavior is not only typical for humans. Also in the animal kingdom we find patterns that reflect insecurity. B. F. Skinner's article on "superstition in the pigeon" (1948) is a classical description of how birds react in situations which to understand transcends their cognitive capabilities and thus become uncontrollable for them. Skinner presented food at regular intervals to hungry pigeons with no reference whatsoever to their current behavior. Soon the birds started to display certain rituals between the reinforcements, such as turning two or three times about the cage, bobbing their head, and incomplete pecking movements. As Skinner remarked, the birds happened to be executing some response as the food appeared the first time, and they tended to repeat this response if the feeding interval was only short enough. In some

---

[2] In (first-order) cybernetics, Ross Ashby's Law of Requisite Variety expresses straightforwardly what it takes for a system to remain in control over another: "Only variety can destroy variety" (Ashby 1956), i.e, the variety of actions available to a control system must be at least as large as the variety of actions in the system to be controlled.

sense, pigeons associated their action with receiving food and started to inductively believe that it causes the food to appear.

At first glance, Skinner's conclusion to liken the pigeons' behavior to superstitions in humans seems far-fetched. In the case of pigeons the superstitious behavior is the result of some unconscious cognitive processes. In the case of humans, rituals are developed due to reflections about the current, possibly threatening situation and a desired goal. Professional athletes who carry out some superstitious activities—eating a certain meal, wearing certain clothes, running in certain patterns over the playground to alter the probabilities—seem to be aware of the fact that they consciously assemble their ritual patterns. However, as I will argue later on, this distinction blurs easily away in the light of some neurophysiological insights which were intended to disprove the independency of a free will but which say probably more about who or what constructs rituals and their inherent anticipations.

Before we can turn to this question, we need to look more closely at some basic cognitive mechanisms as employed, for example, by Malinowski's fishermen. Every day they are exposed to experiences, many of them are familiar, some of them new. What does it mean for an experience to be familiar or unfamiliar? In his works (e.g., the 1954 book "The construction of reality in the child"), Jean Piaget proposed two basic principles when it comes to cope with perception and experience. He argued that in the beginning, a newborn knows little about how to cope with the perceptive impressions around her. Faces might be funny or threatening colorful spots and voices unknown sounds. In fact, she doesn't even know that these are colors and sounds. Only by assimilation and accommodation the child constructs a collection of—as Piaget called it—schemata during her ontogeny. Schemata serve as point of reference when it comes to classify (assimilate) new impressions. If impressions are too alien to be aligned to an older, already assimilated impression, they are either not perceived at all or accommodated, i.e., existing schemata are adjusted in order to include the new "exotic" impression. With each of these assimilating or accommodating steps the child constructs another piece of her reality. This means, only what can be formulated within schemata, can be perceived or expressed in actions.[3] Ulric Neisser's (1975) characterization of perception as a schemata controlled "information pickup" describes this perspective best. An organism's schemata determine the way it is looking at the environment, and are therefore anticipatory. The schemata construct anticipations of what to expect, and thus enable the organism to actually perceive the expected information. If a situation gets out of control—because it is unknown, threatening, and uncertain—assimilation and accommodation have reached their limits and humans are more likely to turn to magical or occult powers.

As experiences are made subsequently, they are connected with each other in a historical manner and form a network of hierarchical interdependencies (Riegler 2001a, b). The components of such a network are mutually dependent; removing one component may change the context of another component. In this sense they impose constraints on each other.

---

[3] Ernst von Glasersfeld has framed this fundamental principle of piecemeal erection of reality as follows. "Knowledge is not passively received but actively built up by the cognizing subject", and the "function of cognition is adaptive and serves the organization of the experiential world, not the discovery of ontological reality" (Glasersfeld 1989).

Whose Constructions?

The picture sketched so far is the following. The individual constructs reality out of the experiences he or she makes. Whether these experiences and constructed reality mirror any outside reality cannot be easily decided, nor can be determined whether the outside reality exists as such. The cognitive apparatus that is doing the constructive work has only unspecific nervous signals at its disposal, i.e., signals which decode the intensity of a stimulus[4] but not its nature or origin. From the perspective of the apparatus, it is therefore of no significance where the signals come from and what entity caused them. However, no rationale speaks against the *construction* that assumes the existence of a reality for practical reasons.[5]

Furthermore, as sketched above, constructions are entrenched in a hierarchical network whose components are mutually dependent. The resulting canalization of future linking possibilities renders arbitrary constructions impossible. That's why we cannot walk through closed doors. It is this ramification of construction details that inherently imposes anticipation, as I have argued in Riegler (2001a, b). Since we construct our own world we limit the degrees of freedom of our constructions at the same time. This apparent paradox could also be read another way, namely that constructs and their unavoidable limitations are *imposed* on us, and all we do is to choose among a few possibilities. Sverre Sjölander (1995) suggests a similar picture. He assumes the existence of an inner "probierbühne" (trial stage) upon which anticipations are formulated, i.e., imaginations about the future in qualitatively (but not necessarily quantitatively) arbitrary ways. (Alternatively, we could think of it as some virtual reality scenario that, detached from "reality", makes it possible to build future scenarios). But who formulates them? Do we have reasons to believe that it is not the I? Who is it then?

The answer to this question might be found by taking a closer look at neurophysiological phenomena, especially at neural correlates of consciousness.

Bridging the gap between subjective experience and objective quantities has been a focus of research ever since. Of specific interest are questions such as: Are acts of free will initialized by conscious decisions? Can physiological insights be reconciled with the view that a free will is responsible for our doings?

These were also the questions Benjamin Libet put to himself (Libet 1985; Nørretranders 1998). In the 1960s he had the opportunity to conduct experiments with patients of the neurosurgeon Bertram Feinstein. Their skull cover had been removed and they remained fully conscious during the surgery. Libet's experiments drew on the well-known insight that by stimulating the motor cortex with electrical impulses one can trigger sensations and even motor movements. It is crucial which area is

---

[4] The term is used synonymously with 'perturbation', i.e., the disturbance caused by an entity on another entity.

[5] Actually, as the philosophy of *radical constructivism* claims, this question cannot be decided at all without recurring to the same perceptive processes which are used in making the experiences in the first place (Glasersfeld 1995; Riegler 2001b). To verify the assumption of the existence (or non-existence) of an outside reality, we need an independent vehicle. Means used so far are, for example, religious (*believing* in reality) or social (*authoritatively claiming* its existence). None of both complies with the scientific method.

stimulated. Certain associative motor areas of the cortex trigger movement together with the subjective sensation that the movement was of one's own volition. On the other hand, if subcortical areas are stimulated the triggered movement appears to be unintentional. Those subcortical areas seem to be beyond the control of the consciousness.

Libet found out that the stimuli have to last at least half a second in order to be registered by the patient. Below that threshold they remain subliminal, i.e., unnoticed by the consciousness. Interestingly, stimulating the skin is already perceived after 20ms. What causes the big difference of awareness between stimulating the cortex and the skin? Libet designed an experimental setup that allowed him to directly compare both sensations. The cortex of a patient was stimulated in a way such that she would feel a light tingle in one hand while the skin of the other hand was directly stimulated to evoke the same sensation. The surprising result of this experiment was that one has to wait half a second between stimulating the cortex and the hand in order to make the two events subjectively happen at the same time. Stimulating the hand earlier caused the sensation in this hand before the other although the skin stimulus happened after the one of the associated cortex area of the other hand. Assuming that it takes about half a second for a stimulus to become aware, Libet concluded that conscious experiences of events are projected backwards in time. This explains why the stimulus is registered immediately. Processing both the artificial stimulation of the cortex and the natural one of the skin takes about the same time. However, the cortex stimulus is not projected back in time as it is no natural sensation but rather a direct intervention in the electrical circuits of the brain which are not subject to the usual "censorship" of nervous pathways through other brain areas.

After Feinstein's death, Libet continued his research in a different way that borrowed from the pioneering work of Hans Kornhuber and Lüder Deeke. In the mid-1960s they had found out that volitional actions are accompanied by a negative electrical potential that arises shortly before in the cortex. This "readiness potential" starts about half up to 1.5 seconds before the actual cortical motor signal and can be made visible in the electroencephalogram (EEG). The readiness potential appears also when the movement is only intended rather than executed, i.e., the motor cortex is not activated. Therefore, the readiness potential reflects the decision to carry out a movement rather than the actual control of the movement by the cortex. If the preparations for a movement take such a relatively long time, when is the decision taken to start it in the first place? And who decides?

Libet chose the following setup in order to correlate three essential points in time with each other. The start RP of the readiness potential, the moment D at which the subject decides to carry out a conscious action, and the time A that marks the begin of the action, registered by an electromyogramm (EMG). In order to maximize the probability that the action is indeed a spontaneous volitional act it has to be as simple as possible. Therefore, Libet asked the subjects to spontaneously bend a finger or bend an arm. This moment A can be measured by the electrical activity of the hand. The parameter RP can be read on an EEG. In order to determine time D, however, it was necessary to fall back on (a modern version of) Wilhelm Wundt's oscilloscope clock ("Komplikationsuhr") which had become a standard instrument in experimental psychology. This clock consists of a screen on which a dot is rotating around its center in

2.56 seconds. All that the subjects had to do was to memorize the relative position of the dot when they spontaneously decided to move a finger.[6]

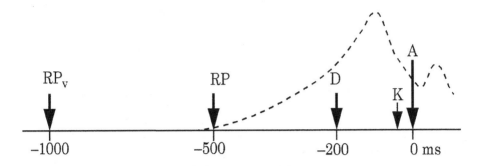

Fig. 1: Sequence of readiness potential (RP), volitional decision (D), and onset of action (A), as well as the control stimulus on the skin (K). If the action is planned ahead, the readiness potential starts already at time RP_v. After Libet (1985).

After statistically averaging the data, Libet obtained the following correlation among the parameters. $A - D = 200$ ms, but $A - RP = 550$ ms. This means that the decision to act starts, as expected, before the action but after the occurrence of the readiness potential (cf. Figure 1). In other words, the consciousness notices only after 350 ms that the unconsciously working part of the brain has started to prepare the "volitional" act.[7] Wolfgang Prinz (1996) has framed this remarkable result as follows: "We don't do what we want, but we want what we do".[8]

Martian Tennis Player

Despite the surprising nature of his results, Libet thought to rescue free will due to the following observation. If subjects interrupt an already decided action, the EEG shows nevertheless a readiness potential. This means that the consciousness—albeit informed belatedly—can still veto an action that has already started. It seems that there is an independently working brain machinery that eludes conscious control and which constantly initializes new actions. The role of the consciousness is then to choose from these actions before they get executed.

---

[6] In order to show that this method allows for precise results, Libet carried out control stimulations with the skin that yielded correct measurements.

[7] In order to eliminate the possibility that the projection back in time is the reason for the readiness potential RP to occur before the volitional decision D, Libet made a control stimulation K on the skin, which subjectively takes place 50 ms before the stimulus. Later on, the results of Libet's have been confirmed both directly and indirectly by others such as Keller & Heckhausen (1990) and Haggard & Eimer (1999).

[8] "Wir tun nicht, was wir wollen, sondern wir wollen, was wir tun."

At first glance this scenario reminds us of Sigmund Freud's concept of the subconscious according to which the human mind is no longer the "master in its own house". Following his horse–rider analogy, it is the horse (the "id") which determines where to move with the rider (the "I"). Similarly, in Libet's interpretation consciousness and free will seem to be at the mercy of the horse "unconsciousness".

But what is more important than finding similarities in psychoanalysis is the fact that Libet's scenario implements Sjölander's inner trial stage. The independent unconscious brain machinery constructs a hierarchy of schemata out of components of experience. The consciousness merely selects the way these components are put together and carried out.[9]

There is a wide variety of experimental results and insights that support this picture. Quite evidently, unconscious processes play a major role in sports where it is of crucial importance to be able to anticipate the opponent's next action. Studies show that the difference between expert and amateur players is based on how they perceive movement. Skilled players read their opponent's game: they look at the right cues and make the proper anticipations from these cues. Looking and making anticipations, however, are no conscious processes. As shown by a many researchers (e.g., Kourtzi & Shiffrar 1999), the perception of motion of a human body is constrained by its anatomical and biochemical properties. So the perception of an opponent's actions is influenced by the unconscious knowledge of the constraints caused by these properties. As Karl Verfaillie and Anja Daems (2002) have argued, this implicit knowledge can be used to anticipate sequences of action. When playing tennis or squash, expert players anticipate the trajectory of the ball from the opponent's body posture and its dynamic aspect before the opponent even hits the ball. Such implicit knowledge is the result of a long and continuous training, or "habit learning", as Ann Graybiel (1998) calls it. This becomes possible by chunking standard perception and action sequences, which can be recalled and replayed without an interfering consciousness.[10] The brain has created a template that can produce the learned behavior as if it was still under conscious control. As a result, a human professional would appear quite inept when playing against a Martian player with a different and therefore unpredictable physiology.

There are two indications that anticipation—whether in sports or other activity—is unconscious. Firstly, the time for conscious responses is, as Libet (see above) has shown, with 500ms much too long in order to react swiftly enough. Secondly, the activity pattern in the brain is much more spread out in unskilled beginners than in experts, indicating that handling the task still needs full attention rather than running smoothly and stereotyped through a small unconsciously working part of the brain. However, if test subjects are asked to pay close attention while carrying out the learned task the symphony of brain activity in the frontal parts of the brain starts again. At the same time, the behavior of the subjects becomes less smooth as if the

---

[9] One might feel tempted to ask for the underlying basis on which the consciousness makes its decision. Unfortunately, such questions lead directly to the qualia problem when trying to reduce the selection criteria to a mere algorithm. Fortunately, we don't need to investigate this dilemma as we are interested in the construction process rather than in judging these constructions.

[10] For a cognitive architecture that features anticipatory chunking, see Riegler (1994).

presence of the consciousness interrupted the execution of the task in an unconstructive manner. (Taking up many areas of the brain while learning the reply to a new challenging situation is also the reason why we can consciously focus only on one single task while in the background a great number of unconscious activities can be carried out in parallel.)

The part of the brain that seems to be responsible for stereotyped habit learning are the basal ganglia (Graybiel 1998). Interestingly, due to their central position they form a bottleneck, which affects all sorts of cognitive activity. Knowlton, Mangels, and Squire (1996) describe the "weather prediction" experiment where subjects are asked to figure out the link between a configuration of cards with geometrical figures and a weather situation. As the authors point out, the mapping between the cues and the outcomes is too complex for the hippocampus, which is usually responsible for learning such propositional representations. Therefore the success of the subjects in this experiment must be attributed to unconscious mechanisms.

We find chunking anticipations on even higher levels of cognition, such as the abstract problem of recalling positions in the game of chess. Already in 1927, Djakow, Rudik, and Petrovsky demonstrated that masters are able to recall the position more accurately than non-players. In his 1946 book, Adriaan de Groot (1978) presented a study according to which a grandmaster can remember up to 93% of the positions of the pieces, while a beginner gets typically only about 50% right. Since the board was presented to the subjects for only a few seconds the propositionally working hippocampus couldn't be made responsible for this surprising performance either. It has been argued that while the novice sees only a random assemble of pieces, the master recognizes well-ordered sets of possibilities. In other words, like the tennis player, the chess master draws on the unconscious ability to learn nonrepresentational constraints and canalizations that are the result of the rules of chess. As Herbert Simon and William Chase (1973) proved, strong players do not recall random positions better than beginners, i.e., configurations that are not the result of the applying the rules of chess.

These results strongly suggest that unconscious anticipations based on the ability to exploit constraints are used in a wide range of cognitive activities, from the sensorimotor level to highly abstract tasks. This seems in agreement with Libet's results. A free consciousness that chooses from what the unconscious suggests.

The Helpless Spectator

According to Gerhard Roth (Haynes et al. 1998; Roth 2001), the situation could be more threatening for the free will than Libet claims. The part of the prefrontal cortex, which is considered the highest authority for planning actions, is under influence of those subcortical areas that elude conscious control. Only through basal ganglia it can access the cerebral cortex which is responsible for motor control. The basal ganglia exert a censorship function and are for their part again under the influence of certain parts of the limbic system that are beyond conscious access. Since the basal ganglia are ultimately under control of the limbic system, the veto-option of the consciousness is considerably reduced if not rendered impossible. Roth attributes an evaluative function to the limbic system, which—based on experiences—is ready with quick reflex-like yet inflexible solutions to many problems. Therefore, it is suggested to be responsible for emotions and can swiftly react in well-known and well-practiced situations. These problem solutions are implemented in form of compact neuronal systems, which are the result of repeated practicing. What presents itself as a new problem for which the brain machinery cannot find a ready-made recipe, will be dealt with by the integrative and flexible yet slowly working consciousness. As the problem occurs repeatedly, new cortical networks are created which transform the solution into an established routine case, which is taken care of without an interfering consciousness. This way, consciousness becomes the deputy sheriff of the unconsciously working evaluation system. It seems to be used in complicated new situations only.

Whether the Libet-Roth picture is valid can't be confirmed yet as the original experiments of Libet are met with vehement criticism (e.g., Gomes 1998; van de Grind & Lokhorst 2000). However, these critics refer mainly to the conclusions Libet draws from his experiments, and to the assumptions that gave rise to his experimental setup. But despite his disapproval even Gilberto Gomes must commit, "I believe we can agree with Libet's conclusion that voluntary acts are nonconsciously initiated" (Gomes 1999) And this is everything we need from Libet's experiments as empirical evidence. After all, the original question was: Who is constructing?

In the late 90s of the 19th century, T. H. Huxely wrote that consciousness is just watching behavior and isn't able to do anything. Julian Jaynes (1976) refers to it as Huxley's Helpless Spectator Theory, and refers to it as follows. "Consciousness can no more modify the working mechanism of the body or its behavior than can the whistle of a train modify its machinery or where it goes. Moan as it will, the tracks have long ago decided where the train will go." Jaynes himself developed a controversial theory about the origin of consciousness. After studying the work of Homer intensively, Jaynes arrived at a remarkable conclusion. He maintained that there was no such thing as consciousness 3000 years ago. The ancient Greeks (the Myceneans of Homer's "Iliad") simply did not have it. He stressed that they heard auditory hallucinations, "voices of gods" instead. There was no sense of subjectivity, no introspection in the modern sense. Rather, the voices of the gods told them what to do and which decisions to take, similar to what schizophrenic experience. In some sense their brain was divided into two—the disassociated hemispheres with different working modes, as Jaynes proposes—resulting in a "bicameral mind". Only later Greeks, like Homer's Odysseus, developed a new "worldview" where the voices of the gods have gone and the mental world of people is enriched by a consciousness instead. Jaynes (1986) wrote: "In his everyday life [bicameral man] was a creature of habit, but when some problem arose that needed a new decision or a more complicated solution than habit could provide, that decision stress was sufficient to instigate an auditory hallucination." Instead of holding an internal monologue in which a person considers different alternatives by making anticipations about the results of a certain action, planning and decision making seem to have happened at an unconscious level and then proclaimed to the person.

We feel immediately reminded to the scenario of an unconscious part in the brain that according to the Libet-Roth model runs the body and informs the consciousness only afterwards. The consequences are evident. When Butz (2002) refers to a "decision-maker" who takes predicted future states into account, the nature of that decision-maker is not the conscious mind, as implicitly assumed. Anticipations are constructed at a level that obviously eludes conscious access. If this view holds water, my claim that "anticipations are the result of internal canalizations which *inevitably 'force'* a particular path" (Riegler 2001a, my italics here) applies to all kinds of anticipatory systems, and not just to implicit ones (in Butz's terminology).

Conclusion

The goal of this paper was to go one step further than my original examination of the nature of anticipation (Riegler 2001a) where I connected anticipations with canalizations in the physical and abstract realms of behavior and reasoning. However, I owed an explanation how these canalizations come about, who or what is responsible for assembling the elements in our hierarchical network of schemata (to use Piaget's terminology).

The answer to this paper's central question is challenging. The interlocking of elements in the hierarchy of schemata, which originate in our experiences, results in interdependencies among these elements and thus in canalizing forces. This puts limits to the accessibility of arbitrary future states. So although the subjective worldview of an individual is the result of a construction process by which elements of experience are linked together, the construction process itself is not arbitrary. In our cognitive repertoire we have constructs like "wall" and "walking" but no "walking through walls" as a valid option. Do we consciously construct concepts like "hard objects" that populate our cognitive space, or does their construction happen "somewhere else"? The evidence presented in this paper speaks in favor of the latter. As our cognitive space forms a hierarchy of interdependent elements, any anticipation we develop is also necessarily subject to canalizations. This is what I referred to as "being firmly rooted in the system rather than being dependent on (deliberately) built internal models" (Riegler 2001a). What follows from the arguments in this paper is that the "generator" of anticipations is buried in layers inaccessible for the conscious experience. While we can a posteriori reflect upon these anticipations we don't produce them in the first place. Whether this restricts volition and free will depends on whether the "spectator consciousness" has a Sjölanderian selection and/or Libetian veto option that prevent us from becoming marionettes in the sense of Huxley.

References

Ashby, W. R. (1956) *An introduction to cybernetics.* Chapman & Hall: London.

Butz, M. V. (2002) Anticipations in natural and artificial systems. Unpublished manuscript. http://www-advancedgec.ge.uiuc.edu/literature%20reviews/anticipation.ps

de Groot, A. D. (1978) *Thought and choice in chess.* The Hague: Mouton Publishers. Dutch original "Het denken van den schaker" published in 1946.

Djakow, I. N., Petrowski, N. W. & Rudik, P. A. (1927) *Psychologie des Schachspiels* [Psychology of the game of chess]. Berlin: de Gruyter.

Dubois, D. M. (2000) Review of incursive, hyperincursive and anticipatory systems. foundation of anticipation in electromagnetism. In: D. M. Dubois (ed.) *Computing Anticipatory*

Systems: CASYS'99—Third International Conference. AIP Conference Proceedings 517. The American Institute of Physics: Woodbury, pp. 3–30.

Glasersfeld, E. von (1989) Constructivism in education. In: Husen, T. & Postlethwaite, T. N. (eds) *The international encyclopaedia of education,* 1st edn, Supplementary Volume 1. Oxford: Pergamon Press, pp. 162–163.

Glasersfeld, E. von (1995) *Radical constructivism.* Falmer Press: London.

Gomes, G. (1998) The timing of conscious experience: A critical review and reinterpretation of Libet's research. *Consciousness & Cognition* 7:559–595.

Gomes, G. (1999) Volition and the readiness potential. *Journal of Consciousness Studies* 6 (8–9): 59–76.

Graybiel, A. (1998) The basal ganglia and chunking of action repertoires. *Neurobiology of Learning and Memory* 70 (1–2): 119–136.

Haggard, P. & Eimer, M. (1999) On the relation between brain potentials and the awareness of voluntary movements. *Experimental Brain Research* 126: 128–133.

Haynes, J.-D., Roth, G., Schwegler, H. & Stadler, M. (1998) Die funktionale Rolle des bewußt Erlebten. [The functional role of conscious experience]. *Gestalt Theory* 20: 186–213. http://www-neuro.physik.uni-bremen.de/~schwegler/Gestalt.html

Jaynes, J. (1976) The origin of consciousness in the breakdown of the bicameral mind. Boston: Houghton Mifflin.

Jaynes, J. (1986) Consciousness and the voices of the mind. *Canadian Psychology* 27(2): 128–148. http://julianjaynessociety.tripod.com/mind.pdf

Keller, J. & Heckhausen, H. (1990) Readiness potentials preceding spontaneous motor acts: Voluntary vs. involuntary control. *Electroencephalography and Clinical Neuropsycholgy* 76:351–361.

Knowlton, B., Mangels, J. & Squire, L. (1996) A neostriatal habit learning system in humans. *Science* 273 (5280): 1399–1402.

Kourtzi, Z., & Shiffrar, M. (1999) Dynamic representation of human body movement. *Perception* 28: 49–62.

Libet, B. (1985) Unconscious cerebral initiative and the role of conscious will in voluntary action. *Behavioral and Brain Sciences* 8: 529-566

Malinowski, B. (1948) *Magic, science and religion and other essays.* Free Press: Glencoe IL. Originally pubished in 1925.

Neisser, U. (1975) *Cognition and reality.* Freeman: San Francisco.

Nørretranders, T. (1998) *The user illusion: Cutting consciousness down to size.* Viking: New York. Originally published in 1991.

Piaget, J. (1954) *The construction of reality in the child.* Ballantine: New York. Originally published as: Piaget, J. (1937) *La construction du réel chez l'enfant.* Délachaux & Niestlé: Neuchâtel.

Prinz, W. (1996) Freiheit oder Wissenschaft? [Freedom or science?] In: Cranach, M. v. & Foppa, K. (eds.) *Freiheit des Entscheidens und Handelns. Das Problem der nomologischen Psychologie.* Asanger: Heidelberg, pp. 86–103.

Riegler, A. (1994) Constructivist artificial life: The constructivist–anticipatory principle and functional coupling. In: Hopf, J. (ed.) *Proceedings of the 18th German Conference on Artificial Intelligence (KI-94) Workshop on Genetic Algorithms within the Framework of Evolutionary Computation.* Max-Planck-Institute Report No. MPI-I-94-241, pp. 73–83.

Riegler, A. (2001a) The role of anticipation in cognition. In: Dubois, D. M. (ed) *Computing Anticipatory Systems. Proceedings of the American Institute of Physics 573.* American Institute of Physics: Melville, New York, pp. 534–541.
http://pcp.vub.ac.be/riegler/papers/riegler01anticipation.pdf

Riegler, A. (2001b) Towards a radical constructivist understanding of science. *Foundations of Science,* special issue on "The Impact of Radical Constructivism on Science" 6 (1–3): 1–30. http://www.univie.ac.at/constructivism/books/fos/riegler/

Rosen, R. (1985) *Anticipatory systems.* Pergamon Press, Oxford.

Roth, G. (2001) Die neurobiologischen Grundlagen von Geist und Bewusstsein. [The neurobiological foundations of mind and consciousness]. In: Pauen, M. & Roth, G. (eds) *Neurowissenschaft und Philosophie. Eine Einführung.* Paderborn/München: Fink/UTB, pp. 155–209.

Simon, H. A. & Chase, W. G. (1973) Skill in chess. *American Scientist* 61: 393-403.

Sjölander, S. (1995) Some cognitive breakthroughs in the evolution of cognition and consciousness, and their impact on the biology of language. *Evolution & Cognition* 1: 3–11.

Skinner, B. F. (1948) 'Superstition' in the pigeon. *Journal of Experimental Psychology* 38: 168–172. http://psychclassics.yorku.ca/Skinner/Pigeon/

Van de Grind, W. N. A. & Lokhorst, G. J. C. (2001) Hersenen en bewustzijn: van pneuma tot grijze massa. [Brains and consciousness: From pneuma to grey mass]. In: Wijnen, F. & Verstraten, F. (eds) *Het brein te kijk: verkenning van de cognitieve neurowetenschappen.* Swets en Zeitlinger: Lisse, pp. 217–246. http://www.eur.nl/fw/staff/lokhorst/grindlokhorst.html

Verfaillie, K. & Daems, A. (2002) Representing and anticipating human actions in vision. *Visual Cognition* 9(1/2): 217–232.

Not Everything We Know We Learned

Mihai Nadin

Program in Computational Design,
University of Wuppertal, Germany;
President, MINDesign, USA/Germany
nadin@acm.org

Abstract. This is foremost a methodological contribution. It focuses on the foundation of anticipation and the pertinent implications that anticipation has on learning (theory and experiments). By definition, anticipation does not exhaust all the forms through which the future affects human activity. Accordingly, guessing, expectation, prediction, forecast, and planning will be defined in counter-distinction to anticipation. The background against which these distinctions are made is explicit in the operational thesis advanced: Anticipation and reaction can be considered only in their unity. The interrelation of anticipation and reaction corresponds to the integrated nature of the physical and the living. Finally, an agent architecture for a hybrid control mechanism is suggested as a possible implementation.

Context and Reference

Einstein [1] observed that, "No problem can be solved from the same consciousness that created it. We must learn to see the world anew." As my own work in anticipatory computing evolved, I have constantly faced attitudes varying between skepticism and sheer enmity from outside the small community of researchers dedicated to the study of anticipation. Every example of anticipation my colleagues or I advanced was whittled down to the reactive explanations of the deterministic cause-and-effect sequence, that is, to a particular form of causality. Even the reviewers of this paper could not agree among themselves on a line of argument that in the final analysis suggests that there is more to causality than what the Cartesian rationality that we learned in school and have practiced since then preaches. (A good source of information on this topic is http://www.culture.com.au/brain_proj/Descartes.htm.) Given this difficult, but not surprising situation, I shall proceed in a more didactic manner than I would otherwise be inclined. The intention is to clarify terminology before working with the concepts—a practice that Charles S. Peirce [2] defined as the "ethics of terminology," a prerequisite of any scientific endeavor.

From among the many definitions of anticipation advanced since the pioneering work of Robert Rosen [3, 4, 5] and my own early work [6], I would like to focus on the following operational definitions:

1. An anticipatory system is a system whose current state is defined by a future state. Eventually, a change was introduced in this definition [7] so to imply that in

M. Butz et al. (Eds.): Anticipatory Behavior ..., LNAI 2684, pp. 23-43, 2003.

addition to the future state, a current or even past state could affect the state of the system (cf. Definition 9 in [8]).

2.   An anticipatory system is a system containing a predictive model of itself and/or of its environment, which allows it to change state at an instant in accord with the model's predictions pertaining to a later instant, in faster than real time (cf. Definition 3 in [8]).

3.   Anticipation is the result of the competition among a number of mind models. Reward mechanisms explain the dynamics of this competition [cf. Definition 2 in [8]; see also [9]).

In the end, I adopted the viewpoint according to which anticipation is a characteristic of the living. At the foundation of this perspective lie the work of Rosen, especially [4] and [5] and Elsasser [10]. Moreover, in discussing the neural basis of deciding, choosing and acting, Jeffrey D. Schall [11] distinguishes between external forces (explaining the "movements of physical bodies, such as rocks") and "reasons" explaining "many human movements" (actions directed towards a goal). Anticipatory behavior—such as the recognition of the prey's trajectory—is not the only expression of anticipation. We can mention design, creative activities, and conceptual elaborations as pertaining to the rich forms expressing anticipation.

In respect to learning, it is rather difficult to limit oneself to a reduced number of definitions. The richness of the forms of learning and of the expressions of learning makes the attempt tenuous at best. However, given the focus of this volume—adaptive learning systems—a first pruning of the rich tree of learning definitions becomes possible.

Computational learning theory (COLT) focuses on "the design and analysis of algorithms for making predictions about the future based on past experiences" (cf. Freund and Schapire, www.learningtheory.org). This definition falls within the reactive paradigm. An applied definition, originating in studies in time series prediction [12] sees learning as an emulation of the structure of time series, i.e., of the dynamics of processes we intend to understand, control, and automate. In this case, the knowledge about the unfolding in time of a process is eventually used in order to produce some desired behavior or to avoid undesired behavior.

Lastly, for the purpose of this paper, learning as the pursuit of regularities or patterns in processes involving the physical/inanimate as well as the living is of special interest to me. This extraction (in forms such as data-mining, perception, knowledge mapping, etc.), by humans, by machines, or by hybrid entities, of regularities or patterns can be subjected to further analysis, i.e., further learning. Learning is a necessary condition of adaptive processes in the living, in machines, and in hybrid systems. In what follows, this methodological section remains the only reference.

Not everything we know was learned. This holds true whether "knowledge is seen as socially situated or whether it is considered to be an individual extraction (cf. Ernest [13]). Taken at face value, it maintains that the process of knowledge acquisition is complemented by knowledge production. Moreover, in addition to what we learn, there is innate knowledge—that space that "has to exist before data," as it was defined by some researchers (cf. Novak *et al* [14]) focusing on language and its active role in learning. There are genetically defined processes (such as language acquisition, (cf. Nadin [15] pp. 77ff) or seeing [15] pp. 321ff). But there are also anticipatory processes through which not only *what is* is acknowledged, but also what *might possibly be* is generated (cf. [10] p. 5), i.e., the principle of creative selection. My focus in this study is on particular processes, which can be defined as anticipatory, through which the complementarity between learning and the activity of knowledge production is accomplished. With this focus in mind, I shall refer to experimental evidence. (More scientific reports based on such evidence are available at www.anticipation.info) Evidence from experiments does not in itself replace a broader understanding. What examples can do, especially when associated with data resulting from experiments, is to make us aware of an unusual or unexpected outcome of some process.

Rich Data

Take something as simple as fluctuations in the heartbeat. Data collected in endless physiological observations made doctors aware of such fluctuations. Patients experience them as discomfort. Sometimes we know what they mean. A good diagnostician can distinguish between "normal" irregularities and others related to a condition requiring treatment. E.F. Adolph [16], following Walter B. Cannon [17], deals with self-regulation mechanisms. Change in posture—standing up from a seated position, for example—would cause changes in blood pressure. This is the "physics" of the body: a liquid (blood), pipes (the various blood vessels), a pump (the heart). That this image is an oversimplification, rooted in the rationality of "the human being as machine," will soon become clear. But it serves as a good point of departure. We can understand the mechanism by taking a rubber tube with liquid inside and manipulating it. The pressure of the fluid varies as we change its position from horizontal to vertical. If human beings were subjected to blood pressure variations as often as we change posture, we all would have a terrible time. Dizziness is only one of the symptoms accompanying variations in blood pressure. It appears that a certain reflex, called the *baroreceptor reflex* [18], affects the heart rate via the nervous system in order to maintain the blood pressure within relatively constant and safe limits. How does the heart "know" about a change in position? Obviously, here the analogy to the simple physical system of a machine, with liquid, pipes, and pump ends. The question asked—how does the heart, or whatever controls the heart, know what the body will do—goes beyond the physics of pumping a liquid through pipes (cf. [12] p. 107).

Almost all of us have been educated in the rationalist spirit of a scientific perspective based on the analogy between an organism and a machine. The most sophisticated adepts of this view will argue that the machine to which the human being can be reduced is probably more complicated than the pump, pipes, and perhaps some controller—but ultimately still a machine. This is the reductionist viewpoint. It maintains that the human being can be reduced to something easier to understand and describe; moreover, that there is some equivalence between the human being and the machine. Reductionism has been practiced along this line of thinking for centuries and, as such, appears to be a valid explanatory model. What animates scientists is the rational thought that what applies to a short time sequence might apply to longer sequences. Beneath this optimistic outlook lie the accepted premises of physical homogeneity, the repetitive nature of physical change (described through the laws of physics), and, of course, reductionist thought.

One of the goals of this text is to show that taking this assertion as a premise can help us satisfactorily explain how humans react to the world in which they live. It also explains, partially, what we can learn in association with experiencing the world as one of action and reaction. However, it cannot explain the example I started out with—why does change in human posture not result in a change in blood pressure. More precisely, it cannot explain the anticipation on whose basis the heart rate compensates for the change in posture *in advance* in order to avoid variations in blood pressure.

The reason for this is that in accepting the reduction of the human being to a machine—no matter how complicated and complex this machine—we simultaneously accept the deterministic premise of the sequential link between cause and effect. And once this is accepted, a time sequence beginning with the cause and continuing to the effect is established. The example of anticipation given above does not fit into this scheme, although it does not preclude causality. The same example suggests that reaction-based learning—learning from something we experienced—is complemented by non-reactive learning, which we should in some ways account for. The simplest example is learning through acknowledging correlations, some of them rather difficult to describe [19] [20].

More examples: Someone approaches with the intention of tickling you. Before the person even touches you, you start laughing. How come? In this case, effect and cause do not seem to follow the deterministic sequence. The effect takes place before the cause (cf. [21], [22]). At work in such situations is the intrinsic dynamics of previous experiences, the learned reflex triggered through association. Speech, a very complicated motoric act (ca. 100 muscles in the face, chest, and throat need to be coordinated [23]) takes place on account of an internal model that "knows" the word before it has been uttered [24]. That we learn part of speaking through imitation is well known. But we also discover speaking as an act of self-definition of the speaking person beyond the learning component: We speak the same language but in so many different ways! Motoric acts not rooted in the cause-and-effect sequence ([25], [26]), i.e., in anticipation of a possible action, are not learned but are rather expressions of choices leading to eventual learning.

In various situations—such as crossing a street or catching a falling object—or when we so often "know" (or do we guess?) what words will follow as someone speaks to us, a joke's punch-line, someone's age, whose footsteps we hear long after we last had contact with the person—something unusual takes place. The event (a car approaches, an object we were unaware of falls, something we surmised would happen, does) to take place in the future—i.e., a *future state*, as it is known in science—affects our immediate behavior—which scientists call *current state*. To be adventurous—a definite means of survival, that is, an evolutionary characteristic—means to be in anticipation of the possible outcome, of the adventure. This means to be in anticipation of what one might eventually learn in the adventure [27], [28].

In regard to all these examples, we can raise the issue of relevance: Is anticipation learned? Can our descriptions of anticipatory processes form a valid model for a theory of learning? Or is learning, as learning theorists would probably be inclined to think, a prerequisite for anticipation? A dogmatic leaning in either direction will actually prevent us from building a coherent explanatory framework, and thus affect our attempts to apply the knowledge gained to practical purposes.

The introductory methodological part advanced, as reference, the notion—by no means unanimously accepted—that to learn means to have an object of learning, an entity subject to our interest (no matter in which form the interest is expressed), and a learner (be this a person, a program, a procedure, etc.). Anticipation is defined as a dynamic characteristic unfolding in a different timeframe, that is, before the object of our interest, and sometimes (as in illusory anticipation) instead of it. This difficult idea of a cause that lies in the future (cf. von Foerster [29]), in the absence of which anticipation becomes meaningless, will be further explained here.

The "Ante" in Anticipation

What all the abovementioned examples have in common is the fact that they question the dominant explanatory model to which we are accustomed, and which we automatically apply in our thinking: There is a cause (in the past or present) that leads to an effect (now or later). This sequence describes determinism in its simplest form. It is also a simple description of causality in its simplest form. (For more on causality, see [30].) But the heartbeat mechanism given as an example changes (the effect) in advance of any modification of someone's in posture (the cause) within a complex dynamic system of interlocking variables. No one has yet tickled us (cause), but we start laughing (effect). Learning takes care of some of the abilities needed, anticipation takes care of others.

We do not compute as we speak (cf. Guenther [24]). We ourselves do not compute data as we drive through heavy traffic (cf. Zadeh [31]). The hand catches the falling object before the person realizes that it's coming. Good reflexes, some say, always amazed that the hand seems to have a mind of its own. And again: there is no computation involved, rather an action corresponding to a possible situation that was not triggered by sensory data. We "guess" so many things (just as we fail to guess others) without realizing how we do it and without being aware that we do it. The outcome precedes or predates the action! The next word in a sentence (spoken or written), the punch-line of a joke, the suspect in a crime novel or movie—in each of these cases there is a realization before what we would be inclined or biased to expect. The logic behind this looks like deductive thinking: from something we know to something else that is its consequence. In some cases we might indeed find that a deductive step is part of the anticipation. Reading a crime novel, not unlike solving a puzzle, involves deduction. However, if a merely deductive process lay behind anticipation, machines would be able to read a crime novel with the same pleasure—a qualifier of high complexity—a human being does, and to produce a summary of the novel. The emphasis is on the production of a summary, a task that any machine is anywhere near accomplishing, although we know by now how to program machines for executing deductive tasks. The ingredients of experiencing pleasure might be several—to take a logical argument—but all of them have to rely on anticipation, since the only source of tension is between what we read vs. what is anticipated. Creativity (cf. [10] p. 146) is intrinsic in the dynamics of the living. Such machines would listen to a concert in which performers and listeners were challenging each other with their respective anticipations. The word guessed in a dialog might be deducted from the flow of succeeding words. At times it might make no logical sense, but rather result from a context in which many factors unrelated to what is uttered are involved. Whether deductions are possible or not, what we deal with is always a temporal sequence from the future (as one possibility among many) to the present. This "before" interval, this *ante* (the Latin prefix that denotes "before") can be and has been repeatedly measured, documented, discussed, and questioned.[1]

The description of the cause-and-effect sequence, often understood as action-reaction, is relatively easy to understand. It corresponds to our intuitive sense of time as this was shaped by the view of the world projected upon us by our own experiences as living creatures: from past through present towards the future. But the opposite sequence description, from the future (goal, state, action) to the present (means, methods, choices) seems to run counter to common sense. Think only of love—is it reaction or is it anticipation? Or is it both? Even making love follows anticipations that need not be spelled out here. Is it of the same nature as logical deduction? If it were, machines endowed with deductive capabilities could fall in love and make love. It is not even fun any more to argue with the machine-obsessed followers of Marvin Minsky [32], who predicted,

In from three to eight years, we will have a machine with the general intelligence of an average human being. I mean a machine that will be able to read Shakespeare, grease a car, play office politics, tell a joke, have a fight. At that point, the machine will begin to educate itself with fantastic speed. In a few months, it will be at genius level, and a few months after that, its power will be incalculable.

Here we experience the fundamental need to overcome the bizarre Cartesian optimism embodied in the cult of the machine and to realize the complementary nature of reaction and anticipation. A self-learning machine ("the machine will begin to educate itself with fantastic speed") is as far from us today as is a good theory of learning that incorporates reactive and anticipatory mechanisms!

---

[1] One of the older versions of the word "anticipation" is the Latin *antecapere*, a "pre-understanding," which eventually became *anticipare*.

As stated at the very beginning, examples do not constitute theories. They are not even hypotheses. Rather, they inform us that among the things we can observe and among the experiments we can, to a certain extent, replicate, there are some which do not fit the pattern of what is already known or accepted. Obviously, before looking for explanations outside the knowledge already acquired and tested, we are inclined to apply what belongs to the shared knowledge of the world. Learning comes immediately to mind.

Maybe a cause does exist—even if we are not yet able to point to it—behind the mechanism controlling the heartbeat in anticipation of a change in posture. Or maybe laughter, the effect of tickling, is after all caused by something else, let us say located in the brain or associated with what is called the mind, or genetic mechanisms. Some scientists even believe that they have found the location of the genetic code leading to adventure. But adventure, which makes learning possible, is also in anticipation of the experience of learning; it is a promise, which might turn out empty.

The realization that "ahead of time" (*ante*) is the legitimate prefix of *anti-cipation* presupposes a different description of time, the implicit dimension of all learning experiences.

To Be Is to Act... Possibly to Venture

Neuroscience, as part of the science of the living (biological science), took it upon itself to examine the relation between sensory data—what we smell, taste, see, hear, and experience through our sense of touch (the haptic sense)—and our actions. There are many reasons for this undertaking. One of them is the increased realization by the scientific community of the inadequacy of the deterministic-reductionist approach. Researchers are trying to understand the ever-changing (dynamic) relation between the living organism and the environment. Questions are focused on the connection—if any could be established—between what organisms do and how the action is embodied in a change in their condition. This change is an expression of learning, especially of adaptive (not only reactive) learning. But this is also how we evaluate—after the fact—the knowledge involved (or, as some would say, "successful learning," which results in acquired knowledge).

What eventually becomes clear is that living processes are characterized by their integrated nature: the whole (organism, being) and its parts are related and reciprocally interdependent. To understand learning means to understand how interdependence—of actions, agents, individuals, groups, etc.—is instantiated in what we do. Sensory data analysis (also called perception) cannot be meaningful unless we correlate sensory information to action.

Expectation, Guessing, Prediction

We become aware of anticipation through the intermediary of some of its aspects: from guessing and expectation to prediction and planning. Each of these various aspects—ultimately a combination of reaction and anticipation—can be pragmatically described. To guess is to choose from what might happen on account of various experiences: one's own, in the case of a similar endeavor or one that is likely to happen; the experience of others; or experience based on unrelated patterns—the so-called "lucky throw" of a coin or dice, for example. The reaction component and the anticipation are combined according to a formula that qualifies some of us as better than others in guessing. Reactions are based on the evaluation of the information pertinent to the situation—different when one visits a casino than when one guesses the correct answer in a multiple-choice test. Anticipations result from the self: From all possible games in the casino, some are more "favorable" at a certain time. In the multiple-choice situation, one infers from the known to the unknown. There is learning in guessing as patterns emerge: the next attempt integrates the observation of related or unrelated information. This associative action is the cognitive ingredient most connected to guessing and, at the same time, to learning. The rest is often statistics at work, combined with *ad hoc* associative schemes pertinent to what is possible. Let us acknowledge that guessing (as well as learning) is reduced to nil in predictable situations. Only surprise justifies the effort even when the result is negative. Recent research of responses of the human frontal cortex to surprising events (cf. Fletcher *et al* [33]) points to the relation to learning that I mentioned above. The dorsolateral prefrontal cortex seems to perform a role in the adjustment of inferential learning. Associative relationships leading to associative learning are based on the action of discriminating the degree (strength) of interrelation.

In comparison to guessing, expectation does not entail choosing ("Heads or tails?"), but rather an evaluation of the outcome of some open-ended process. Look at your child's expression and guess what might happen when he or she will "hang out" with friends. (For instance, facial expression reveals a lot more about coming actions than we actually see.) This evaluation might be difficult, if at all possible, to describe (e.g., "I know what you guys plan to do"). In the act of forming an expectation, the focus changes from the probable to the desired and/or expected. These pertain to what is possible. Several sources of information pertinent to forming an expectation are weighed against each other. What appears most probable out of all that is possible gets the higher evaluation, especially if its outcome is desirable. If the outcome is judged to be negative, then avoiding it is the basis for action. Again, anticipation—reflected in what is perceived as possible—meets reaction, and information is associated with probable cause. Weather is often expected, not guessed. So are the outcomes of activities that weather might influence, such as agriculture practiced prior to the integration of digital information in agricultural praxis. A cornfield is not equally fertile in every spot. We can learn how to increase production by extracting data (through GPS-based measurements) pertinent to fertility and applying fertilizers or planting more seeds in certain spots. Based on the evaluation of the outcome (see introductory definition) we form new expectations. Events with a certain regularity prompt patterns of expectation: a wife awaits her husband, a child awaits his/her parent, a dog awaits its owner, who usually returns from work at a certain time. We encounter these regular events on many occasions and in many activities. And we try to use them to our benefit.

Prediction results from connecting cause and effect, by associating the data describing their connection. Causality, as the primary, but not exclusive, source of our predictive power is rarely explicit, and even more rarely easy to understand. Accordingly, prediction—explicit or implicit—expresses the limits of our own understanding of whatever we want to predict. In some cases, the prediction is fed back into what we want to predict: how a certain political decision will affect society; how an economic mechanism will affect the market; how technological innovation (let's say multimedia) will affect education. As a result, a self-referential loop is created. The outcome is none other than what we input as prediction, although we are not fully aware of the circularity inherent in the process. The impossibility of disconnecting the observer (subject, in learning) from the observed (the object of learning) is an inherent condition of learning, whether human or machine learning. The constructivist perspective demonstrated the point quite convincingly (cf. Ernest von Glasersfeld [34]).

But there are also predictions driven, to a certain extent, by anticipatory mechanisms. Falling in love at first sight is a prediction almost impossible to make explicit. There is no cause-and-effect connection to establish. The future state (the romantic ideal of a great love, or the calculated outcome of an arranged marriage) affects current states as these succeed each other in a sequence of a time often described as "out of this world." Facial expression as a predictor is yet another example. In very sophisticated studies (cf. Ekman *et al* [35]), it was shown that the "language" of facial expression speaks of facts to happen before they are even initiated—which is anticipation in pure form. For those who "read" the face's expression, i.e., for those who learned the language of facial expression, predictions based on their own anticipation guides their action. (A convincing case is that of a Los Angeles policeman who reads on the face of the criminal holding a gun on him that he will not shoot, leading the officer to spare his life, cf. Gladwell [36].) The expectation—criminal pulls out gun and points it at the policeman pursuing him—and the prediction—this person with the particular facial expression, as studied by the interpreter, will not shoot—collide.

Descriptions of the relation between expectation and prediction are informative in respect to the mechanisms on which both are based, and to the various levels at which learning takes place. Expectations pertain to more patterned situations. An acceptable description is that the learner extracts regularities or uses innate knowledge (such as in speech). At times, they are an expression of what in ordinary language is described as stereotype or, in some cases, as wishful thinking. However, when the individuals become involved in the activity of predicting (literally, *to say beforehand,* i.e., before something happens), they not infrequently expect the prediction to actually take place. It is no longer a wish, but rather the human desire, expressed in some action, to succeed.

The attempt to understand change (macro-level, i.e., behavior) without looking at the processes through which it takes place (micro-level processes, cognitive processes) also leaves anticipation out of the picture that this understanding of change eventually provides. Change means modification over time. Predictive efforts are focused on understanding sequences: how one step in time is followed by another. However, these efforts focus on what ultimately anticipatory processes are—a modeling of the entity for which they are an agency, and the execution of the model in faster than real time speed. The limited deterministic perspective, mechanic in nature—i.e., which cause leads to which ensuing effect—affects the understanding of anticipation through a description of predictive mechanisms. Predictions following known methods (such as time series analysis and linear predictors theory) capture the reaction component of human action. (For more information and bibliography, see, respectively, http://www.ubmail.ubalt.edu/~harsham/stat-data/opre330Forecast.htm, and http://www.neurocolt.com/abs/1996/abs96037.html).The anticipatory component is left out most of the time, as a matter of definition and convenience: complexity is difficult to handle. Once a predictive hypothesis—let's say every minute the clock mechanism engages the minute hand—is adopted, it defines the cognitive frame of reference. Should the predicted behavior of the mechanism somehow not take place, the expectation is tested. However, mechanisms, as embodiments of determinism, rarely fail. And when they do, it is always for reasons independent of the mechanism's structure.

Predictions concerning the living are less obliging. It happens at all levels of the living that predictions—what will happen next (immediate or less immediate future)—are either partially correct or not at all. In studying learning and selective attention, Peter Dayan *et al* [37] refer to reward mechanisms in the Kalman filter model (more experience leads to higher certainty). In such cases, expectations turn out to be a measure of how much the deterministic instinct (culture, if you prefer) takes over the more complex model that accounts for both reaction and anticipation in the dynamics of the living.

Predictors reflect our desire to understand how change takes place. They express our practical need to deal with change: However, they leave our own change out of the equation. Actions from thoughts, as Nicolelis [38] calls them, account for the self-awareness of change. What is learned supports inferences (statistical or possibilistic); uncertainty results as the competitive resources engaged in the inference are overwritten by unrelated factors. Predictions also try to capture the interconnectedness of all elements involved in the dynamics of the observed. And not last by any means, they are an instrument of learning. In this sense, they open access to ways in which we could emulate (or imitate, to use another word) change. It is important to realize that expectations have no direct learning component: one cannot learn explicitly how to expect, even if we accept that there might be structure in the learning process and in the representation. Expectations only occasionally produce knowledge: a series of expectations with a certain pattern of success, or failure for that matter. Predictions, even when only marginally successful, support activities such as forecasting—for short or less than short sequences of change—of modeling, and of inference to the characteristics of the observed dynamic entities. A good prediction says quite a bit about the complexity of the change observed. Even through what it does not capture, it says a lot about complexities we tend to leave aside.

In this respect, predictions regarding the living, although inappropriate for systematically capturing their anticipatory dimension, are a good indicator of what we miss when we ignore anticipation. An example: In focusing only on human beings, predictions of physiological data remain at a primitive stage at best, despite the spectacular progress in technology and in the scientific theory of prediction. Albeit, if we could improve such predictions by accounting for the role of anticipation, we would be in a better position to deal with life-threatening occurrences (strokes, sudden cardiac death, diabetic shock, epileptic seizure, etc. [cf. 12]). This means that we would learn about such occurrences in ways transcending their appearance and probability. Things are not different in the many and varied attempts we undertake in predictions concerning the environment, education, health, and the functioning of markets. Unless and until anticipation is acknowledged and appropriate forms of accounting for it are established, the situation will not change drastically. This becomes even clearer when we look at the very important experiences of forecasting and planning.

Forecasting and Planning

Not fundamentally different from predictions are forecasts. Actually, predictions are a prerequisite of forecasting. The etymology points to a different pragmatics, one that involves randomness (as in casting). Under certain circumstances, predictions can refer to the past: Take a sequence in time—let's say the San Francisco earthquake of 1906—and try to describe the event (after the fact). In order to do so, the data, as registered by many devices (some local, some remote) and the theory are subjected to interpretations. The so-called Heat-Flow Paradox is a good example. If tectonic plates grind against one another, there should be friction and consequently heat. This is the result of learning from other physical phenomena. Along the well-known San Andreas Fault, geologists (and others) have measured (and keep measuring) every conceivable phenomenon. No heat has been detected. The generalization of the learned in respect to friction proved doubtful at best. Accordingly, in order to maintain the heat dissipation hypothesis as a basis for forecasting, scientists started to consider the composition of the fault. This new learning—extraction of regularities other than those pertaining to friction and heat dissipation—was focused on a different aspect of friction. A strong fault and a weak fault behave differently under stress, and therefore release different quantities of heat. Here is a case in which data is fitted to a hypothesis—heat release resulting from friction. This is an attempt to adapt what was learned to a different context. Therefore adaptive learning and forecast offer a different context for interpretation.

In other cases, as researchers eventually learned, what was measured as "noise" was treated as data. Learning noise patterns is a subject we rarely approach, because we do not yet know of procedures for effectively distinguishing between noise and data. In medicine, where the qualifiers "symptomatic" vs. "non-symptomatic" are applied in order to distinguish between data and noise, this occurs to the detriment of predictive performance.

In general, theories are advanced and tested against the description given in the form of data. Regardless, predictions pertinent to previous change (i.e., descriptions of the change) are not unlike descriptions geared to future change. In regard to the past, one can continue to improve the description (fitting the data to a theory) until some pattern is eventually discerned and false knowledge discarded.

To ascertain that something will happen in advance of the actual occurrence—prediction (the weather will change, it will rain)—and to cast in advance—forecast—(tomorrow it will rain) might at first glance seem more similar than they are. A computer program for predicting weather could process historic data: weather patterns over a long time period. It could associate them with the most recent sequence. And in the end, it could come up with an acceptable global prediction for a season, year, or decade. In contrast, a forecasting model would be local and specific. The prediction based on "measuring" the "physical state" of a person (how the "pump," i.e., heart, and "pipes," i.e., blood vessels, are doing, the state of tissue and bone) can be well expressed in such terms as "clean bill of health" or "worrisome heart symptoms." But it can almost never become a forecast: "351 days later, you will have a heart attack;" or "In one year and seven hours, you will fall and break your jaw."

Predictions do not involve interpretations. They result from whatever explanatory model (expressed or not) is adopted. Forecasts, even when delivered without explanation, are interpretive. They contain an answer to the question behind the forecasted. "The price of oil will change due to...." You can fill in the blank as the situation prompts: cold winter, pipeline failure, war. "Tomorrow at 11:30 am it will rain...." because of whatever brings on rain. "There will be a change in government...." "Your baby will be born in ... hours." A good predictive model can be turned into a machine—something we do quite often, turning into a device the physics or chemistry behind a good prediction: "If you don't watch the heat under the frying pan, the oil in it will catch fire."

Forecasts are not reducible to the machine structure. They involve data we can harvest outside our own system (the sensorial, in the broadest sense). In addition, they involve data we ourselves generate. The interplay of initial conditions (internal and external dynamics, linearity and non-linearity, to name a few factors), that is, the interplay of reaction and anticipation, is what makes or breaks a forecast.

Our own existence is one of never-ending change. Implicit in this dynamic condition of the living are a) the impossibility of accurate forecasting, and b) the possibility of improving the prediction of physical phenomena, to the extent that we can separate the physical from the living.

Our guesses, expectations, predictions, and forecasts—in other words, our learning in a broad sense, as I defined learning in the methodological introduction—co-affect our actions and affect our pragmatics. Each of them, in a different way, partakes in shaping actions. Their interplay makes up a very difficult array of factors impossible to escape, but even more difficult to account for in detail.

Mutually reinforcing guesses, expectations, predictions, and forecasts corresponding to a course of events for which we have effective descriptions allow us to proceed successfully in some of our actions. In other cases, they appear to cancel each other out, and thus undermine the action, or negatively affect its outcome. Learning and unlearning (which is different from forgetting) probably need to be approached together. Indeterminacy can be experienced as well. It corresponds to descriptions of events for which we have insufficient information and experience, or lack of knowledge. They can also correspond to events which by their nature seem to be ill defined. We react *and* anticipate. This conjunction defines how effective we can be.

Self-awareness, Intentionality, and Planning

We human beings are what we do (the pragmatic foundation of identity, cf. [15] pp. 258ff). The only identifier of our actions is their outcome. This is an instantiation of our identity at the same time. The question, "What do you do?" cannot be answered with "I anticipate," followed, or not, by an object, such as "I anticipate that an object will fall," or "I anticipate my wife's arrival," or "I anticipate smelling something that I never experienced before." Anticipation is a characteristic of the living, but not a specific action or activity. We do not undertake anticipation. It is not a specific task. Anticipation is the result of a variety of processes. We are *in* anticipation. As an outcome, anticipation is expressed through consequences: increased performance—an anticipated tennis serve is returned; danger—such as a passing car—is avoided; an opportunity—in the stock market, for instance—is used to advantage. Implicit in the functioning of the living, anticipatory processes result in the proactive dimension of life. This is where identity originates. Anticipatory processes are defined in contrast to reaction. Characteristic of the deterministic sequence of action-reaction defined in physics, reaction is the expression of our physical nature. Identity is expressed in the unity of the reactive and proactive dimensions of the human being. It appears as a stable expression, but actually defines change. It is the difference between what we seem to be and what we are becoming as our existence unfolds over time. Identity is affected by, but is not the outcome of, learning.

No matter what we do, the doing itself—to which explicit and implicit learning belongs—is what defines our unfolding identity. The outcome is the expression of our physical and intellectual abilities. It also reflects our knowledge and experience. The expression of goals, whether they are specifically spelled out or implicitly assumed, affects the outcome of our actions as well. We identify here the process through which our existence is preserved at the lowest level—not unlike the phototropic mono-cell and progressing all the way up to the human being. But at a certain level of life organization and complexity, the preservation drive assumes new forms through which it is realized. Anticipation is the common denominator. However, the concrete aspect of how it is eventually expressed—i.e., through self-awareness, intentionality, or in the activity we call "planning"—changes as the complexity of the processes through which the living unfolds increases.

Anticipation at the level of preserving existence is unreflected. Facial expression in anticipation of an action is a good example here, too. It seems that facial expression is not defined on a cultural level but is species wide (cf. Ekman [35], Gladwell [36]). It is not a learned expression. We can control our facial expression to an extent, but there is always that one second or less in which control is out of the question. Intentionality is always entangled with awareness—one cannot intend something without awareness, even in vague forms. But this awareness does not automatically make human expressions carry anticipations more than the expression of the rest of the living. The difference is evident on a different level. Humans reach self-awareness; the mind is the subject of knowledge of the mind itself. As such, we eventually recognize that our faces "speak" before we act. They are our forecasts, the majority of these involuntary. Those intent on deciphering them obtain access to some intriguing anticipatory mechanisms, or at least to their expression.

Planning, expressed through policymaking, management, prevention, logistics, and even design, implies the *ante* element—giving advance thought, directing towards something, looking forward, engaging resources (including the self). Moreover, it implies understanding, which resonates with the initial form of the word denoting anticipation: *antecapere*. As such, the activity through which human beings identify themselves as authors of the blueprint of their actions takes place no longer at the object level, but on a meta-level. It is an activity of abstracting future actions from their object. It is also their definition in a cognitive domain not directly associated with sensory input, but rather with what we understand, with knowledge. Plans synthesize predictive abilities, forecasting, and modeling. A plan is the understanding of what is planned, expressed in goals and means to attain these goals, as well as the time sequence for executing the plan. A plan is a timeline and a script for interactions indexed to the timeline. To what we call *understanding* belong goals, the means, the underlying structure of the endeavor (tasks assumed by one person, by several, the nature of their relation, etc.), a sense of progression in time. As such, they report upon the physical determination of everything we do and of the anticipatory framework. In every plan, from the most primitive to the utmost complex, the goal is associated with the reality for which the plan provides a description (a theory), which is called *configuration space*. If it is a scientific plan, such as the exploration of the moon or the genome project, the plan describes where the "science" actually resides, where those equations we call descriptions are "located." If it is a political plan, or an education plan, the configuration space is made up of the people that the plan intends to engage. Our own description of the people, like the mathematical equations of science, is relative. They "shape" the configuration space, and, within that space, the interactions through which people learn from each other.

The plan also has to describe the time-space in which the goal pursued will eventually be embodied. This is a manifold, towards which the dynamics of our own actions and the actions of those with whom we interact (social context) will move us. Again, in science, this is the landing on the moon, the map of the human gene, a new educational strategy or, in politics, the outcome of equal opportunity policies—to name very few examples. All of these are anticipations projected against the background of our understanding the world as one that unites the physical and the living. Plans spell out the variables to be affected through our actions, and the nature of their interrelationships. Quite often, plans infer from the past (the reactive component) to the future (proactive component). They also project how the future will eventually affect the sequence of ensuing current states. Planning cannot be disassociated from self-regulation, i.e., from the inner dynamics of phenomena and their attractors. These attractors are the states into which they will settle. They are the descriptions of self-organizing processes, their eventual destination, if we can understand it as a dynamic entity, not a static finality. Planning sets the limits within which adaptive processes are allowed. Each plan is in effect an expression of learning in action, the very goal pursued having the nature of an attractor in a dynamic system.

Anticipation and Learning

Learning is an active pursuit of regularities in whatever we do, for instance: meeting a bear, falling in a lake, walking in the dark (to remain with the realm of fear), or smelling something that never existed before (and for which we accordingly have no genetic information or predisposition). But learning is also the realization of time cycles (day and night, seasons), natural cycles (related to our environment), patterns in space, patterns in human interactions, and in the interactions among various organisms. Such regularities appear as independent of our own activity, that is, independent of our self-constitution. They seem to be necessary in nature. For example, night must necessarily follow day; we cannot avoid this. Accordingly, the anticipation of night (in the form of rhythm changes in the organism), or of seasons (the thickening of animal fur, falling leaves, changes in metabolism), takes the appearance of something unavoidable. A change in the space through which we move exemplifies the same in respect to our space-based experiences. Our pace changes when a hill is followed by a long level pasture, or even when the "tactility" of the road changes from a smooth, dry surface to a rough, uneven surface. The degree of necessity is expressed in the regularities we extracted and became aware of, that is, in our learning. This learning can be further transferred to machines (e.g., cars, to remain in the domain of mobility).

Here a major question has to be posed. If we want to understand how learning is expressed in the form of anticipation, we need to be able to describe what partakes in the "forecast" leading to anticipatory behavior. The hypotheses to entertain are obvious: Anticipation might be associated with conditioning through previous responses—we do not put our hands into fire in order to find out that they can get burned. Anticipation might result from learning experiences. Or, not to be too hastily discarded, even after accounting for Elsasser's [10] storage-free memory model (based on Bergson's [40] considerations of memory), anticipation might be embryonically predefined and influenced by the subsequent dynamics of the organism unfolding in a given environment. Maturing and ageing, with all that these entail in bodily and cognitive changes, that is, in new learning patterns, is the clearest example (cf. Haylick [41]). Here we have a very powerful illustration of the dynamics affecting the relation between anticipation and learning as the body increasingly returns to its "physicality."

Anticipation as Agent

The agency of anticipation is ultimately about supporting the life of an organism. Decreased anticipation corresponds to a decreased living dynamics. As we have just seen, the living returns to the merely physical, which is the substratum of all there is. In some convoluted way, and by focusing on anticipation, we've transcended the question of modality ("How?")—in particular "How does anticipation take place?" It is progressively replaced by the fundamental question, existential in nature, of "Why?"—indeed, "Why does the living individual lose its 'living' characteristics?" By its own condition, learning is driven by this very "Why?" question. To repeat: the answer to "Why?" is the finality called life. We also integrated conditioning—a characteristic of the living very well documented in science (psychology, cognitive science, neuroscience, etc.)—learning, and the predefined characteristics (embryonically and otherwise) into anticipation. The living learns, but some of its anticipatory characteristics are inherited. Better yet, they are implicit in the condition of being alive, or implicit in survival. Adventure, as an inclination, is telling more of survival than of curiosity—unless curiosity itself is related to survival.

It is really difficult to distinguish between what seems to be learned and what is expressed as a new hypothesis. The variability in such activities is the result of a delicate interplay between conditioning through previous responses and learning, from others or from experience. For example, "When the pension funds sell a stock, it is time to sell!"—the participants in the market "guess," "expect," "forecast," or "predict" a future state that becomes the future state of the market through which they identify themselves. In many instances, variability and the anticipation of the sense and magnitude of change can be the result of noise. Rumors have their own dynamics, as does an instable political situation.

Example: A Hybrid Anticipatory Control Mechanism. The "Learning" Automobile

Driving a car involves reaction-based knowledge (acquired through experience, instructions, interaction with other drivers, etc.) and anticipatory skills (sometimes manifested as anticipatory behavior). The hybrid anticipatory control mechanism integrates automatic car supervision and control and human performance. A simple architecture for implementation in the form of intelligent agent technology was suggested by Davidsson et al [42].

In this architecture, the model of the car operating in faster than real time (the module described as model) allows predictions to be turned into specific actions. The model affects the system's performance: the prediction of a curve results in lowering speed, for instance.

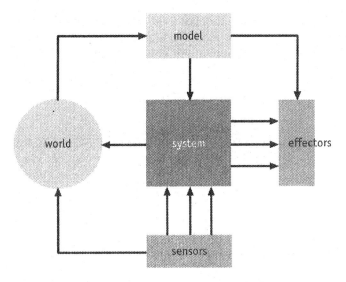

Fig. 1. Anticipation-based architecture for an agent-controlled mechanism.

Here we have two concurrent processes: 1) a reactive process at the object level (the system controlled is the car); b) a predictive process, with anticipatory characteristics, at a meta-level (the model). The integration of the two processes is not trivial. Once we implement an improved architecture of selections corresponding to progressive learning (slow down in a curve, if the road is icy, the slow-down has to be even greater; when driving in heavy traffic on an icy road, deceleration has to be accompanied by a different placement of the car in respect to the median; and so on),

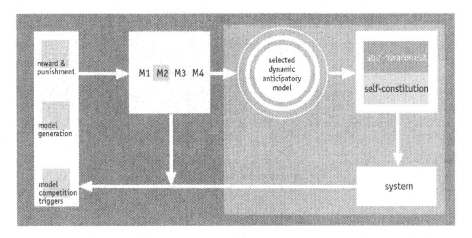

Fig. 2. Anticipative mechanism based on competition among models and the reward mechanism.

complexity increases. If we introduce competition among models and a reward mechanism, we can augment the adaptive qualities of the integrated control mechanism. In this case, alternatives are generated—some probabilistic (integrating knowledge extracted from statistics), such as "informed guesses," expectations (driving up a hill, one can expect a slower speed), forecasts (after rain, the system can forecast a longer brake-path), and predictions.

This architecture makes the following available:
- a space of possibilities (in Zadeh's [43] sense) in the model generation module
- conflicting possibilities: for each situation, a certain possibility is better than the others
- a selection and reward mechanism: learning is stimulated by allocating resources to the "winning" mechanism.

The processes with anticipatory, predictive, and forecasting characteristics are described through

Control = function of (past state, current state, future state) system

Knowledge of future states is a matter of possibilistic distributions. The anticipated performance (cf. [44] p. 281) and the actual performance are related. Their difference, together with the reward mechanism, guides the learning component.

Functioning under continuously changing conditions means that control mechanisms will have to reflect the dynamics of the activity. Without learning, this is not possible. If we finally combine the automated part and human performance, expressed in driver behavior features, we arrive at an improved architecture:

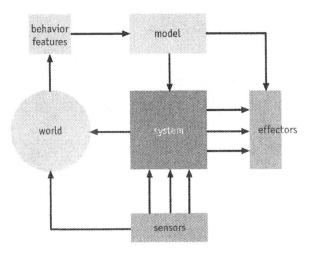

Fig. 3 . Hybrid control mechanism in which driver's behavior features are a source of "learning" for the car

Essential in this new architecture are the indexed behavior features and the mechanism for extracting regularities from them, i.e., the learning. The "learning" car is thus one which combines its own dynamic (modified, evolving knowledge) and that of the driver (anticipation included). This example is only an indication of ways through which implementations of anticipatory characteristics can be achieved. A more detailed analysis guides current research efforts and cooperative projects with industry.

In Lieu of Concluding Remarks

For those who hope that by better understanding anticipation we could improve our prediction and forecast performance, a *caveat:* Anticipatory processes are not reducible to deterministic sequences. Once we accept this, we automatically also accept that although the past remains important, we can no longer act on the assumption that the past contains the present and the future. Moreover, we can no longer rely exclusively on a scientific description according to which the past alone determines the present and the future. We can, however, improve prediction and forecasting by facilitating improved anticipation. This can result from a better knowledge of the possibility space: all that is possible in a given context. This knowledge extends from better methods for evaluating the weight (significance) of possibilities, for dynamically pruning what is of lesser impact from the possibility space, to learning how to handle the imprecise knowledge that we have, instead of searching for more data and higher precision. This aspect, of the relation between anticipation, human performance, and soft computing, will require our attention beyond the usual inclination to improve the performance of our programs, or to improve the hardware in order to handle more data with higher precision.

Only one more word at this juncture: Planning, ideally a practical activity of integrating reaction and anticipation, can benefit from improved prediction and forecasting. It can benefit more, however, from learning, especially learning associated with anticipatory generalization. Based on this understanding, we have to pursue the more specific levels at which anticipation—the proactive path—and all the physical processes described in relation to it—the reactive path—eventually merge. This is a task of an order of magnitude well beyond a context of understanding for the subject of anticipation.

Acknowledgments

Research supported by the DFG (German Science Foundation) and ANTE-Institute for the Research of Anticipatory Systems. The author wishes to express gratitude to the reviewers and to Professor Martin Butz, one of the editors of this volume, for the frank evaluations and for the suggestions that eventually helped me improve the structure of this text. Obviously, I bear responsibility for the viewpoint expressed and for the scientific line of arguments herewith advanced.

References

1. Clark, R.W.: Einstein. The Life and Times. Avon, New York (1972)
2. Peirce, Charles S.: CP 2.222. In: Hartshorne, C., Weiss, P. (eds): The Collected Papers of Charles Sanders Peirce, Vol. I-VI. Harvard University Press, Cambridge (1931). Per convention, Peirce is cited by volume number:article number.
3. Rosen, Robert: Anticipatory Systems. Pergamon Press, New York (1985)
4. Rosen, Robert: Life Itself: A Comprehensive Inquiry into the Nature, Origin, and Fabrication of Life. Columbia University Press, New York (1991)
5. Rosen, Robert: Essays on Life Itself. Complexity in Ecological Systems. Columbia University Press, New York (2000)
6. Nadin, Mihai: Mind—Anticipation and Chaos. Belser Presse, Zürich Stuttgart (1991), based on lecture: Mind—Intelligence is Process, Ohio State University (1988)
7. Dubois, Daniel, Resconi, G.: Hyperincursivity: A New Mathematical Theory. Presses Universitaires de Liège, Liège (1992)
8. Nadin Mihai: Anticipation: The Cause Lies in the Future, Lars Mueller Publishers, Baden Switzerland (2003)
9. Schultz, Wolfram: Multiple reward signals in the brain. In: Nature Reviews/Neuroscience, vol. 1, December (2001) 199-207
10. Elsasser, Walter M.: Reflections on a Theory of Organisms. Johns Hopkins University Press, Baltimore (1998). Originally published as Reflections on a Theory of Organisms. Holism in Biology. Orbis Publishing, Frelighsberg, Quebec (1987)
11. Schall, Jeffrey: Neural Basis of Deciding, Choosing, and Acting. In: Nature Reiews Neuroscience, vol. 2 January (2001) 33-42
12. Rigney, David R., Goldberger, Ary L., Ocasio, Wendell C., Ichimaru, Yuhei, Moody, George B., and Mark, Roger G.: Multi-Channel Physiological Data: Description and Analysis (Data Set B). In: Weigend, Andreas S. and Gershenfeld, Neil A. (eds.): Time Series Prediction: Forecasting the Future and Understanding the Past. Addison Wesley Longman (1994)
13. Ernest, P.: The one and the many. In: Steffe, L. and Gale, J. (eds.): Constructivism in Education. Lawrence Erlbaum Associates, Inc., New Jersey: (1995) 459-486
14. Novak, Joseph and Gowin, D. Bob: Learning How to Learn. Cambridge University Press, Cambridge, UK and New York (1984, 1999)
15. Nadin, Mihai: The Civilization of Illiteracy. Dresden University Press, Dresden (1997)
16. Adolph, Edward F.: Origins of Physiological Relations. Academic Press, New York (1961)
17. Cannon, Walter B.: The Wisdom of the Body (2nd ed.). Peter Smith Publisher, New York (1939)
18. Trzebski, Andrzej: Baroreceptor Reflex Revisited (editorial comment). in: Polish Heart Journal, vol. L, 5, November (1999) 405
19. Nadin, Mihai: Anticipation—A Spooky Computation. In Dubois, D. (ed.): CASYS, International Journal of Computing Anticipatory Systems, Partial Proceedings of CASYS 99, Liege: CHAOS , Vol. 6 (1999) 3-47
20. Meegan, Daniel V., R.N. Aslin, and R.A. Jacobs. Motor timing learned without motor training. In: Nature/Neuroscience, vol. 3, No. 9 September (2000) 860-862
21. Carlsson, Katrina, Petrovic, Skane, S., Petersson, K.M., Ingvar, M.: Tickling Expectations: Neural Processing in Anticipation of a Sensory Stimulus. In: Journal of Cognitive Neuroscience 12, (2000) 691-703
22. Serrien, Deborah J., Mario Wiesendanger. Role of the Cerebellum in Tickling. Anticipatory and Reactive Free Responses. Journal of Cognitive Neuroscience 11, 1999, pp. 672-681.
23. Braun, A.R., H. Guillemen, L. Hosey, M. Varga. The neural organization of discourse, Brain, Vol. 124, No. 10, October 2001, pp. 2028-2044.

24. Guenther, F.H.: Neural modeling of speech production. In: Proceedings of the 4th International Nijmegen Speech Motor Conference, Nijmegen (2001)

25. Max, Ludo: Stuttering and Internal Models for Sensorimotor Control: Deriving Testable Hypotheses for a Theoretical Perspective. in: Maassen, B., Kent, R., Peters, H.F.M, Lieshout, P. van, and Hulstijn eds): Speech Motor Control in Normal and Disorderd Speech. Oxford University Press, Oxford (forthcoming)

26. Gallese, Vittorio: The Inner Sense of Action. Agency and Motor Representations. In: Journal of Consciousness Studies, 7:10 (2000) 23-40

27. Wiesendanger, Mario and Serrien, Deborah J.: Toward a physiological understanding of human dexterity. In: News in Physiological Sciences, vol. 16, no. 5, October (2001) 228-233

28. Wolpert, Daniel M. and Ghahramani, Youben: Computational principles of movement neuroscience. In: Nature/Neuroscience Supplement, vol. 3, November (2000) 1212-1217

29. Foerster, Heinz von: Der Anfang von Himmel und Erde hat keinen Namen (2nd ed.). Döcker Verlag, Vienna (1999)

30. Pearl, Judea: Causality. Models, Reasoning and Inference. Cambridge University Press, Cambridge (2000)

31. Zadeh, Lotfi A.: An Outline of Computational Theory of Perceptions Based on Computing with Words. In: Sinha, N.K., Gupta, M.N., and Zadeh, L.A. (eds.): Soft Computing & Intelligent Systems. Academic Press, New York (2000) 3-22

32. Minsky, Marvin: The Virtual Duck and the Endangered Nightingale. In: Digital Media, June 5, (1995) 68-74

33. Fletcher, P.C., Anderson, J.M., Shanks, D.R., Honey, R., Carpenter, T.A., Donovan, T., Papadakis, N., Bullmore, E.T.: Responses of human frontal cortex to surprising events are predicted by formal associative learning theory. In: Nature Neuroscience, 4:10, (October, 2001) 1043-1048

34. Glasersfeld, Ernest von: Radical Constructivism: A Way of Knowing and Learning. The Falmer Press, London Washington (1995)

35. Ekman, Paul, Rosenberg, E.L. (eds.): What the Face Reveals: Basic and Applied Studies of Spontaneous Expression Using the Facial Action Coding System (FACS). Oxford University Press, New York (1997)

36. Gladwell, Malcolm: The Naked Face. Can experts really read your thoughts? In: The New Yorker (August 5, 2002) 38-49

37. Dayan, Peter, Kakade Sham, and Montague, P. Read: Learning and selective attention. In: Nature Neuroscience, 3 (2000) 1218-1223

38. Nicolelis, Miguel A.: Actions from thoughts. In: Nature 409 (2001) 403-407

39. Ekman, Paul: Universals and cultural differences in facial expressions of emotion. In: Cole, J. (ed.): Nebraska Symposium on Motivation 1971, Vol. 19, University of Nebraska Press, Lincoln (1972) 207-283

40. Bergson, Henri: La matière et la mémoire (1896) (The English edition, cited by Elsasser, was published as Matter and Memory, Allen and Unwin, London)

41. Haylick, Leonard: The future of ageing. In: Nature 408, November (2000) 267-269

42. Davidsson, Paul, Astor, Eric, and Ekdahl, Bertil: A framework for autonomous agents based on the concept of anticipatory systems. In: Cybernetics and Systems. World Scientific, Singapore (1994) 1427-1434

43. Zadeh, Lotfi A.: Fuzzy sets as a basis for a theory of possibility. In: Fuzzy Sets and Systems 1 (1978) 3-28

44. Ragheb, M., Tsoukalas,L.: A coupled probability-possibility method for decision-making in knowledge-based systems. In: S. C-Y. Lu, S.C-Y, Commanduri, R. (eds.): *Knowledge-Based Expert Systems for Manufacturing* New York: ASME, New York (1986)

Anticipatory Behavioral Control

Joachim Hoffmann

Department of Cognitive Psychology,
University of Würzburg, Germany
hoffmann@psychologie.uni-wuerzburg.de

Abstract. This contribution introduces the psychological learning framework of anticipatory behavioral control (ABC). Departing from the premise that almost all behavior of humans and higher animals is goal oriented, the framework proposes that (1) a voluntary action is preceded by a representation of the to-be-attained effect(s), (2) learning of such effect representations is triggered by the comparison of predicted and actual effects resulting in the primary learning of action-effect relations, (3) situational context is integrated secondarily, (4) action-effect representations are activated by the need or desire of an effect-related goal, and (5) conditionalized action-effect relations can also be activated by contingent stimuli. The framework is supported by a variety of experimental studies and an extensive literature survey.

1 Control of Voluntary Behavior: The Challenge

Almost all behavior is purposive or goal oriented. People behave, for example, in order to cross the street, to open a door, to ring a bell, to switch on a radio, to fill a cup with coffee etc. Animals likewise behave to attain various goals as for example to escape from an enemy, to catch a prey, to feed their descendents, etc. If one is willing to accept that behavior is basically goal oriented, and I think one should do so, the question arises how behavior is determined in such a way that the various goals, organisms strive for are really attained. Already more than 150 years ago, Hermann Lotze acknowledged that this question indeed addresses an intricate problem:

> We understand that we can will but not perform. Rather, our will is connected with processes that 'mechanically' determine certain movements of our limbs. We can do nothing else than to evoke such states in our mind from which autonomous processes subsequently depart that let our body move in agreement with our purposes, and that are completely impenetrable to conscious experience [37, p.288, shortened translation by the author].

Thus, in the saying of Lotze, science is challenged to designate the states from which the movements of the body are finally determined and to elucidate the mechanisms by which these states are "mechanically" transformed to appropriate movements, i.e. to the required muscle contractions.

M. Butz et al. (Eds.): Anticipatory Behavior ..., LNAI 2684, pp. 44–65, 2003.

1.1   The Ideo-Motor Principle

At the time of Lotze the only access to this issue was introspection which consistently led the leading figures at that time to the assumption that anticipations of the desired sensory consequences precede voluntary acts (e.g. [18, 27, 28, 37, 39]). For example, William James stated:

> An anticipatory image, then, of the sensorial consequences of a movement, plus (on certain occasions) the fiat that these consequences shall become actual, is the only psychic state which introspection lets us discern as the forerunner of our voluntary acts.[28, p.1112]

The additional assumption, that anticipations of sensory effects do not only precede but also determine voluntary movements was the basic tenet of the Ideo-Motor Principle (IMP). Such a determination necessarily presupposes that movements become connected to their contingent effects; otherwise effect-anticipations couldn't have the power to address the movements they were effects of. How such action-effect learning may take place was already proposed by Johann Friedrich Herbart [21] as early as 65 years before William James propagated the IMP:

> Right after the birth of a human being or an animal, certain movements in the joints develop, for merely organic reasons each of these movements elicits a certain feeling ... In the same instance, the outside senses perceive what change has come about... If, at a later time, a desire for the change observed earlier arises, the feeling associated with the observation reproduces itself. This feeling corresponds to all the inner and outer states in nerves and muscles through which the intended change in the sphere of sensual perception can be brought about. Hence, what has been desired actually happens; and the success is perceived. Through this, the association is reinforced: the action, once performed, makes the following one easier and so on [21, p. 464, rough translation by the author].

In short, Herbart assumed that bidirectional connections between representations of movements and representations of their concrete sensory effects are formed so that the desire or anticipation of formerly experienced sensory consequences attain the power to address the movements they were effects of. Furthermore, he assumed that producing a desired or anticipated concrete effect acts like a reinforcement of the corresponding action-effect association, irrespective of whether the effect is a rewarding stimulus or not.

1.2   Behavioral Control by Stimuli

The assumptions of the IMP are contrary to what Behaviorism has claimed and what later became a tacit consensus in Cognitive Psychology, namely that behavior and psychic processes in general are mainly determined by stimuli.

According to the behaviorist doctrine, instrumental behavior is to be attributed to stimulus-response (S-R) instead of response-effect (R-E) associations, whereby, according to Thorndike's "Law of Effect", S-R associations were assumed to result from a reward of the response in the presence of the stimulus [60, 67, 66]. Thus, behavioral effects were assumed to be of importance merely with respect to their rewarding function (reinforcement learning of S-R associations, cf. [59]) but it was not assumed that the concrete effects become integrated into the emerging behavior-determining representations. Consequently, instrumental behavior was considered as habitual reflexes to training situations which ensure the most expected gain but not as purposefully willed actions in order to reach a particular outcome.

When in the late 50ies of the last century cognitive psychology displaced Behaviorism as the leading paradigm of general psychology, research interests changed from the formation of S-R associations to the formerly neglected processes which take place between the stimulus and the response (cf. [38]). From the beginning on, cognition was defined as referring "[...]to all the processes by which the sensory input is transformed, reduced, elaborated, stored, recovered, and used" [40, p.4].

Thus, mental states like a "desire for a change with associated feelings" or an "anticipatory image of sensorial consequences" are neither from a behaviorist nor from cognitive perspective respectable candidates for a determining cause of overt behavior. Instead, stimuli are more or less explicitly considered as the driving force behind all examined processes. In view of the very fact that almost all behavior is oriented not only to receive a reward but also to reach particular outcomes, it is at least astonishing that 'what behavior concretely leads to' has received so little scientific attention for so long time. Anthony Dickinson recently called it even perverse that knowledge about the instrumental contingency between action and outcome should play no role in the performance of the action [12, p.48].

1.3   The Primacy of Action-Effect Learning

According to the IMP, it is consistently reinforced R-E relations instead of rewarded S-R relations what primarily counts in the acquisition of behavioral competence. In animal learning the impact of R-E associations on behavior is indeed well established:

If animals are trained with only one reinforcer (E), it is difficult to distinguish the contribution of S-R and R-E associations on the resulting instrumental behavior. If, for example, a rat is trained to press a lever for food, it is likewise possible that lever pressing is triggered by the situation (S) as well as by a desire for the food pellets (E). Thus, in order to disentangle the influence of S-R and R-E associations, multiple R-E contingencies have been trained and subsequently, the value of a certain outcome has been changed. If during the initial training only S-R habits would have been established, the propensity to perform the formerly reinforced responses should not be changed by the manipulation of outcome values. By contrast, if training has led also to the formation of R-E associations, the propensity to perform the different responses should be altered in correspondence to the changed value of their outcomes.

Consider, for example, a study by Colwill and Rescorla [9]: Rats were first separately reinforced with food pellets after performing R1 and with a sucrose solution after R2. Once instrumental training had occurred, one of the two reinforcers (outcomes/effects) was devalued by associating it with a mild nausea. Finally, the rats were given a choice between the two responses, but with all outcomes omitted. In this test-phase rats showed a clear suppression of performing the response the outcome of which had been devalued. Obviously, the rats had not only associated two responses with a situation wherein these were reinforced (S-R1 and S-R2), but they had also learned which response leads to which outcome (R1-food pellets, R2-sucrose solution). Consequently, the rats avoided in each situation respectively that response that they know to be followed in this context by the devalued outcome.

There are numerous studies which have used the devaluation technique under yet more sophisticated settings (cf. [10, 12, 42, 47, 48, 49]). Although there is evidence also for the contribution of S-R and S-E associations on the resulting behavior, the evidence for the formation of response-effect associations is by far the strongest. Thus, it seems appropriate to conclude that associations between responses and their concrete effects are a central part of the representations which are formed by animals in instrumental learning.

Since the late seventies response-effect or response-outcome learning has also been examined in humans [2, 3, 13, 51, 52, 64, 65]. Typically, participants were presented with a number of trials on each of which they could perform or not perform a particular response (e.g. a key-press) and subsequently observe whether or not a certain effect (e.g. a light) appeared. The critical variation concerns the contingency between the response and the effect, i.e. the probability $p(E/R)$ of an effect (E) given the performance of the response, and the probability $p(E/R)$ of an effect given that the response is not performed. Participants were to judge the effectiveness of the action in controlling the effect. Although judgments have been shown to be biased by several factors they are generally found to be highly correlated with the difference between both probabilities $p = p(E/R) - p(E/R)$, i.e. with one of the normative measures of the contingency between two variables.

Although this finding clearly indicates that humans are sensitive to response-effect contingencies, it hardly says much about the contribution of R-E learning to instrumental behavior, as the experimental settings differ too much from the conditions of ordinary goal oriented behavior. First, subjects continuously watch the contingency between only one response and one effect, whereas humans usually aim for steadily changing goals facing multiple R-E contingencies. Second and more important, subjects are explicitly instructed to judge the strength of the experienced R-E contingencies, and it is by no means surprising that they adopt appropriate strategies in order to do so. In contrast, in ordinary behavioral settings the task is not to judge R-E contingencies but to learn what to do in order to attain various goals. Thus, it is questionable whether participants learn in the same way about multiple R-E contingencies as they learn if required to judge a single R-E contingency.

In a recent series of experiments, Stock and Hoffmann [55] explicitly compared the impact of S-R and R-E relations on the acquisition of behavioral competence in humans. Participants accomplished a computer-controlled learning task which required them to learn which action achieves which goal under which start condition, similar to what people permanently have to learn under natural circumstances. In each trial first, a start symbol and after a short delay a goal symbol was presented, both taken from the same set of four different symbols (A, B, C, and D). Furthermore, four different actions were available, consisting of pressing one of four keys. Pressing a key resulted in the presentation of another effect-symbol (see Figure 1). Participants could freely choose among the four keys in each trial but they were requested to select respectively that key that would re-produce the actual goal symbol in the presence of the actual start symbol. If the selected key indeed resulted to the current goal symbol, an uprising melody was played, as a reinforcement signal so to say

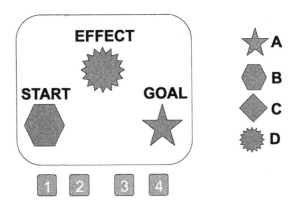

Fig. 1. Illustration of settings in an experiment by Stock and Hoffmann [55]: Participants were requested to press the one of the four keys which would produce an effect-symbol that corresponds to the actual goal-symbol.

In the main experiment, two reinforcement schedules were applied. In one schedule each key was always reinforced in the presence of a particular start symbol. For example, in the presence of Start Symbol A, pressing Key 1 always resulted in the presentation of the actual goal symbol whatever it may be, whereas Key 1 never resulted in the presentation of the required goal symbol if another start symbol was present. In the other schedule each key always was reinforced in the presence of a particular goal symbol. For example, in the presence of Goal Symbol A pressing Key 1 always resulted in the presentation of Symbol A whatever start symbol was presented, whereas Key 1 never resulted in a success if other goal symbols were required. Thus, in both cases participants could accomplish the given task simply by mapping the four keys to either the four start or to the four goal symbols.

Figure 2 presents the main result in terms of the increasing proportion of hits (left) as well as in the mean number of trials to reach a criterion of 16 consecutive hits. All participants with a consistent reinforcement of goal-key relations managed the task rather fast. The mean number of trials to criterion amounted to 120. In contrast, in the group with an equally consistent reinforcement of start-key relations only 3 out of 15 subjects accomplished the task rather late in the session so that the mean number of trials to reach the criterion amounted to 243 trials (the maximum number of trials was 256). The remaining 12 subjects neither reached the criterion nor did they report any insight into the reinforcement rule. Instead, they reported that they despaired more and more and most of them were convinced that there was no solution at all and that they had been tricked.

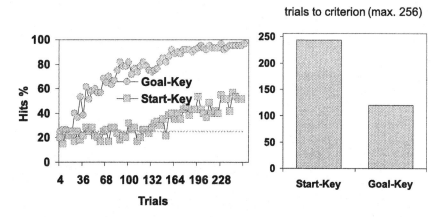

Fig. 2. Proportion of hits (left) and the mean number of trials to criterion (right) in dependence on whether start-key or goal-key relations were consistently reinforced (after [55]).

The basic finding that consistently reinforced relations between the goals and the keys, or more precisely, between the keys and successfully produced effects are learned much faster than equally reinforced Start-Key relations was confirmed in several further experiments. Especially, if the reinforcement schedule was arranged in such a way that each participant received both, 100% reinforcements of certain start-key relations as well as only 75% reinforcements of certain key-effect relations, the selection of keys nevertheless showed to be increasingly stronger dependent on successful key-effect relations than on reinforced start-key relations, despite the fact that the former were only reinforced with a rate of 75%. These finding confirms the primacy of action-effect learning against stimulus-response learning if participants strive for attaining various goals under various start-conditions, i.e. under conditions which are almost always given under natural circumstances.

1.4   The Conditionalization of Action-Effect Relations

As convincing the evidence for the primacy of action-effect learning might be, it would be silly to deny the impact also situations have on behavior. For example, if a bus driver who goes by his private car home, stops at a bus stop, his behavior obviously is not determined by his goal to drive home but rather by perceiving the bus stop, which immediately evokes the habit to stop there [20]. Indeed, several theoretical conceptions in psychology acknowledged the fact that situations may attain the power to evoke associated behavior. For example, Kurt Lewin [36] spoke in this context of the "Aufforderungscharakter" of objects, Narziß Ach [1] coined the term "voluntive Objektion", and James Gibson [15] argued that objects are not only to be characterized by their physical features but also by their "affordances". All these terms refer to the fact that suitable objects often afford us to do the things we mostly do with them and that they immediately trigger habitual behavior if one is already ready to do it. For example, if one intends to post a letter, the sight of a mailbox immediately triggers the act of posting and in driving a car, flashing stop lights of the car ahead immediately evokes applying the brakes.

Thus, in order to reach a more complete picture of the representations which underlie behavioral control, the integration of situational features are also to be concerned. Given the evidence for the primacy of action-effect learning, the situational context presumably becomes integrated either if a particular context repeatedly accompanies the attainment of a particular effect by a particular action or if situational conditions systematically modify action-effect contingencies. Especially the latter deserves attention as it points to the frequent case that the effects of an action change with the situational context, as, for example, the effect of pressing the left mouse button may dramatically change with the position of the cursor. There is no doubt that people learn to take into account critical situational conditions but the topic has not yet received much experimental examination.

An illustrative example for a corresponding study in animal learning provides an experiment by Colwill and Rescorla [10]: Rats were trained with two different actions leading to two different reinforcers under two different situational contexts. For example, in the presence of noise, rats were trained to press a lever for food pellets and to pull a chain for sucrose solution. In the presence of light however, lever pressing resulted in sucrose and chain pulling resulted in food pellets. After this discrimination training one of the two outcomes, let's say sucrose, was devalued by pairing its consumption with a mild nausea. Finally, the animals were given the choice between the two actions in the presence of either the noise or the light. The rats clearly preferred the action that in the presence of the current stimulus results into the non-devalued outcome. In the given example, the rats preferred lever pressing in the presence of noise whereas they preferred chain pulling in the presence of light. The authors concluded that during the discrimination training the experienced action-effect relations have been conditionalized to the respective context in which they were repeatedly experienced. The animals, they argue, have acquired hierarchical S-(R-E) representations. Consequently, these representations enable the animal to anticipate the outcomes of their actions in dependence of the given situation so that they preferably choose that action that in the present situation leads to the relatively more desirable (non-devalued) effect.

A comparable study with human participants has been conducted by [23]. Participants sat in front of a screen, showing the face of a wheel of fortune with four windows. In each window a playing card was to see. By pressing one of four buttons the wheel behind the face could be rotated, either to the right or to the left, slow or fast, depending on the used button. When the wheel comes to stop, four new cards were to see in the four windows. Thus, by pressing a button, participants release the presentation of four new cards.

Participants were requested to turn the wheel in such a way that in the right upper window a card of fortune, let's say an ace, would appear. If such a hit was produced, a fanfare sounded and a stock of points was raised. If subjects failed, the stock of points was reduced. Moreover, the chance of producing hits by the four buttons was manipulated in dependence on the current situational context. For this reason either a lucky pig or a lucky beetle was presented in the center of the wheel (see Figure 3). Finally, the rate of hits was manipulated in dependence on the current symbol (cf. Table 1). There were two buttons that resulted in a hit rate of 20% irrespective of the central symbol. The other two buttons resulted in a total hit rate of 50% but in a hit rate of 100% if used for rotating a particular wheel: For example, Button 1 resulted always into hits with the pig but never with the beetle, and Button 2 resulted always into hits with the beetle but never with the pig. These conditions were designed in order to explore whether participants would learn to prefer Button 1 if a pig is presented and Button 2 if a beetle is presented, just in the same way as rats learned to prefer pulling the chain in one situation and pressing the lever in another situation.

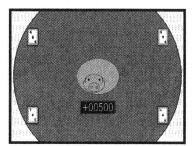 

Fig. 3. Illustration of the two wheels of fortune as they were used in an experiment by Hoffmann and Sebald [23].

Table 1. The percentage of hits in dependence on whether buttons 1 to 4 were used to turn a wheel of fortune with a pig or a beetle as the central symbol.

	button 1	button 2	button 3	button 4
pig	100%	0%	20%	20%
beetle	0%	100%	20%	20%
total	50%	50%	20%	20%

The data reveal that in average the number of choices of respectively that button continuously increased which in the present situation would lead to a hit for sure. Finally, wheels with a pig were almost always rotated with Button 1 and wheels with the beetle were almost always rotated with Button 2. Interestingly, individual data showed that this consideration of the central symbol results from a sudden insight instead being a continuous process. Figure 4 shows the data of a typical participant. In the first blocks the relative frequency continuously increases with which participants use one of the two critical buttons (dotted line). This corresponds to the fact that these buttons result with 50% to a success in comparison to a hit rate of 20% of the two other buttons. Then, in Block 7, the choice of the two already preferred buttons suddenly becomes dependent on the central symbol on the wheel leading to a hit rate of 100%. The data nicely show that the use of the buttons first follows the unconditionalized predictability of a success before the situational context is taken into account. This confirms again the general notion that the formation of action-effect relations is the primary process and that only secondarily action-effect relations become conditionalized to relevant features of the situational context.

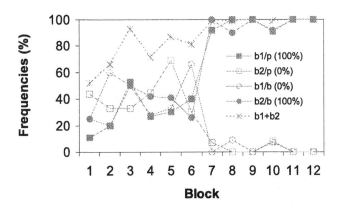

Fig. 4. The relative frequencies with which a typical participant (E.K.) used Button 1 (b1) and Button 2 (b2) to turn a wheel of fortune in dependence on whether a pig (p) or a beetle (b) was the central symbol, plotted against blocks of 25 trials.

2   The ABC Framework: Anticipative Behavioral Control

Hoffmann [22] proposed a tentative framework for the acquisition and use of behavioral knowledge that takes the primacy of R-E learning as well as the secondary conditionalization of R-E relations on relevant situational contexts into account. The framework departs from the following basic assumptions (cf. Figure 5):

1. A voluntary action (A_{volunt}) is defined as performing an act to attain some desired outcome or effect. Thus, a desired outcome, as general and imprecise as it might be specified in the first place, has to be represented in some way before a voluntary action can be performed. Consequently, it is supposed that any voluntary act is preceded by an anticipation of a to-be-attained effect (E_{ant}).
2. The actual effect is compared with the anticipated effect. If there is sufficient coincidence between what was desired and what really happened, representations of the just-performed action and of the experienced effect become interlinked, or an already existing link is strengthened. If there is no sufficient coincidence, no link is formed, or an already existing link is weakened. This formation of integrated action-effect representations is considered as being the primary learning process in the acquisition of behavioral competence.
3. It is assumed that situational contexts (S) become integrated into action-effect representations, either if a particular action-effect episode is repeatedly experienced in an invariant context or if the context systematically modifies the contingencies between actions and effects. This conditionalization of action-effect relations is considered as being a secondary learning process.
4. A "awakening" need or a concrete desire activates action-effect representations whose outcomes sufficiently coincide with what is needed or desired. Thus, anticipations of effects address actions that are represented as being appropriate to produce them. If the activated action-effect representations are conditionalized, the coincidence between the stored conditions and the present situation is checked. In general, an action will be performed that in the actual situational context most likely produces the anticipated effect.
5. Conditionalized action-effect representations can also be addressed by stimuli that correspond to the represented conditions. Thus, a certain situational context in which a certain outcome has been repeatedly produced by a certain action can elicit the readiness to produce this outcome by that action again.

The sketched framework integrates, still rather roughly, important aspects of the acquisition of behavioral competence: First, it considers the commonly accepted fact that behavior is almost always goal oriented instead of being stimulus driven. Second, it is assumed that any effect which meets an anticipated outcome will act as a reinforcer. Consequently, learning is not only driven by a satisfaction of needs but by anticipations which can flexibly refer to any goal.

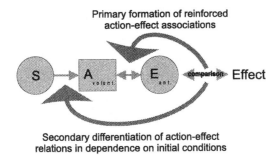

Fig. 5. Illustration of the ABC framework: The acquisition of anticipative structures for the control of voluntary behavior.

Third, the framework considers the given evidence that voluntary behavior is primarily determined by action-effect instead by stimulus-response associations. Finally, also stimulus driven habitual behavior is covered, as it is assumed that action-effect relations become conditionalized and can be evoked by the typical contexts in which they are experienced.

Certainly, all the presumed mechanisms need specification, preferentially by designing a concrete computational model and by comparing the model with data of animal or human learning. The first steps in this direction are already taken. For example, Butz and Hoffmann [6] designed an anticipatory classifier system according to the basic ideas of the sketched ABC framework and used the model to simulate data of action-effect learning in animals. The contributions of Butz and Goldberg in the present volume presents further examples of applications of the ABC framework (cf. also [57, 58]). The given evidence confirms the ABC framework as a fruitful departure point for a further elaboration of the basic processes which underlie the acquisition of behavioral competence in animals and humans.

3   Latent Formation of Effect Anticipations

The ABC framework so far considers the control of goal oriented behavior. However, to address a certain act in order to produce a certain outcome presupposes the existence of action-effect relations as already William James noted:

> When a particular movement, having once occurred in a random, reflex, or involuntary way, has left an image of itself in the memory, then the movement can be desired again, proposed as an end, and deliberately willed. But it is impossible to see how it could be willed before.[28, p.1099]

Remember that also Herbart [21] has argued that actions initially become autonomously associated with their sensory effects, i.e. without the effects being neither desired nor presenting any kind of reward. Thus, the "latent" formation of associations between actions and their correlated effects is to be assumed a necessary precursor for an anticipatory control of voluntary behavior.

In animals, the latent formation of action-effect expectancies has its acknowledged place since the work of Edward Tolman and his collaborators [63, 61, 62]. In humans, the latent or incidental adaptation of behavior to environmental structures became nowadays examined in numerous studies on so called "implicit learning" (cf. for recent overviews [7, 50, 54]). For example, in memorizing letter strings, participants incidentally adapt to grammatical structures of the to-be-memorized strings without being able to report any of the grammatical rules [16, 45, 46]. Likewise, participants respond to structured stimulus sequences increasingly faster than to random sequences although they are apparently unaware of any structural constraints (e.g. [8, 29, 41]). However, almost all studies on implicit learning explored stimulus-stimulus or response-response learning, and only a few studies so far considered incidental action-effect learning.

One instructive example of latent action-effect learning is provided by a recent study of Elsner and Hommel [14]. Participants were first requested to respond to a unique Go-Signal either by pressing a left key with the left index finger or a right key with the right index finger. Both responses were to be selected approximately equally often and in an unsystematic order. The left response resulted in a high-pitch effect-tone and the right response into a low-pitch effect-tone (or vice versa). After this acquisition phase the task switched from a free choice to an ordinary choice reaction time task in which the former effect-tones served as the imperative response signals. In a "nonreversal group" the effect-tones required respectively those responses which they formerly were effects of. In a "reversal group" the stimulus-response assignments were reversed. The data show that responses in the non-reversal group are approximately 50 ms faster than in the reversal group, i.e. participants responded faster if the imperative stimuli required respectively that action they were effects of in the preceding acquisition phase. The authors conclude: "The present experiments suggest an automatic acquisition of action-effect knowledge and the impact of this knowledge on action control." [14, p.239]

Hoffmann et al. [24] applied a similar technique to demonstrate latent action-effect learning in a standard serial-reaction-time task (SRT). Participants responded to asterisks presented at one of four horizontally aligned locations by pressing one of four response keys which were also horizontally aligned. The keys were assigned to the asterisk locations in a spatially compatible fashion, i.e. the response keys from left to right were respectively assigned to the asterisk locations from left to right. Each keystroke triggered the presentation of the next asterisk, such that participants performed a sequence of keystrokes to a sequence of self-triggered asterisks. In the first two blocks the sequence of asterisks and so the sequence of the required responses was random. In six following blocks a fixed sequence of stimuli was cyclically repeated resulting in a cyclic repetition of a fixed response sequence. In a subsequent test block the sequence of stimuli and responses switched back to random before in a last block the fixed sequence was again presented. Typically, reaction times (RTs) and errors continuously decreased with repetitions of the fixed sequence and they increased in the test block. This increase indicates serial learning, as it reveals that the preceding decrease of RTs were due to acquired knowledge about the serial structure of the fixed sequence, which becomes useless if a random sequence is presented again.

Hoffmann et al. [24] argued that serial learning should be improved if each keystroke would produce a contingent tone-effect so that an anticipation of the to-be-produced fixed tone-sequence could be used in order to control the fixed keystroke sequence, just like a pianist may control his strokes by an anticipation of the melody to be played. In the corresponding experiment the four tones of a C Major chord were assigned to the keys from left to right in an ascending order (condition CT). Consequently, if participants responded to the fixed sequence of asterisks they cyclically produced a fixed sequence of tones that could be integrated into a to-be-produced "melody". There were a first control condition in which no tones were presented (condition NT) and a second control condition in which the fixed tone sequence was presented one serial position ahead to the fixed asterisks/response sequence. This manipulation resulted to the same tone-sequence but with the tones no longer contingently mapped to the keys (condition NCT).

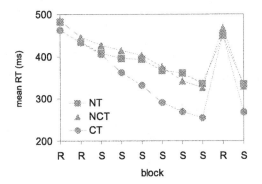

Fig. 6. Mean reaction times (RT) in an SRT task in dependence on whether the responses produced contingent tone-effects (CT), non-contingent tone-effects (NCT), or no tones (NT), plotted against random (R) and structured (S) blocks (after [24]).

Figure 6 presents the mean RTs for the three groups plotted against the blocks. The results show that the mere presentation of an additional tone-sequence does not affect RTs as there were no differences between the conditions with no-tones (NT) and with non-contingent tones (NCT). In contrast, RTs decrease substantially more if the tones were contingently mapped to the keystrokes (CT). As RTs increase in the test block to the same level as in both control groups, the data confirm that the contingent tone-effects improved the acquisition and use of serial knowledge about the fixed sequence. Obviously, key presses and tone-effects became associated so that anticipations of the to-be-produced tones gained control over the to-be-executed fixed sequence of key presses, otherwise their impact on response speed would be hard to be explained (cf. also [17, 56]).

Note, that in both mentioned experiments the tone-effects were completely irrelevant to accomplish the respective task demands. The fact that the effects nevertheless influenced the behavior strongly supports the notion that effects need not to be intended in order to become associated with the actions which they are results of. Rather, it seems that attending the effects and temporal overlap of code activation suffices in order to integrate action and effect representations in a bidirectional connection.

4   From Anticipations to Behavior

The theoretically most challenging claim of anticipative behavioral control regards the notion that actions, i.e. motor commands, are assumed to be addressed by anticipations of their sensory feedback. Thus, it is argued that the causal relationship between actions and their sensory effects is reversed as actions are thought to be caused by anticipations of their sensory effects.

An experimental examination of this basic assumption is an intricate matter as anticipations are difficult to control. However, if any selection of a voluntary movement does indeed require an anticipation of its sensory effects, manipulations of the to-be-expected effects should have an impact on the action which produces these effects. Following this logic, Wilfried Kunde recently provided convincing evidence for the impact, effect anticipations have not only on the selection and initiation but also on the execution of voluntary acts [35, 32, 33, 34, 30].

Kunde started from the well established stimulus-response compatibility effect: If in a choice reaction time experiment the imperative stimuli and the required responses vary on a common dimension (dimensional overlap) compatible S-R assignments are faster accomplished than incompatible assignments (cf. [31]). Consider for example, spatial compatibility: if participants have to respond to left and right stimuli with the left and right hand, they respond faster with the left hand to a left stimulus and with the right hand to a right stimulus than vice versa (e.g. [53]). Ongoing from S-R compatibility Kunde [35] proceeded to argue that if selecting and initiating a response does indeed require the anticipation of its sensory effects, the same compatibility phenomena should appear between effects and responses as between stimuli and responses.

Figure 7 illustrates an example for spatial action-effect compatibility as it was used by Kunde [35]. In the compatible condition each keystroke triggered a spatially compatible effect on the screen, whereas in the non-compatible condition the assignments of the effect locations to the key locations were scrambled. Participants were required to press the keys to imperative color signals, i.e. there was no dimensional overlap between the stimulus set and the response set. The results show that participants responded significantly faster if their responses triggered spatially compatible effects than if they triggered spatially incompatible effects.

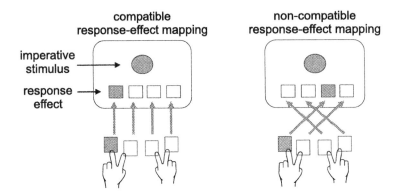

Fig. 7. Illustration of spatially compatible as well as incompatible visual effects of key presses as used by Kunde [35].

Further examples of action-effect compatibility confirm the generalizability of the phenomenon. Consider for example an experimental setting in which participants are requested to press a key either short (dit) or long (dah) in response to a corresponding color signal. Pressing the key triggers either a long tone or a short tone. In the compatible condition the short key press triggers a short tone and the long key press triggers a long tone. In the non compatible condition the response effect assignments were reversed. Response latencies were again significantly reduced in the compatible compared to the non-compatible condition (Kunde, in press). In another experiment responses and effects varied both with respect to intensity. Participants were required to press a key either softly or strongly in response to a corresponding color signal. Pressing the key triggered either a quiet or a loud tone. In the compatible condition the soft key press triggered a quiet tone and the strong key press triggered a loud tone. In the non-compatible condition, the response-effect assignments were reversed. Response latencies were again significantly reduced in the compatible compared to the non compatible condition [34].

Thus, the response-effect compatibility phenomenon is a very robust one. It occurs in the dimensions of space, time, and intensity. As in all these experiments the effects were exclusively presented after the execution of the response, their impact on response latencies proves that representations of the effects were activated before the responses were selected and initiated. The impact, the subsequently appearing effects have on RTs, i.e. on processes that precede the release of the response would be otherwise hard to explain (see [32] for experiments which rule out possible alternative accounts).

The use of response alternatives that differ in intensity also allows a qualification of response execution. Thus, if participants are required to complete a soft or a strong key press the slope with which the force on the key increases or the peak force that is reached provide highly correlated measures of response execution. Kunde [32] used these data in order to explore whether the compatibility of the to-be-expected effect of a key press would not only influence its latency but also its execution. This was indeed the case. The intensity of the effect-tones uniquely affected the peak forces of soft as well as of strong key presses in a contrast like fashion. As Figure 8 illustrates, loud effect-tones reduced and accordingly quiet effect tones intensified the peak forces of intended soft as well as of intended strong key presses.

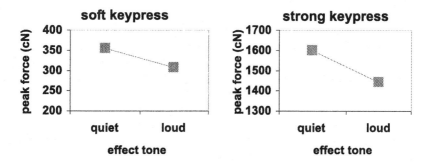

Fig. 8. The impact of quiet and loud effect-tones on the peak forces of intended soft and strong key presses (after [34]).

For an appropriate account of the found contrast it is to be considered that peak forces assesses the intensity of the tactile feedback by which participants start to reduce the force of their hand because they feel the intended force (strong or soft) to be reached. In this view, the data show that participants need less strong tactile feedback to feel a soft key press completed if a loud effect-tone is expected and accordingly that participants need stronger tactile feedback in order to feel a strong key press completed if a quiet effect-tone is expected. Thus, it seems that during the execution of a particular force the anticipated intensity of the effect-tone is somehow allocated with the intensity of the tactile feedback. For example, simply averaging the intensities of the given tactile with the anticipated auditory feedback would suffice in order to account for the found contrast (cf. also [4, 5, 32] for further evidence in support of this speculation).

However it might be, the present data already provides profound evidence that anticipations of the concrete response effects are not only involved in the selection and initiation of voluntary actions but take also part in the control of their execution. The assumption that anticipations of behavioral effects are involved in the online feedback control of movements corresponds well to the recent literature on feed forward mechanisms in behavioral control (e.g. [11]).

5 Review and Outlook

In the introduction to this chapter Hermann Lotze [37] was cited who claimed more than 150 years ago that voluntary behavior is determined by internal states which autonomously let our body move in agreement with our purposes. From this perspective two basic issues are to be treated: First, the internal states from which the appropriate movements result are to be designated and second, the mechanisms by which these states determine the appropriate movements are to be elucidated.

We started our discussion with a reference to the ideo-motor principle (IMP), which was, at the time of Lotze, a common plausible speculation on the determination of voluntary behavior (e.g. [18, 21, 27, 28]). The IMP designated the internal states from which goal oriented movements of the body autonomously depart as being "ideas" of the sensory consequences which these movements shall bring about. "Ideas" attain the power to trigger body movements, it was assumed, by the formation of bidirectional associations between representations of movements and representations of their contingent sensory effects, so that the "idea" of a particular effect, i.e. its anticipation, immediately addresses that movement that formerly brought this effect about.

Ongoing from the IMP, we proposed a general framework of anticipative behavioral control (ABC) which takes also into account the impact, the situational context has on voluntary behavior. According to the ABC framework two learning processes are to be distinguished: A primary process that associates actions, i.e. motor representations with representations of their contingent sensory effects and a secondary process by which action-effect associations become conditionalized to relevant features of the situational context. In performing goal oriented behavior, it is assumed that attaining a desired effect reinforces the experienced action-effect relation. However, action-effect associations are also assumed to become, at least initially, latently formed, i.e. without the effects being explicitly aimed at or presenting a reward. Regarding the secondary process of conditionalization, it was assumed that action-effect relations become associated with situational features that invariantly accompany their realization. Furthermore, conditionalization should take place if the particular outcomes of a particular action change in dependence on the situational context, so that the act sometimes produces the desired outcome and sometimes fails to produce it.

Furthermore, according to the ABC framework an intended voluntary act starts with the anticipation of a desired outcome. If the actor disposes about some experiences in producing this particular outcome (or at least similar ones), the anticipation immediately addresses that act with which it is most strongly associated. If the corresponding action-effect association is conditionalized, anticipations of the stored situational features are also evoked and the addressed action is released if the anticipated conditions are actually met. In performing the act, the anticipated effects are compared with the actual ones and the intention is completed if the circle is closed, i.e. if the actual outcomes sufficiently correspond to the anticipated ones.

From a behaviorist perspective, the notion that un-observable internal states like anticipations of a desired outcome may determine behavior was a sacrilege and so undebatable. When later behaviorism was replaced by the cognitive perspective, internal states became a preferred research object but they still were seen as primarily stimulus driven. Thus it happened that the proper idea of an anticipative behavioral control was not scientifically elaborated for decades (cf. Stock and Stock, in press, for a comprehensive overview of the origins and fate of the IMP). No more than recently the idea attracts the scientific debate anew (e.g. [22, 26, 43, 44]) and inspires substantial experimental research (cf. also [19, 25, 68]).

The reported experimental findings provide profound evidence in support of some of the basic claims of the ABC framework: The primacy of action-effect over stimulus-response learning has been convincingly demonstrated with animals as well as with humans. Also the latent formation of action-effect associations is now a well established fact. Likewise there are data which support the notion that action-effect units become secondarily conditionalized if necessary. Recent experiments also confirm the assumption that voluntary acts are preceded by anticipations of their sensory effects. Moreover, the data show that effect anticipations affect the selection and initiation as well as the execution of voluntary acts what confirms that they not only precede the behavior but also take part in its determination. Altogether, the given evidence encourages for a further theoretical and empirical elaboration of anticipatory behavioral control.

Among the numerous issues which are still to be resolved, the conditions under which action-effect relations are latently formed deserves especially attention. Is it the case that actions are inevitably associated to all sensory effects that contingently accompany their execution or are certain conditions to be fulfilled for the latent formation of action-effect associations? The reported experiments showed that actions become associated with irrelevant effect-tones which participants did not strive for. However, the effect-tones were salient as they were the only effects of the to-be-performed actions which could be hardly ignored. Thus, it might well be that only attended effects become associated. Furthermore, it is to explore to what extent also delayed effects become associated and what degree of contingency is needed for a latent formation of an action-effect association.

Another open issue regards the conditions which provoke an actor to consider the situational context as behaviorally relevant. It is certainly plausible to assume that the situational context is attended if a certain act sometimes is and sometimes is not successful in attaining a certain outcome. However, the conditions under which a failure to attain a particular outcome by a particular act either leads to an attenuation or to a conditionalization of the corresponding action-effect association are still to be specified.

Finally and perhaps most important, the processes by which anticipations are transformed into actions are not yet appropriately understood. Obviously, not every anticipation results into acting. For example, one can very precisely anticipate raising the right hand without doing so. Remember in this context the quotation of William James where he argued that "on certain occasions" it needs not only anticipations but also a "fiat" that the anticipated consequences shall become actual in order to bring the body into move. However, what is a "fiat" and whereof it depends whether it is needed or not? What opens the gate that let a fully prepared action run and what prevents its execution despite full preparation? Likewise, any certain outcome can be attained by mostly numerous actions, so that action-effect relations are almost always ambiguous. By what mechanisms the numerous alternatives by which a certain act can be realized are restricted to the movements which are finally performed? Furthermore, we have not yet distinguished so far between the proprioceptive and the distal exteroceptive consequences of an action although it might well be that anticipations of both types play a different role in the preparation and execution of an intended act.

Thus, we finish as usual with more unsettled issues than we started with. Nevertheless, the given evidence displays the ABC framework as a proper groundwork for a further exploration of what is still a mystery: the learning dependent establishment of structures and processes by which the mind controls the body in such a way that what is desired really happens.

Acknowledgments

The author is grateful for support of the reported work by grants from the German Research Council (DFG).

References

[1] Ach, N.: Über den Willensakt und das Temperament: Eine experimentelle Untersuchung. Verlag von Quelle & Meyer, Leipzig (1910)
[2] Allan, L.G.: A note on measurements of contingency between two binary variables in judgment tasks. Bulletin of the Psychonomic Society **15** (1980) 147–149
[3] Allan, L.G., Jenkins, H.M.: The judgment of contingency and the nature of the response alternatives. Canadian Journal of Psychology **34** (1980) 1–11
[4] Aschersleben, G., Prinz, W.: Synchronizing actions with events: The role of sensory information. Perception and Psychophysics **57** (1995) 305–317
[5] Aschersleben, G., Prinz, W.: Delayed auditory feedback in synchronization. Journal of Motor Behavior **29** (1997) 35–46
[6] Butz, M.V., Hoffmann, J.: Anticipations control behavior: Animal behavior in an anticipatory learning classifier system. Adaptive Behavior (in press, 2003)
[7] Cleeremans, A.: Mechanisms of implicit learning. MIT Press, Cambridge, M (1993)
[8] Cohen, A., Ivry, R., Keele, S.W.: Attention and structure in sequence learning. Journal of Experimental Psychology: Learning, Memory, and Cognition **16** (1990) 17–30
[9] Colwill, R.M., Rescorla, R.A.: Postconditioning devaluation of a reinforcer affects instrumental learning. Journal of Experimental Psychology: Animal Behavior Processes **11** (1985) 120–132
[10] Colwill, R.M., Rescorla, R.A.: Evidence for the hierarchical structure of instrumental learning. Animal Learning & Behavior **18** (1990) 71–82

[11] Desmurget, M., Grafton, S.: Forward modeling allows feedback control for fast reaching movements. Trends in Cognitive Scienc **4** (2000) 423–431

[12] Dickinson, A.: Instrumental conditioning. In Mackintosh, N., ed.: Animal learning and cognition. Academic Press, San Diego, CA (1994) 45–79

[13] Dickinson, A., Shanks, D.R., Evenden, J.: Judgment of act-outcome contingency: The role of selective attribution. Quarterly Journal of Experimental Psychology **36A** (1984) 29–50

[14] Elsner, B., Hommel, B.: Effect anticipation and action control. Journal of Experimental Psychology: Human Perception and Performance **27** (2001) 229–240

[15] Gibson, J.J.: The ecological approach to visual perception. Houghton, Boston, MA (1979)

[16] Gomez, R.L., Schvaneveldt, R.W.: What is learned from artificial grammars? Transfer tests of simple associations. Journal of Experimental Psychology: Learning, Memory, and Cognition **20** (1994) 396–410

[17] Greenwald, A.G.: Sensory feedback mechanisms in performance control: With special reference to the ideo-motor mechanism. Psychological Review **77** (1970) 73–99

[18] Harle, E.: Der Apparat des Willens. Zeitschrift für Philosophie und philosophische Kritik **38** (1861) 50–73

[19] Hazeltine, E.: The representational nature of sequence learning: Evidence for goal-based codes. In Prinz, W., Hommel, B., eds.: Common mechanisms in perception and action. Volume XIX of Attention and Performance. Oxford University Press, Oxford (2002) 673–689

[20] Heckhausen, H., Beckmann, J.: Intentional action and action slips. Psychological Review **97** (1990) 36–48

[21] Herbart, J.: Psychologie als Wissenschaft neu gegründet auf Erfahrung, Metaphysik und Mathematik. Zweiter, analytischer Teil. August Wilhelm Unzer, Königsberg, Germany (1825)

[22] Hoffmann, J.: Vorhersage und Erkenntnis: Die Funktion von Antizipationen in der menschlichen Verhaltenssteuerung und Wahrnehmung. [Anticipation and cognition: The function of anticipations in human behavioral control and perception.]. Hogrefe, Göttingen, Germany (1993)

[23] Hoffmann, J., Sebald, A.: Lernmechanismen zum Erwerb verhaltenssteuernden Wissens [Learning mechanisms for the acquisation of knowledge for behavioral control]. Psychologische Rundschau **51** (2000) 1–9

[24] Hoffmann, J., Sebald, A., Stöcker, C.: Irrelevant response effects improve serial learning in serial reaction time tasks. Journal of Experimental Psychology: Learning, Memory, and Cognition **27** (2001) 470–482

[25] Hommel, B.: The cognitive representation of action: Automatic integration of perceived action effects. Psychological Research **59** (1996) 176–186

[26] Hommel, B.: Perceiving one's own action — and what it leads to. In Jordan, J.S., ed.: Systems theory and apriori aspects of perception. Elesevier Science B.V., Amsterdam (1998) 143–179

[27] James, W.: The Principles of Psychology. Volume 1. Harvard University Press, Cambridge, MA (1981 (orig.1890))

[28] James, W.: The Principles of Psychology. Volume 2. Harvard University Press, Cambridge, MA (1981 (orig.1890))

[29] Koch, I., Hoffmann, J.: Patterns, chunks, and hierarchies in serial reaction time tasks. Psychological Research **63** (2000) 22–35

[30] Koch, I., Kunde, W.: Verbal response-effect compatibility. Memory and Cognition (in press, 2003)

[31] Kornblum, S., Hasbroucq, T., Osman, A.: Dimensional overlap: Cognitive basis for stimulus response compatibility: A model and taxonomy. Psychological Review **97** (1990) 253–270

[32] Kunde, W.: Temporal response-effect compatibility. Psychological Research (in press, 2003)

[33] Kunde, W., Hoffmann, J., Zellmann, P.: The impact of anticipated action effects on action planning. Acta Psychologica **109** (2002) 137–155

[34] Kunde, W., Koch, I., Hoffmann, J.: Anticipated action effects affect the selection, initiation and execution of actions. The Quarterly Journal of Experimental Psychology. Section A: Human Experimental Psychology (in press, 2003)

[35] Kunde, W.: Response-effect compatibility in manual choice reaction tasks. Journal of Experimental Psychology: Human Perception and Performance **27** (2001) 387–394

[36] Lewin, K.: Wille, Vorsatz und Bedürfnis. Psychologische Forschung **7** (1928) 330–385

[37] Lotze, H.: Medizinische Psychologie oder Physiologie der Seele. Weidmann'sche Buchhandlung, Leipzig (1852)

[38] Mandler, G.: Cognitive Psychology. An essay in cognitive science. Erlbaum, Hillsdale, NJ (1985)

[39] Münsterberg, H.: Beiträge zur Experimentalpsychologie. Heft 1. J.C.B. Mohr, Freiburg i.B. (1889)

[40] Neisser, U.: Cognitive psychology. Appleton, New York (1967)

[41] Nissen, M.J., Bullemer, P.: Attentional requirements of learning: Evidence from performance measures. Cognitive Psychology **19** (1987) 1–32

[42] Pearce, J.M.: Animal learning and cognition. 2 edn. Psychology Press, Hove (1997)

[43] Prinz, W.: Ideomotor action. In Heuer, H., Sanders, A.F., eds.: Perspectives on perception and action. Erlbaum, Hillsdale, NJ (1987) 47–76

[44] Prinz, W.: Perception and action planning. European Journal of Cognitive Psychology **9** (1997) 129–154

[45] Reber, A.S.: Implicit learning of artificial grammars. Journal of Verbal Learning and Verbal Behavior **6** (1967) 855–863

[46] Reber, A.S.: Implicit learning and tacit knowledge. Journal of Experimental Psychology: General **118** (1989) 219–235

[47] Rescorla, R.A.: Evidence for an association between the discriminative stimulus and the response-outcome association in instrumental learning. Journal of Experimental Psychology: Adaptive Behavior Processes **16** (1990) 326–334

[48] Rescorla, R.A.: Associative relations in instrumental learning: The eighteenth Bartlett memorial lecture. Quarterly Journal of Experimental Psychology **43** (1991) 1–23

[49] Rescorla, R.A.: Full preservation of a response-outcome association through training with a second outcome. Quarterly Journal of Experimental Psychology **48** (1995) 252–261

[50] Seger, C.A.: Implicit learning. Psychological Bulletin **115** (1994) 163–196

[51] Shanks, D.R.: Continuous monitoring of human contingency judgment across trials. Memory and Cognition **13** (1985) 158–167

[52] Shanks, D.R., Dickinson, A.: Associative accounts of causality judgment. In Bower, G.H., ed.: The psychology of learning and motivation. Volume 21. Academic Press, San Diego, CA (1987) 229–261

[53] Simon, J.R., Rudel, A.P.: Auditory S-R compatibility: The effect of an irrelevant cue on information processing. Journal of Applied Psychology **51** (1967) 300–304

[54] Stadler, M.A., Frensch, P.A.: Handbook of implicit learning. Sage Publications, Thousand Oaks, CA (1998)

[55] Stock, A., Hoffmann, J.: Intentional fixation of behavioral learning or how R-E learning blocks S-R learning. European Journal of Cognitive Psychology **14** (2002) 127–153

[56] Stöcker, C., Sebald, A., Hoffmann, J.: The influence of response effect compatibility in a serial reaction time task. Quarterly Journal of Experimental Psychology (in press, 2003)

[57] Stolzmann, W.: Antizipative Classifier Systems [Anticipatory classifier systems]. Shaker Verlag, Aachen, Germany (1997)

[58] Stolzmann, W.: An introduction to anticipatory classifier systems. In Lanzi, P.L., Stolzmann, W., Wilson, S.W., eds.: Learning Classifier Systems: From Foundations to Applications. Springer-Verlag, Berlin Heidelberg (2000) 175–194

[59] Sutton, R.S., Barto, A.G.: Reinforcement learning: An introduction. MIT Press, Cambridge, MA (1998)

[60] Thorndike, E.L.: Animal intelligence: Experimental studies. Macmillan, New York (1911)

[61] Tolman, E.C.: Purposive Behavior in animals and men. Appleton, New York (1932)

[62] Tolman, E.C.: There is more than one kind of learning. Psychological Review **5b** (1949) 144–155

[63] Tolman, E.C., Honzik, C.: Introduction and removal of reward, and maze performance in rats. University of California, Publications in Psychology **4** (1930) 257–275

[64] Wasserman, E.A., Chatlosh, D.L., Neunaber, D.J.: Perception of causal relations in humans: Factors affecting judgments of interevent contingencies under free-operant procedures. Learning and Motivation **14** (1993) 406–432

[65] Wasserman, E.A., Elek, S.M., Chatlosh, D.L., Baker, A.G.: Rating causal relations: Role of probability in judgments of response-outcome contingency. Journal of Experimental Psychology: Learning, Memory, and Cognition **19** (1993) 174–188

[66] Watson, J.B.: Psychology from the standpoint of a behaviorist. 2 edn. Lippincott, Philadelphia (1924)

[67] Watson, J.: Psychology as a behaviorist views it. Psychological Review **20** (1913) 158–177

[68] Ziessler, M.: Response-effect learning as a major component of implicit serial learning. Journal of Experimental Psychology: Learning, Memory, and Cognition **24** (1998) 962–978

Towards a Four Factor Theory of Anticipatory Learning

Mark Witkowski

Interactive and Intelligent Systems Section
Department of Electrical & Electronic Engineering
Imperial College
Exhibition Road
London SW7 2BT
United Kingdom
{m.witkowski@imperial.ac.uk}

Abstract. This paper takes an overtly anticipatory stance to the understanding of animat learning and behavior. It analyses four major animal learning theories and attempts to identify the anticipatory and predictive elements inherent to them, and to provide a new unifying approach based on the anticipatory nature of those elements based on five simple predictive "rules". These rules encapsulate all the principal properties of the four diverse theories (the four factors) and provide a simple framework for understanding how an individual animat may appear to operate according to different principles under varying circumstances. The paper then indicates how these anticipatory principles can be used to define a more detailed set of postulates for the Dynamic Expectancy Model of animat learning and behavior, and to construct its computer implementation SRS/E. Some of the issues discussed are illustrated with an example experimental procedure using SRS/E.

1 Introduction

This paper takes a particular stance on animat behavior generation and learning. At the heart of this problem is how an animat should select actions to perform, under what conditions and to what purpose. It will argue that the generators of animat behavior have a strong anticipatory or predictive quality, and that learning, and our animal models of learning, should exploit the anticipatory and predictive properties inherent in an animat's structure. The ability of entities, including living organisms and machines, to anticipate future events and be in a position to react to them in a timely manner has long been recognised as a key attribute of intelligence For instance, the discussions between Charles Babbage and Italian scientists in 1840, where the meeting concluded that "*... intelligence would be measured by the capacity for anticipation*" [15].

Recently, a growing number of researchers have identified and emphasized the importance of anticipation as the basis of models of animal learning and behavior. Butz, Sigaud and Gerard [9] categorize four main distinctions between different forms of anticipatory learning and behavior: implicit anticipation, payoff anticipation, sensorial anticipation and state anticipation. This paper focuses on the last of these, systems that exploit the properties of anticipation between states or partial state

M. Butz et al. (Eds.): Anticipatory Behavior ..., LNAI 2684, pp. 66-85, 2003.

descriptions. Stolzmann *et al.* [18] describe a classifier system model based on anticipatory principles, Tani and Nolfi [20] an Artificial Neural Network approach and Witkowski the Dynamic Expectancy Model (DEM, [25], [26], [27]), which places anticipation and prediction at the center of the learning and behavioral process. The anticipatory stance imbues an animat with several important properties. First, the ability to determine possible future situations following on from the current one, thereby anticipating those situations that might be advantageous (or harmful) in anticipation of them occurring. Second, to determine the possible outcome of actions made by the animat, leading directly to the ability to establish chains or plans of actions to satisfy some desired outcome. Third, the ability to rank the effectiveness of each action in its immediate context, independently of any particular goal or task specific reward or reinforcement (by "corroboration"). Fourth, to determine when structural learning should take place, by detecting when unpredicted events occur.

From time to time goals, activities required of the animat, will arise. By constructing a Dynamic Policy Map (DPM), a graph like structure derived from the predictions it has discovered during its lifespan and then determining an intersection of this graph with the goals and current circumstances, the Dynamic Expectancy Model may determine appropriate actions to satisfy those goals. Part of the structure of the DEM provides the animat with rules by which this discovery process proceeds. Part imbues the animat with sufficient behavior to set goals and to initiate and continue all these activities until learned behavior may take over from the innate.

Section two of this paper reviews four well-established theories of animal learning: behaviorist, classical conditioning, operant (or instrumental) conditioning and cognitive or 'expectancy' models. During the 20th century each attracted strong proponents and equally strong opponents, and each was the dominant theory for a time. Each position is supported by (typically large numbers of) detailed and fully repeatable experiments. However, none of these stances could be made to explain the full range of observable behaviors, and none was able to gain an overall dominance of the others. Yet the fact remains that each regime can be shown to be present in a single animal (though not all animal species will necessarily demonstrate every attribute). Each is made manifest in the animal according to the experimental procedures to which it is subjected. Examples will be drawn from both the animal and artificial animat research domains.

Section three analyses (selected) data from each school with the express purpose of generating a new, unifying, set of principles or "rules" of prediction and propagation, specifically related to the anticipatory properties that can be extracted from the observations leading to the four models of learning and behavior. This section also reviews a number of computer models inspired by each of these the four stances.

These rules are presented and discussed in section four. The purpose of this section is to consider the anticipatory role of prediction as a unifying factor between these approaches to learning, where previously differences may have been emphasized. This section introduces the primary contribution of this paper. In developing the unifying, anticipatory, framework, this paper does not suggest that any of these theories are in any sense incorrect, only that they each need to be viewed in the context of the whole animal, and of each other, to provide a satisfying explanation of the role of each part. Equally, no assertion is made that, either taken individually or as a whole, these theories represent a complete description of the perceptual, behavioral or learning capabilities of any individual or type of animal.

Section five develops these arguments to show how they have influenced the development of the Dynamic Expectancy Model. Dynamic Expectancy Model animats may be seen as machines for devising hypotheses that make predictions about future events, conducting experiments to corroborate them and subsequently using the knowledge they have gained to perform useful behaviors. A critical feature is the creation and corroboration of these self-testing experiments, each derived from simple "micro-hypotheses", which are in turn created directly from observations in the environment. Each hypothesis will be viewed as describing and encapsulating a simple experiment. Each "micro-experiment" takes the form of an expectancy or prediction that is either fulfilled, so corroborating the effectiveness of the hypothesis, or is not fulfilled, weakening or denying the hypothesis. Anticipatory principles seem interesting in this context as they define a continuing process of discovery and refinement. This allows an animat to progress throughout its lifetime; incrementally developing is structures, and so match its behavior patterns to its environment.

Section six briefly describes the control architecture for SRS/E, an implementation of the Dynamic Expectancy Model. Section seven presents some illustrative results from an experimental procedure with the model. Further results using this model have been previously reported elsewhere ([25], [26], [27]).

2   Prediction and Theories of Behavior

We continue with the view that behavior generation ("action selection") is properly described by the direct or indirect interaction of sensed conditions, Sign-stimuli (S) and response, action or behavior (R) generators. This section will outline four major theoretical stances relating to animal behavior and learning, and will particularly focus on those issues relating to predictive ability, which will be considered in further detail later.

2.1   The Behaviorist Approach

It has been a long established and widely held truism that much of the behavior observed in natural animals can be described in terms of actions initiated by the current conditions in which the animal finds itself. This approach has a long tradition in the form of stimulus-response (S-R) behaviorism, and, although proposed over a century ago ([22]), continues to find proponents, for instance in the behavior based models of Maes [11], the reactive or situated models of Agre [1] and Bryson [8], and was a position vigorously upheld by Brooks [7] in his "intelligence without reason" arguments.

All argue that the majority of observed and apparently intelligent behavior may be ascribed to an innate, pre-programmed, stimulus response mechanism available to the individual. Innate intelligence is not, however, defined by degree. Complex, essentially reactive, models have been developed to comprehensively describe and (so largely) explain the behavioral repertoire of several non-primate vertebrate species, including small mammals, birds and fish. Tyrrell [24] provides a useful summary of a variety of action selection mechanisms drawn from natural and artificial examples.

Behaviorist learning is considered to be "reinforcement", or strengthening of the activating bond between stimulus and response. That is the occurrence of a desirable event concurrently (or immediately following) an application of the S-R pair enhances the likelihood that the pairing will be invoked again over other, alternative pairings, conversely, with a reduced likelihood for undesirable events. New pairings may be established by creating an S-R link between a stimulus and a response that were active concurrently with (or immediately preceding) the desired event.

2.2 Classical Conditioning

A second, deeply influential, approach to animal learning developed during the 1920's as a result of the work of Ivan Pavlov (1849-1936), now usually referred to as classical conditioning. The procedure is well known and highly repeatable. It is neatly encapsulated by one of the earliest descriptions provided by Pavlov. Dogs naturally salivate in response to the smell or taste of meat powder. Salivation is the *unconditioned reflex* (UR), instigated by the *unconditioned stimulus* (US), the meat powder. Normally the sound of a bell does not cause the animal to salivate. If the bell is sounded almost simultaneously with the presentation of meat powder over a number of trials, it is subsequently found that the sound of the bell alone will cause salivation. The sound has become a *conditioned stimulus* (CS). The phenomena is widespread, leading Bower and Hilgard ([6], p. 58) to comment *"almost anything that moves, squirts or wiggles could be conditioned if a response from it can be reliably and repeatably evoked by a controllable unconditioned stimulus"*.

The conditioned response develops with a characteristic sigmoid curve with repeated CS/US pairings. Once established the CS/UR pairing will diminish if the CS/US pairing is not regularly maintained (*extinction*). We may note that the scope of the US may be manipulated over a number of trials to either be highly differentiated to a specific signal, or conversely gradually generalized to respond to a range of similar signals (for instance, a tone of particular frequency, versus a range of frequencies about a center). *Higher-order conditioning* ([3]; [6], p. 62) allows a second neutral CS' (say, a light) to be conditioned to an existing CS (the bell), using the standard procedure. CS' then elicits the CR.

2.3 Operant Conditioning

An radically alternative view of learning was proposed by B.F. Skinner (1904-1990), that of instrumental or operant conditioning. In this model, responses are not "elicited" by impinging sensory conditions, but "emitted" by the animal in anticipation of a reward outcome. Reinforcement strengthening is therefore considered to be between response (R) and rewarding outcome (O), the R-O model, not between sensation and action, as in the S-R model.

The approach is illustrated by reference to an experimental apparatus developed by Skinner to test the paradigm, now universally referred to as the "Skinner box". In a typical Skinner box the subject animal (typically a rat) operates a lever to obtain a reward, say a small food pellet. The subject must be prepared by the experimenter to associate operating the lever with the food reward. However, once the subject is conditioned in this manner the apparatus may be used to establish various regimes to investigate effects such as stimulus differentiation, experimental extinction, the effects of adverse stimuli ("punishment schedules") and the effects of different schedules of reinforcement (such as varying the frequency of reward). As the apparatus may be set up to automatically record the activities of the subject animal (lever pressing), long and/or complicated schedules are easily established.

Operant conditioning has found application in behavior "shaping" techniques, where an experimenter wishes to directly manipulate the overt behavioral activities of a subject, animal or human. In the simplest case the experimenter waits for the subject to emit the desired behavior, and immediately afterwards presents a reward (before a rat may be used in a Skinner box it must be prepared in this way). Importantly, it is to be noted that the R-O activity may be easily manipulated so as to occur only in the presence of a specific stimulus, which may in turn be differentiated or generalized by careful presentation or withholding of reward in the required circumstances.

This has lead to the assertion that operant conditioning is properly described by as three-part association, S-R-O. It is also interesting to note that the stimulus itself now appears to act as a conditioned reinforcer, where it had no inherent reinforcing properties before. In turn, then, a new response in the context of another stimulus (Sy) and response (Ry) may be conditioned to the existing triple (Sx-Rx-O):

Sy-Ry-Sx-Rx-O

Chains of considerable length and complexity have been generated in this way, and have been used, for instance, in the film industry to prepare performing animals. It is, of course, a given that the rewarding outcome is itself a sensory event with direct (innate) association with some condition the subject wants (or in the case of aversive condition, does not want). If the subject animal is not, for instance, hungry when offered food, the connection will not be manifest, and might not be formed. It is also the case that an apparently non-reinforcing sensory condition can attain reinforcing properties if presented in conjunction with an innately reinforcing (positive or negative) one, the *secondary* or *derived reinforcement* effect ([6], p. 184). Derived reinforcers will also condition responses unrelated to the original one.

2.4   The "Cognitive" Model

In the final model to be considered, derived from Tolman's [23] notion of a *Sign-Gestalt Expectancy*, that is a three part "basic cognitive unit" of the form S1-R-S2, in which the occurrence of the stimulus S1 in conjunction with the activity R, leads to the expectation or prediction of the outcome S2 (which may or may not be "rewarding"). This is largely equivalent to Catania's [10] description of the fully discriminated operant connection as a *three-part contingency* of "stimulus – response – consequence", but with the essential difference that it is the identity of the outcome that is to be recorded, rather than just a measure of the desirability or quality of the connection as assumed in purely behaviorist S-R or operant conditioning approaches. Tolman's means-ends approach inspired, and remains one of the central techniques of, Artificial Intelligence problem solving and planning techniques.

3   Interpreting Behavior as Prediction

It is clear that the standard S-R formulation makes no explicit prediction as to the outcome of performing the action part. But there is nevertheless an implicit prediction that responding in this way will place the animal in a "better" situation than the current one, and that the animal will be driven forward to a situation where further behaviors are triggered. Maes' model [11] makes this explicit. The S-R model is an effective one, and explains much about innate behavior generation. However the implicit prediction is one shared with the species as a whole (actually with the forebears of the individual).

Modern reinforcement learning techniques ([19], for a recent review) have revitalized our view of how this implicit prediction should be viewed. They provide robust and analytically tractable ways to guarantee the prioritization of multiple S-R connections to achieve optimized performance. Such policy maps, while finding many important applications, tend to be "over stable" with respect to sources of reward. In contrast, when reward states change, animals respond quickly to these changing needs.

The anticipatory attributes of the classical conditioning paradigm have long been noted, not least because it is almost impossible to establish the effect when the CS occurs after the US. Indeed for best results the CS must be presented (a short time) before the US, implying that there is a predictive effect. It remains an open question as to whether classical conditioning should be interpreted as a general predictive principle, or if it is just a highly specific phenomenon only associated with autonomic reflexes. This paper tends on the side of generality. Classical conditioning has been extensively and accurately modeled by computer simulation ([4], for review). Barto and Sutton [5] comment in particular on the anticipatory nature of the process.

Even though they arise from profoundly different points of view, i.e. "behaviorist" vs. "cognitivist", there are many similarities between the operant conditioning and "cognitive" approaches. A key issue that separates them is the role of overt reward as a driver for learning. Is reward necessary for learning, as would be suggested by the operant conditioning approach? Clearly not, as indicated by the *latent learning* procedure ([21] for a review of the animal literature, and, e.g., [25], for a simulation using the DEM), in which rats (for instance) may be demonstrated to learn mazes in the absence of any externally applied reward. It is not until some rewarding condition is introduced into the maze that the same rats are observed to act in an obviously purposive manner within the maze. This, and similar observations, would suggest that learning and the motivation to utilize what is learnt are generally separate. It may, of course, be the case that an animal is partially or highly pre-disposed to learn combinations that are, have been, or might be "rewarding" (Witkowski, [25] models such an effect using the DEM).

Saksida *et al.* [14] present a computer model of operant conditioning for robot behavior shaping tasks. The Associative Control Process (*ACP*) model ([2]) develops the two-factor theorem of Mowrer ([12]). The ACP model reproduces a variety of animal learning results from both classical and operant conditioning. Schmajuk [16] presents a two-part model combining both classical and operant conditioning modules emulating escape and avoidance learning behavior. Mowrer's work, combining aspects of classical conditioning and operant conditioning (the "two factors"), provides the inspiration for the title of this paper.

Several anticipatory and predictive three-part models have recently appeared in the Animat literature. Stolzmann *et al.* [18] describe an Anticipatory Classifier System (ACS), Witkowski ([25], [26], [27]) describes the Dynamic Expectancy Model (DEM). Developed independently, both are overtly predictive three-part systems, with a number of significant parallels and differences.

4   The Anticipatory Framework

This section proposes a framework of the three fundamental kinds of connection between stimulus Signs and Action response, and five basic rules relating their behavioral and predictive activities. The purpose of this section is to show that each of the four apparently disparate learning theories introduced in the previous sections can be unified from a single anticipatory or predictive viewpoint, and so how they might each serve a purpose within the individual animat.

Henceforth, the term sign-stimulus or simply *Sign* will be used to refer to an identifiably distinct conjunction of sensory conditions, all of which must be individually present for the Sign as a whole to be deemed *active*. A Sign that is predicted is referred to as *sub-active*, a status distinct from full activation as there are circumstances where anticipated activations must be treated differently from actual activation. The component parts of a Sign may be sensitive to a broad or narrow range of phenomena, and the Sign is active whenever each component is detecting anywhere in its range. The range of these components may be altered marginally at any given time. In principle, a Sign may detect external phenomena (as from a sensor or perceptual system), the activity status of an Action or a variety of other, internal, conditions. The total set of Signs currently known to the animat will be indicated by the calligraphic capital letter S, an individual Sign by the lower case letter γ and the active sub-set of Signs by S^*.

The term *Action* (used from now on in preference to the pejorative, but largely synonymous term "response") will refer to recognizable units of activity performed by the animat, taken from the set of actions available to the animat. The animat will have a fixed repertoire of such action patterns (which may be simple or complex). Any action being currently expressed (performed) is deemed *active*. Actions may be overt, causing physical change to the animat's effector system or covert, specifically changing the status of a Sign's valence level or forming a connection between other Signs and Actions. The total set of Actions available to the animat will be indicated by the letter A, and individual Actions by α; the active sub-set of Actions by A^*.

The generally neutral term *valence* (after Tolman [23]) will be adopted to indicate that a Sign has goal like properties, in that it may give the appearance of driving or motivating the animat to activity. In this framework any Sign may have valence, which is separate property from activation or sub-activation. Like sub-activation it may be propagated to other Signs; but does not, in itself, give rise to overt behavior. It remains unclear how motivation is derived in the brain, although its observable effects are clear enough.

We will assume that the animat has memory, conventionally, of past occurrences, but also a temporary ordered memory of predicted future occurrences. The extent of this memory (in terms of what may be recalled and the time period over which it is defined) will limit what may be learned and predicted. First we recognize three types of connection:

1) SA-Connection: Signs can be connected to Actions, either innately or as a consequence of learning.
2) SS-Connection: Signs may predict other Signs, where a predictive link has been established.
3) SAS-Connection: Signs may be attached to an Action and a second Sign, as prediction.

Next consider the following five "rules of propagation", which define (a) when an Action becomes a candidate for activation, (b) when a prediction will be made, (c) when a Sign will become sub-activated, and (d) when a Sign will become valenced:

1. When the Sign in an SA-Connection is active or sub-active the associated action becomes a candidate for activation (expression).
2. When the stimulus-sign in an SS-Connection is active or sub-active the consequent Sign becomes sub-active. Where the prediction implies a time delay, a future "memory" may be made of the predicted activation.
3. Any Sign that predicts another Sign (either SS or SAS) that has valence, itself becomes immediately valenced.
4. An SAS-Connection where the antecedent Sign and Action are both active is itself active and predicts its Consequent Sign, taking into account any delay.
5. The Action in an SAS-Connection where the antecedent Sign is both active and has valence (because its consequent Sign does, by rule 3) becomes a candidate for expression.

Rule 1: This is the standard behaviorist Stimulus-Response model. It may be applied to SA connections both in the sense of an Unconditioned Reflex in the classical conditioning domain, and in the sense of an action pattern releaser/trigger for a more complex behavior module, for instance using a "winner takes all strategy". As many Signs may be active at any one time and are not assumed to be mutually exclusive. It will also be assumed (in the absence of data to the contrary) that several UR may be initiated concurrently.

As action patterns become more complex the activation strategy becomes more critical. It is largely assumed that such complex activities are mutually exclusive (even though several Signs may be active), such that the activated behavior patterns will be in a priority order. The description of this process as a simple S-R activity belies the potential, and typical, complexity of the behaviors than can be initiated. Bryson's [8] EDMUND model, for instance, extends Rosenblatt and Payton's [13] feed-forward network model with elements of parallel activation and hierarchical control structures in order to explain the range of phenomena noted in nature.

Rule 2: Describes a simple predictive step, the occurrence of one Sign leading to the expectation or anticipation that a second will follow within a specified period. This rule accounts for the observations of classical (and higher order) conditioning phenomena (section 2.2) when in conjunction with rule one. Note that rule one only expresses the expression criteria in conditional terms, that sub-activation (the result of the predictive connection) may (or may not) activate the SA-connection. Despite an assumption of *equivalence of associability* (i.e. that any two Signs may act as either predictor or predictee, [6], p. 67), it is clear that not all stimuli are equally amenable to act as the CS in conditioning experiments. Shettleworth [17] found (in the case of golden hamsters) that it was easy to associate certain UR behaviors, such as "digging", "scrabbling" and rising up on the hind legs with a food outcome, and almost impossible to condition others, such as washing or scratching. Shettleworth also noted that the behaviors that could be conditioned were in any case those that the animal tended to emit ordinarily in anticipation of feeding, where the ones that could not be conditioned were not.

By rule 2, sub-activation is defined as self-propagating; sub-activation of the antecedent will in turn sub-activate the consequent. This defines the mechanism for longer chains, as would be the case in, for example, higher order conditioning. In some examples of second order conditioning schedules (e.g. Rizley and Rescorla, cited in [6]) it is possible to extinguish the initial (directly predicting) CS, without affecting the second-order CS. This appears consistent with the notion of propagating sub-activation, rather that full activation, which would indeed sever the chain.

A question remains as to the degree to which sub-activation should propagate in this manner. Given the reported difficulties of sustaining higher order conditioning schedules, it would seem plausible to suggest that propagated sub-activation in this sense will typically be a highly attenuating process in most instances. By treating sub-activation also as an anticipatory mechanism (in the Shettleworth [17] sense), that is, priming the animat for other activities, it would seem equally reasonable that the consequences of this predictive effect should remain localized. Without this restriction too many Signs would become sensitized and the effect would be diluted.

Rule 3: This rule describes the reverse effect of propagating valence (back) across a predictive link, from predictee to predictor. We may take the derived reinforcer as an exemplar of this process. Some Signs clearly have innate connection to the source of valence. That is, their occurrence predicts or is associated with a change in the state of the valence source. For a hungry dog, it seems that the taste or smell of meat has just such an effect. This is apparently innate and does not need to be established. By rule 3, the derived reinforcer, otherwise neutral, gains its valence by predicting that smell or taste. Clearly, the prediction link persists after the conditions that lead to its formation are lost.

Rule 3 applies to both SS and SAS type connections, as they are both overtly predictive forms. However they have different properties and should therefore propagate differently. The SAS connection, like a conventional problem-solving operator, implies action by the animat to move across the link. In this form the animat actively initiates the transitions. It is assumed that valence will propagate well across these links, capable of forming long chains of outcome predictions (section 2.3). Applying this rule rigorously, we note, however, that the propagation takes the form of a graph between Signs linked by predictions. The sequence of actions it actually generates, on the other hand, will indeed appear as a linear sequence.

In the SS form the animat must essentially wait and see if the transition occurs. While useful for some schedules ("wait for the bell"), to rely on long chains of such connections would lead to effective behavioral paralysis. It is therefore assumed that valence, as with prediction, propagates poorly across SS connections.

Rule 4: Defines the conditions under which an SAS connection makes its prediction. Note that the prediction is made (and any sub-activations instigated) regardless of how the action was initiated.

Rule 5: Defines the conditions under which the action in an SAS connection itself becomes a candidate for activation. When an antecedent Sign is both active and has valence, it is at a point of intersection in the valence graph forming a plausible "chain of actions" to a source of valence (acting as a goal) from the animat's current situation. The Dynamic Expectancy Model takes into account the total (estimated) effort between each Sign and sources of valence by combining the effort that must be expended at each step with the strength of the prediction across the connection. Consequently the model defers action choice until the graph of Sign connections is completely evaluated, so as to be sure of selecting the action at the start of the most advantageous chain.

5   The DEM Postulates

The Dynamic Expectancy Model defines an animat controller based on the principles of the anticipatory approach described. The five "rules" discussed in section 4 serve to establish a broad framework. This section adds operational detail to those principles as a step to the computer program implementation of the model (SRS/E) in the form of a larger number of "postulates". Where the five rules emphasize the generation of predictions and the activation of behaviors from existing anticipatory connections, the postulates extend the discussion by considering how SS and SAS connections may be formed and maintained.

The anticipatory connections, SS and SAS, constitute a form of overtly predictive hypotheses. In the Dynamic Expectancy Model they will be referred to as μ-*hypotheses* (spoken "micro-hypotheses"). These are encapsulations of the two predictive, and so capable of corroboration "by experiment", forms (SS and SAS). Applications of these forms, where they make their prediction, will be considered as a form of experiment, or μ-*experiments* ("micro-experiments"). Each activation acts as a test to determine their overall effectiveness in representing the animat and its environment. The construction and corroboration of low-level observation based μ-hypotheses would appear a useful pre-cursor to the independent development of any systematic theoretical model, whose structure is not wholly or primarily dependent on an *originator* (the individual or process responsible for the creation of the animat and its ethogram).

5.1   The Hypothesis Postulates

Definition H0: The μ-*hypothesis*. Each of the forms SS and SAS shall be considered as μ-hypotheses, as each type is capable of forming a prediction and so is inherently "testable". Call the set of all μ-hypotheses $\mathcal{H}$, with h indicating an individual μ-

hypothesis. A μ-hypothesis is composed of Signs ($\mathcal{s}$) and Actions ($\mathcal{a}$) from the respective Sign (S) and Action (A) lists. So:

$$\text{SS:} \qquad h_{SS}: \mathcal{s}' \ {}^{0}\leftrightarrows_{t \pm \tau} \ \mathcal{s}'' \tag{eqn. 1}$$

$$\text{SAS:} \qquad h_{SAS}: \mathcal{a} \wedge \mathcal{s}' \ {}^{0}\leftrightarrows_{t \pm \tau} \ \mathcal{s}'' \tag{eqn. 2}$$

Each records a possible transition between two conditions that may be sensed by the animat (signs $\mathcal{s}'$ and $\mathcal{s}''$). In an SAS connection $\mathcal{s}'$ must be concurrent ($\wedge$) with an action $\mathcal{a}$. The double arrow ($\leftrightarrows$) now jointly indicates the left to right prediction (rules 2 and 4), of the consequent, and the instantaneous (rule 3) reverse transfer of valence.

Postulate H1: μ-Experimentation. μ-Experimentation is the mechanism by which predictive self-testability is conducted. μ-Experimentation is a two-part process. (1) making a prediction based on matching a μ-hypothesis' antecedent conditions to current activations, and (2) comparing those predictions, *a posteriori*, with the actual activations that hold true at the time stipulated by the prediction.

Postulate H2: Prediction. Prediction (implementing rules 2 and 4) records the predicted sign in prediction memory whenever a μ-hypothesis is active. Denoting the total set of active predictions made by the animat and currently awaiting confirmation with the letter $\mathcal{P}$, with p indicating an individual prediction. So:

$$\text{SS:} \quad \text{if}\,(h.\mathcal{s}' \in S^*) \text{ then } h.p^{+t} \leftarrow \mathcal{s}'' \tag{eqn. 3}$$

$$\text{SAS:} \ \text{if}\,(h.\mathcal{s}' \in S^* \,\&\, h.\mathcal{a} \in A^*) \text{ then } h.p^{+t} \leftarrow \mathcal{s}'' \tag{eqn. 4}$$

This memory is a property of the predicting μ-hypothesis, not of the sign predicted, as one Sign may be independently predicted by several μ-hypotheses.

In the SRS/E implementation, prediction memories are implemented as shift register like *traces*, the prediction being placed into the register $+t$ units ahead. The register moves one place backwards towards "the current time" with each execution cycle. The dot notation indicates that p is an attribute of h. Thus $h.p^{+t}$ indicates a prediction due at time t, made by the hypothesis h. A different implementation might record individually time stamped predictions, and so have an arbitrary time horizon.

Postulate H3: Corroboration. To match these recorded predictions against immediate sensations at the time the predictions fall due. If a μ-experiment is to be valid it must encapsulate all of the pre-conditions under which it will be judged. The antecedent components in a SS or SAS connection serve exactly as the definition of those pre-conditions. The quality of each μ-hypothesis is determined solely by its ability to accurately predict its consequent sign. This record of the animat's ability is encoded in the *corroboration measure* (C_m).

One might suppose that the corroboration measure is properly defined as the simple ratio of the total number of predictions made by the μ-hypothesis to the number of correct predictions made. This is equivalent to the probability (P_m), thus:

$$P_m = p(\mathcal{s}''|^{t}(\mathcal{s}' \wedge \mathcal{a})) \qquad \text{(SAS form)} \tag{eqn. 5}$$

The use of the "t" symbol here acts as a reminder of the temporal relationship that exists between the predicted outcome and context. However, this measure is highly sensitive to sample size, if a μ-hypothesis were to change from being valid to invalid (the world changed) a long established μ-hypothesis would react slowly.

In practice, a confidence measure related to probability is adopted. Each successful prediction reinforces confidence in a μ-hypothesis. Conversely every unsuccessful prediction extinguishes confidence in that μ-hypothesis. The contributions of past predictions are discounted as further predictions are made and μ-hypotheses remain largely insensitive to their age and experience. The corroboration measure (C_m) is increased by the quantity:

$$\Delta C_m = \alpha(1 - C_m) \quad \text{if } h.p^0 \in S^* \tag{eqn. 6}$$

following each instance of a successful prediction of an active Sign, and

$$\Delta C_m = -\beta(C_m) \text{ if } h.p^0 \notin S^* \tag{eqn. 7}$$

following an unsuccessful prediction. C_m is updated following the widely used delta rule form. Under constant conditions these relationships give rise to the widely observed "negatively accelerating" form of the learning curve. Two proper fractions, the *reinforcement rate* (α) and the *extinction rate* (β) respectively define a "learning rate" for successful and unsuccessful prediction situations. They control the rate at which the influence of past predictions will be discounted. The C_m value of a μ-hypothesis that makes persistently successful predictions tends to 1.0, the C_m value of a μ-hypothesis that persistently makes unsuccessful predictions tends to 0.0. The positive reinforcement rate need not be equal to the negative extinction rate.

DEM μ-hypotheses are not derived from explicit theories, but are instead created from examples and may be thought of as "competing" to attain higher confidence measures and so be incorporated into goal-directed valence sequences and therefore influence the overt behavior of the animat.

Postulate H4: Learning by Creation. μ-Hypotheses may, of course, be innate to the animat, part of the ethogram definition. The prediction and corroboration mechanism will effectively tune them to the animat's actual circumstances. This both pre-disposes the animat to useful and (presumably) appropriate behavior patterns, and allows innate and learned behaviors to be integrated. However, to be a fully-fledged learning entity, the model must define a "Learning by Creation" method by which the animat extends the set of μ-hypotheses. This learning proceeds in two parts, (1) detecting circumstances where a new μ-hypothesis is required, and (2) the actions required to construct the double (SS) or triple (SAS) connection.

μ-Hypotheses exist to predict future occurrences of Signs; it is therefore reasonable to suppose that new μ-hypotheses should be created under two specific circumstances. Potentially, every sign should have at least one μ-hypothesis capable of predicting it, and ideally the Sign would be correctly predicted for every occurrence. Novel signs (ones not previously recognized by the system) can appear in the system as a result of the differentiation process (H5, below) where new, distinct Signs are formulated - **postulate H4-1**, *novel event*. In the second creation circumstance, known signs are detected without a corresponding prediction, **postulate H4-2**, *unexpected event*. Novel and unexpected Signs are recognized within the SRS/E system by detecting the condition:

$y \in S^* \& y \notin P^0$, that is, the Sign y is active, but was not predicted to be so at the current time.

In either case a new μ-hypothesis may be created. The consequence Sign ($\mathcal{S}'$) for this new μ-hypothesis will be the novel or unexpected Sign. The context and action drawn from the set of recent Signs (and Actions for an SAS connection) recorded by the system in the memories associated with individual Signs and Actions (modeled in SRS/E as the shift register like "traces"). The new μ-hypothesis may then be constructed by incorporated the remembered components into the antecedent and shifting the predicted time by an amount equivalent to the depth in the memory trace of the antecedent item(s).

Postulate H5: Refinement. Refinement is the mechanism by which the animat may differentiate or generalize its existing set of μ-hypotheses. Differentiation adds extra conditions to the context of an existing μ-hypothesis, reducing the range of circumstances under which that μ-hypothesis will be applicable. Generalization removes or relaxes existing conditions to the context, increasing the range of circumstances. Differentiation may be appropriate to enhance μ-hypotheses that have stabilized, or stagnated, at some intermediate corroborative measure value. μ-Hypotheses should not be subject to differentiation until they have reached an appropriate level of testing (their "maturity", or extent of corroboration). Maturity is a measure of the degree of corroboration of a μ-hypothesis. It is otherwise independent of the age of a μ-hypothesis. It is expected that the refinement process will create new, separate μ-hypotheses that are derived from the existing ones. Both old and new μ-hypotheses are retained and may then "compete" to determine which offers the best predictive ability. In the specific implementation SRS/E, creation (H4) is heavily biased to formulating over-generalized μ-hypotheses, so differentiation is the primary refinement method. Anticipatory Classifier Systems (ACS), due to their design, tend to emphasize generalization [18] as the primary refinement mechanism.

Postulate H6: Forgetting. Forgetting is the mechanism by which the animat may discard μ-hypotheses found ineffective from the set of μ-hypotheses held, or when the system needs to recover resources. A μ-hypothesis might be deleted when it can be determined that it makes no significant contribution to the abilities of the animat. This point can be difficult to ascertain. Evidence from animal learning studies indicates that learned behaviors may be retained even after considerable periods of extinction. Experimental evidence drawn from the implementation of the Dynamic Expectancy Model points to the value of not prematurely deleting μ-hypothesis, even though their corroborative measures fall to very low levels [27]. Where a Sign is predicted by many μ-hypotheses there may be good cause to remove the least effective. It is presumed that the last remaining μ-hypothesis relating to a specific consequent Sign will not be removed, on the basis that some predictive ability, however poor, is better than none at all. As no record is retained of the forgotten μ-hypothesis, a new μ-hypothesis created later may be the same as one previously removed (by H4-2).

5.2   The Valence Postulates

Definition G0: Goals. A goal establishes a valence condition within the animat causing the animat to select behaviors appropriate to the achievement or *satisfaction* of that goal. Goals (denoted by the letters G/g) are a special condition of a Sign; goals are therefore always drawn from the set of available Signs.

Postulate G1: Goal Valence. From time to time the animat may assert any of the Signs available as a goal. Any Sign asserted to act as a goal in this way is termed as having valence (or be valenced). None, one or many Signs may be valenced at any one time.

Postulate G2: Goal Priority. Each valenced goal is assigned a positive, non-zero priority. This priority value indicates the relative importance to the animat of achieving this particular goal, in the prevailing context of other behaviors and goals. Goal priority is determined within the innate behavioral component of the ethogram. In the current SRS/E implementation only one goal is pursued at any time - the *top-goal*, the goal with the highest priority.

Postulate G3: Valenced Behavior. Whenever a goal is valenced, SRS/E will, by rule 4, propagate valence across existing μ-hypotheses to establish a graph of valenced connections within the system. In the SRS/E implementation each SRS connection will impose a *cost effort* estimate, C_e, proportional to the effort of performing the action and inversely to the current C_m value for the link:

$$C_e \leftarrow (\text{action\_cost}(\alpha) / C_m) \qquad\qquad (\text{eqn. 8})$$

This effort accumulates across the graph, so that each antecedent Sign in the network defines the beginning of a path or chain (of actions) that represents the "best estimate" for the animat forward to the top-goal. This graph is referred to as the *Dynamic Policy Map* (DPM), as it defines a both preference ranking for activation for every Sign reached by rule 3 and indicates which of the actions associated by μ-hypotheses with the Sign should be activated. The DPM is recalculated frequently as goal priorities and confidence measures change due to corroboration, and as μ-hypotheses are added and removed from the system. In the SS connection case, it is convenient to consider a "dummy" action. By assigning it a high (notional) action_cost, propagation across these links is disadvantaged.

Postulate G4: Valenced Action Selection. When a DPM exists the system will apply rule 5 to activate a μ-hypothesis and so select an action. SRS/E selects the μ-hypotheses with the lowest overall cost estimate to the top-goal where several nodes compete for activation under rule 5.

Postulate G5: Goal Satisfaction. A valenced goal is deemed "satisfied" once the conditions defined by the goal are encountered, when the sign that defines the goal becomes activate. The priority of a satisfied goal is reduced to zero and it ceases to be a source of valence. Where goal-seeking behavior is to take the form of sustained maintenance of a goal state, the goal selection process must maintain the valence of the goal Sign following each satisfaction event.

Postulate G6: Goal Extinction. In a situation where all possible paths to a goal are unavailable, continued attempts to satisfy that goal will eventually become a threat to the continued survival of the animat, by blocking out other behaviors and needlessly consuming resources. Such a goal must be forcibly abandoned. This is the *goal extinction point*. Witkowski [27] has modeled goal extinction using the DEM, arguing that is it substantially different from a simple reversal of the development of corroboration and from extinction in classical conditioning.

5.3   The Behavior Postulates

Definition B0: Behaviors. Behaviors (indicated by the letter $\mathcal{B}$) are non-learned activities inherent within the system. Behaviors are explicitly Stimulus-Response (SA) connections and are activated according to the tenets of rule 1. They are defined prior to parturition as part of the ethogram. There is no limit to the complexity (or simplicity) of innate behavior. An animat might be solely dependent on innate behaviors, with no learning component.

Postulate B1: Behavior Priority. Each behavior within the animat is assigned a priority relative to all the other behaviors. This priority is defined by the ethogram. The action associated with the behavior of highest priority is selected for expression.

Postulate B2: Primary Behaviors. Primary behaviors define the vocabulary of behavior patterns available to the animat at parturition. These behaviors provide a repertoire of activities enabling the animat to survive in its environment until learning processes may provide more effective behaviors.

Postulate B3: Goal Setting Behaviors. The ethogram defines the conditions under which the animat will convert to goal seeking behavior. Once a goal is set the animat is obliged to pursue that goal while there is no primary behavior of higher priority. Where no behavior can be selected from the DPM, the animat selects the primary behavior of highest priority that is currently active. Behavior selection from the DPM resumes once there is any match between the set of active signs and the current DPM.

Interruption of goal directed behavior by a higher priority innate behavior turns the animat away from pursuing its current top priority goal. For instance, goal directed food-seeking behavior should be interrupted by high priority predator avoidance activity. Once the threat has passed the goal directed behavior resumes, although the animat's perceived "place" in the DPM will have shifted as a result of the intervening behavior. The structure and corroboration of the DPM may have changed, and it must be re-evaluated as behavior reverts to the goal directed form. Where goal seeking takes the form of a sustained maintenance of the selected goal state, the selection process must re-valence the required goal each time it is satisfied.

Postulate B4: Default (exploratory) Behaviors. Default Behaviors provide a set of behaviors to be pursued by the animat whenever neither a primary nor a goal setting behavior is applicable. Typically these default behaviors will take the form of exploratory actions. Exploratory actions may be either random (*trial and error*), or represent a specific exploration strategy. Selection of this strategy will impact the rate and order in which the μ-hypothesis creation processes occur (H4). Default behaviors have a priority lower than any of the primary (B2) or goal setting (B3) behaviors. The provision of some default behaviors is mandatory within the ethogram.

6   The SRS/E Program Architecture

Figure one illustrates the flow of control within the SRS/E program architecture and the interaction between parts. The flow of control forms a non-terminating loop incorporating each of the eight steps identified in the figure. The computational effort of each cycle is relatively light, each activity being initiated opportunistically according to the prevailing circumstances. It is the cumulative effect over many cycles that gives rise, over time, to a refined set of corroborated μ-hypotheses.

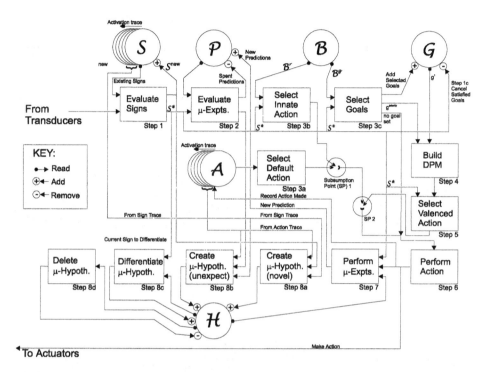

Figure One: The SRS/E Evaluation Cycle

Step 1 evaluates every sign to create the Sign activation list S^* using the current status of the animat's transducers. Step 2 compares past predictions falling due at the current time with the current activations and updates the corroboration measure of the μ-hypotheses responsible for the predictions tested (postulate H3). Step 3a selects a default (exploratory, B4) behavior. If an innate behavior is activated (postulate B2, step 3b) this will override the default behavior on the basis of priority (B1) at the *subsumption point* (SP1). Step 3c determines the valence status of Signs, and updates the Goal List (G), assigning each goal a priority (G2) on the basis of the defined goal setting behaviors (B3). Where at least one goal has valence (G1) step 4 is initiated and a Dynamic Policy Map constructed (G3). Step 5 applies rule 5 to find an intersection between S^* and μ-hypotheses valenced by step 4 (postulate G4). The highest priority action is passed (via subsumption point SP2) to step 6, which causes the animat's actuators to perform that action. Once an Action has been selected every μ-hypothesis in H can be evaluated (postulates H1/H2) to determine the predictions to be made, which will be evaluated by step 2 in future cycles. Steps 8a, 8b, 8c and 8d implement postulates H4-1, H4-2, H5 and H6 respectively. The loop starts again at step 1.

7   An Illustrative Experimental Procedure

Figure two shows key stages from a single experimental run using the SRS/E program. The example illustrates a number of points arising from the use of the anticipatory learning approach described in this paper. First, demonstrating the use of anticipatory learning techniques to create new μ-hypotheses and corroborate them in the absence of explicit reward. Second, showing the effects of motivation on the behavior of the system and third, the effects of failed predictions in causing substantial changes to overt behavior during valenced behavior.

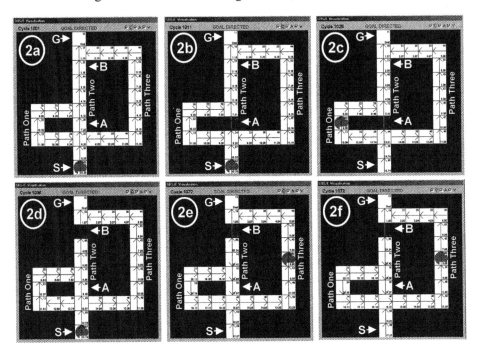

Figure Two: Key Stages in the Experimental Procedure

The set-up represents a single animat (shown as the circular object) in a maze like environment where white squares represent traversable paths and black ones are blocked and cannot be entered. The animat may make one of four actions, moving "north" (up), "south", "east" or "west", taking it into an adjacent location. It does not move if the destination square is blocked or the edge of the environment is encountered. In this simulation the animat only senses a block when it attempts the move into it. The animat may directly and uniquely sense, as a Sign, the identity of the location it is currently occupying. The arrows represent the preferred action in the event that the Sign representing a given location is encountered. The number in the top right corner the total number of times the location has been visited, that in the bottom corner the sum of cost effort estimates (C_e, eqn. 8) along the preferred path to the source of valence. The experimental procedure applied in the following stages (α = 0.5, β = 0.2 throughout):

1) The animat is allowed to explore the maze for 1000 action cycles, but with no source of valence. Cycles 1 – 1000.

2) The animat is returned to the start 'S' and the Sign representing location 'G' is assigned valence by the experimenter. The animat is allowed to run to 'G'. In animal experiments, valence would be attached to the goal location by placing a "reward" (food, for example) there, acting as a secondary or derived reinforcer (section 2.3). Cycles 1001-1010.

3) A block is introduced at location 'A', the animat returned to 'S' and 'G' valenced. The animat is allowed to run to 'G'. Cycles 1011-1035.

4) The block at 'A' is removed and a block is now introduced at location 'B'. The animat is again returned to 'S' and 'G' valenced. The animat is allowed to run to 'G'. Cycles 1036-1080.

5) Stage 4 is repeated, 'B' remains blocked. Cycles 1081-1098.

During stage one of the experiment the animat uses the learning by creation (H4-1, H4-2) methods to formulate many μ-hypotheses relating to the environment and subsequently corroborates them using the methods described in postulates H1, H2 and H3. As there is no motivation or goal setting active during the first stage, all learning that takes place is "latent" (learning in the absence of explicit motivation or reward). The corroborative learning processes described by postulates H1, H2 and H3 depend only on the ability to make a prediction and so anticipate a specific outcome. A successful prediction is taken as its own reward. The system will also learn the maze if motivation is present, but, apparently paradoxically, learning may be less successful than under the unmotivated conditions ([27] for further discussion). This phenomenon has also been observed in the direct animal experiments on latent learning.

In stage two, the behavior of the animat is controlled by the Dynamic Policy Map, constructed by the repeated application of rule 3 as the Sign indicating the goal location 'G' now has valence. On the first run, the animat goes straight to the goal location via path two (figure 2a shows the DPM at cycle 1001). This is as expected. Stage three illustrates the effects of failed predictions, due to the introduction of a block at location 'A'. Figure 2b shows the situation at the start of stage 2 (cycle1011), anticipating a path via 'A'. On reaching the blockage, the animat attempts the action (North) to traverse into the expected location using the μ-hypothesis predicting the move. This fails, causing the corroboration measure for this μ-hypothesis to fall (eqn. 7) on successive attempts. As this individual hypothesis weakens, path two becomes less viable than path one (one may observe the DPM changing with each failed prediction and action). Figure 2c shows the situation at cycle 1026, with the animat now traversing path one. The visualization shows the situation after the route via path one becomes preferred, and the animat's behavior has changed to follow path one to 'G'. Figure 2d shows the DPM at the start of stage four. Due to the repeated failure of the μ-hypothesis entering location 'B', the animat backtracks to 'G' via path three once the blocked hypothesis is sufficiently weakened (figure 2e, cycle 1072). Figure 2f shows the situation at the start of stage five (cycle 1081). Inspection reveals that the DPM indicates a path to 'G' via path three. This is confirmed by the animat's actual path.

On an historical note, this procedure is based on the "place learning" procedure employed by Tolman and Honzik (see [6], p. 337) to demonstrate "inferential expectation" or "insight" in rats. The key question was whether the rat would take path two (as would be the case with a pure reward based reinforcement learning strategy) - the animat having no "insight" that the block at 'B' would also block path two) - or path three (the animat has a "map" strategy) at stage five. However, even this result relies on the animat being given sufficient time to fully explore its environment. Using a random or chance exploration strategy, insufficient exploration time can lead to non-optimal path choice preferences. This level of exploration is easily confirmed with the visualization tool (of figure two), but not so obvious with live rats. Where insufficient exploration is permitted, incomplete DPMs are formed and may not reach to the current location. The animat performs exploratory actions until a location with valence is encountered, when goal-seeking behavior resumes.

8   Summary and Conclusions

This paper has developed a minimal set of five "rules" of prediction and propagation by which an animat may exploit anticipation as a model of intelligence. The five rules are used to place the important attributes of four major learning and behavioral schemes (the "four factors") into a single anticipatory context and to develop a unified approach to modeling them. These are then elaborated with a larger number of "postulates", which act as a bridge to a realizable model (DEM) and the specific implementation (SRS/E). The key strength is the encapsulation of anticipatory prediction into μ-hypotheses, self-contained and capable of corroboration without recourse to any outside agency. Such μ-hypotheses anticipate what might happen (SS) and predict what can be made to happen (SAS), and can be used by the animat to derive appropriate behaviors in its environment as the need arises.

References

1. Agre, P.E.: Computational Research on Interaction and Agency, Artificial Intelligence, **72** (1995) 1-52
2. Baird, L.C. and Klopf, A.H.: Extensions to the Associative Control Process (ACP) Network: Hierarchies and Provable Optimality, 2nd Int. Conf. on Simulation of Adaptive Behavior (SAB-2), (1993) 163-171
3. Balkenius, C.: Natural Intelligence in Artificial Creatures, Lund University Cognitive Studies 37 (1995)
4. Balkenius, C. and Morén J.: Computational Models of Classical Conditioning: A Comparative Study, 5th Int. Conf. on Simulation of Adaptive Behavior (SAB-5), (1998) 348-353
5. Barto, A.G. and Sutton, R.S.: Simulation of Anticipatory Responses in Classical Conditioning by a Neuron-like Adaptive Element", Behavioral Brain Research, **4** (1982) 221-235
6. Bower, G.H. and Hilgard, E.R.: Theories of Learning, Englewood Cliffs: Prentice Hall Inc., fifth edition (1981)
7. Brooks, R.A.: Intelligence Without Reason, MIT AI Laboratory, A.I. Memo No. 1293. (Prepared for Computers and Thought, IJCAI-91, pre-print (April 1991)
8. Bryson, J.: Hierarchy and Sequence vs. Full Parallelism in Action Selection, 6th Int. Conf. on Simulation of Adaptive Behavior (SAB-6), (2000) 147-156

9. Butz, M.V., Sigaud, O. and Gerard, P.: Internal Models and Anticipations in Adaptive Learning Systems, Adaptive Behavior in Anticipatory Learning Systems 2002 Workshop (ABiALS 2002) 23pp.

10. Catania, A.C.: The Operant Behaviorism of B.F. Skinner, in: Catania, A.C. and Harnad, S. (eds.) "The Selection of Behavior", Cambridge: Cambridge University Press (1988) 3-8

11. Maes, P.: Behavior-based Artificial Intelligence, 2nd Int. Conf. on Simulation of Adaptive Behavior (SAB-2), (1993) 2-10

12. Mowrer, O.H.: Two-factor Learning Theory Reconsidered, with Special Reference to Secondary Reinforcement and the Concept of Habit, Psychological Review, 63, (1956) 114-128

13. Rosenblatt, J.K. and Payton, D.W.: A Fine-Grained Alternative to the Subsumption Architecture for Mobile Robot Control, IEEE/INNS Int. Joint Conf. on Neural Networks, Vol. II. (1989) 317-323

14. Saksida, L.M., Raymond, S.M. and Touretzky, D.S.: Shaping Robot Behavior Using Principles from Instrumental Conditioning, Robotics and Autonomous Systems, 22-3/4, (1997) 231-249

15. Schaffer, S.: Babbage's Intelligence: Calculating Engines and the Factory System, hosted at http://cci.wmin.ac.uk/schaffer/schaffer01.html (1998)

16. Schmajuk, N.A. Behavioral Dynamics of Escape and Avoidance: A Neural Network Approach, 3rd Int. Conf. on Simulation of Adaptive Behavior (SAB-3), (1994) 118-127

17. Shettleworth, S.J.: Reinforcement and the Organization of Behavior in Golden Hamsters: Hunger, Environment, and Food Reinforcement, Journal of Experimental Psychology: Animal Behavior Processes, 104-1 (1975) 56-87

18. Stolzmann, W., Butz, M.V., Hoffmann, J. and Goldberg, D.E.: First Cognitive Capabilities in the Anticipatory Classifier System, 6th Int. Conf. on Simulation of Adaptive Behavior (SAB-6), (2000) 287-296

19. Sutton, R.S. and Barto, A.G.: Reinforcement Learning: An Introduction, Cambridge, MA: MIT Press (1998)

20. Tani, J. and Nolfi, S.: Learning to Perceive the World as Articulated: An Approach for Hierarchical Learning in Sensory-Motor Systems, 5th Int. Conf. on Simulation of Adaptive Behavior (SAB-5), (1998) 270-279

21. Thistlethwaite, D.: A Critical Review of Latent Learning and Related Experiments, Psychological Bulletin, 48(2), (1951) 97-129

22. Thorndike, E.L.: Animal Intelligence: An Experimental Study of the Associative Processes in Animals, Psychol. Rev., Monogr. Suppl., 2-8 (1898)

23. Tolman, E.C.: Purposive Behavior in Animals and Men, New York: The Century Co. (1932)

24. Tyrrell, T.: Computational Mechanisms for Action Selection, University of Edinburgh, Ph.D. thesis (1993)

25. Witkowski, M.: Dynamic Expectancy: An Approach to Behaviour Shaping Using a New Method of Reinforcement Learning, 6th Int. Symp. on Intelligent Robotic Systems, (1998) 73-81

26. Witkowski, M.: Integrating Unsupervised Learning, Motivation and Action Selection in an A-life Agent, 5th Euro. Conf. on Artificial Life (ECAL-99), (1999) 355-364

27. Witkowski, M. The Role of Behavioral Extinction in Animat Action Selection, proc. 6th Int. Conf. on Simulation of Adaptive Behavior (SAB-6), (2000) 177-186

Internal Models and Anticipations in Adaptive Learning Systems

Martin V. Butz[2,3], Olivier Sigaud[1], and Pierre Gérard[1]

[1] AnimatLab-LIP6, 8, rue du capitaine Scott, 75015 Paris France
{olivier.sigaud,pierre.gerard}@lip6.fr
[2] Department of Cognitive Psychology, University of Würzburg, Germany
butz@psychologie.uni-wuerzburg.de
[3] Illinois Genetic Algorithms Laboratory (IlliGAL),
University of Illinois at Urbana-Champaign, IL, USA

Abstract. The explicit investigation of anticipations in relation to adaptive behavior is a recent approach. This chapter first provides psychological background that motivates and inspires the study of anticipations in the adaptive behavior field. Next, a basic framework for the study of anticipations in adaptive behavior is suggested. Different anticipatory mechanisms are identified and characterized. First fundamental distinctions are drawn between implicit anticipatory behavior, payoff anticipatory behavior, sensory anticipatory behavior, and state anticipatory behavior. A case study allows further insights into the drawn distinctions. Many future research direction are suggested.

1 Introduction

The idea that *anticipations* influence and guide behavior has been increasingly appreciated over the last decades. Anticipations appear to play a major role in the coordination and realization of adaptive behavior. Various disciplines have explicitly recognized anticipations. For example, philosophy has been addressing our sense of reasoning, generalization, and association for a long time. More recently, experimental psychology confirmed the existence of anticipatory behavior processes in animals and humans over the last decades.

Although it might be true that over all constructible learning problems any learning mechanism will perform as good, or as bad, as any other one [71], the psychological findings suggest that in natural environments and natural problems learning and acting in an anticipatory fashion increases the chance of survival. Thus, in the quest of designing competent artificial animals, the so called *animats* [69], the incorporation of anticipatory mechanisms seems mandatory.

This book addresses two important questions of anticipatory behavior. On the one hand, we are interested in *how* anticipatory mechanisms can be incorporated in animats, that is, which structures and processes are necessary for anticipatory behavior. On the other hand, we are interested in *when* anticipatory mechanisms are actually helpful in animats, that is, which environmental preconditions favor anticipatory behavior.

M. Butz et al. (Eds.): Anticipatory Behavior ..., LNAI 2684, pp. 86–109, 2003.

To approach the *how* and *when*, it is necessary to distinguish first between different anticipatory mechanisms. With respect to the *how*, the question is *which* anticipatory mechanisms need which structure. With respect to the *when*, the question is *which* anticipatory mechanisms cause which learning and behavioral biases. In this chapter, we draw a first distinction between (1) *implicit anticipatory* mechanisms in which no actual predictions are made but the behavioral structure is constructed in an anticipatory fashion, (2) *payoff anticipatory* mechanisms in which the influence of future predictions on behavior is restricted to payoff predictions, (3) *sensory anticipatory* mechanisms in which future predictions influence sensory (pre-)processing, and (4) *state anticipatory* mechanisms in which predictions about future states directly influence current behavioral decision making. The distinctions are introduced and discussed within the general framework of *partially observable Markov decision processes* (POMDPs) and a general animat framework based on the POMDP structure.

The remainder of this chapter is structured as follows. First, psychology's knowledge about anticipations is sketched out. Next, we identify and classify different anticipatory mechanisms in the field of adaptive behavior. A non-exhaustive case study provides further insights into the different mechanisms as well as gives useful background for possible extensions. The conclusions outline many diverse future research directions tied to the study of anticipatory behavior in adaptive learning systems.

2 Background from Psychological Research

In order to motivate the usage of anticipations in adaptive behavior research, this section provides background from cognitive psychology. Starting from the behaviorist movement, we show how the notion of anticipation and its diverse impact on behavior was recognized in psychology research. While behaviorism gave rise to successful experimental psychology it somewhat ignored, and often even denied, anticipatory behavior influences. However, the experimental approach itself eventually revealed inevitable anticipatory influences on behavior. Recent neuron imaging techniques and single-cell recordings provide further proof of anticipatory cognitive processes.

2.1 Behaviorist Approach

Early suggestions of anticipations in behavior date back to Herbart [21]. He proposed that the "feeling" of a certain behavioral act actually triggers the execution of this act once the outcome is desired later.

The early 20th century, though, was dominated by the behaviorist approach that viewed behavior as basically stimulus-response driven. Two of the predominant principles in the behaviorist world are *classical conditioning* and *operant conditioning*.

Pavlov first introduced classical conditioning [39]. Classical conditioning studies how animals learn associations between an unconditioned stimulus (US)

and a conditioned stimulus (CS). In the "Pavlovian dog", for example, the unconditioned stimulus (meat powder) leads to salivation — an unconditioned reflex
(UR). After several experiments in which the sound of a bell (a neutral stimulus
NS) is closely followed by the presentation of the meat powder, the dog starts
salivating when it hears the sound of the bell independent of the meat powder.
Thus the bell becomes a conditioned stimulus (CS) triggering the response of
salivation.

While in classical conditioning the conditioned stimulus may be associated
with the unconditioned stimulus (US) *or* with the unconditioned reflex (UR),
operant conditioning investigates the direct association of behavior with favorable (or unfavorable) outcomes. Thorndike [60] monitored how hungry cats learn
to escape from a cage giving rise to his "law of effect". That is, actions that lead
to desired effects will be, other things being equal, associated with the situation
of occurrence. The strength of the association depends on the degree of satisfaction and/or discomfort. More elaborate experiments of operant conditioning
were later pursued in the well known "Skinner box" [46].

Thus, classical conditioning permits the creation of new CS on the basis of
US, and operant conditioning permits to chain successive behaviors conditioned
on different stimuli. Note that the learning processes take place backwards. To
learn a sequence of behaviors, it is necessary to first learn the contingencies at
the end of the sequence. In addition, the consequences are only learned because
they represent punishments or rewards. Nothing is learned in the absence of any
type of reward or punishment.

While behaviorism allowed cognitive psychology to make significant progress
due to its principled study of behavior phenomena, a persisting drawback of the
approach is the complete ignorance to, or denial of, any sort of mental state.
Skinner's and others' mistake was to disallow future predictions or expectations,
described as intentions, purposes, aims, or goals, to influence behavior.

> No one is surprised to hear it said that a person carrying good news
> walks more rapidly because he feels jubilant, or acts carelessly because of
> his impetuosity, or holds stubbornly to a course of action through sheer
> force of will. Careless references to purpose are still to be found in both
> physics and biology, but good practice has no place for them; yet almost
> everyone attributes human behavior to intentions, purposes, aims, and
> goals. [47, p.6]

Although Skinner is correct that the unscientific reference to e.g. "purpose" might
result in the obstruction of scientific progress in psychology, we will show that
it is possible to formalize future representations and behavior dependent on
future representations. First, however, we present psychological investigations
that clearly show that representations of the future are influencing behavior.

2.2   Expectancy Model

First experimental evidence for anticipatory behavior mechanisms can be found
in Tolman's work [61, 62, 63]. Tolman proposed that, additionally to conditioned

learning, *latent learning* takes place in animals. In latent learning experiments animals show to have learned an environmental representation during an exploration phase once a distinct reinforcer is introduced in the successive test phase (e.g. [61, 58]).

In typical latent learning experiments animals (usually rats) are allowed to explore a particular environment (such as a maze) without the provision of particular reinforcement. After the provision of a distinctive reinforcer, the animals show that they have learned an internal representation of the structure of the environment (by e.g. running straight to the food position).

More technically, the rats must have learned some environmental map (i.e., a predictive model) during exploration. Next, a goal emerges, that is, a certain state in the environment is desired. Finally, without any further active exploration, the rats are able to exploit the learned model and consequently move directly towards the desired state.

The observation of latent learning led Tolman to propose that animals form *expectancies*,

> [...] a condition in the organism which is equivalent to what in ordinary parlance we call a 'belief', a readiness or disposition, to the effect that an instance of this sort of stimulus situation, if reacted to by an instance of that sort of response, will lead to an instance of that sort of further stimulus situation, or else, simply by itself be accompanied, or followed, by an instance of that sort of stimulus situation.[64, p.113]

Essentially, expectancies are formed predicting action effects as well as stimulus effects regardless of actual reinforcement. A whole set of such expectancies, then, gives rise to a *predictive environmental model* which can be exploited for anticipatory behavior.

2.3   More Recent Psychological Evidence

In cognitive psychology anticipations have been experimentally shown to influence behavior ranging from simple reaction time tasks to elaborate reasoning tasks [29, 48]. It becomes more and more obvious that anticipations influence actual behavior as well as memory mechanisms and attention [37]. Neuropsychology gained further insights about the role of anticipatory properties of the brain in attentional mechanisms and, conversely, highlighted the role of attentional mechanisms in e.g. the anticipation of objects [43]. This section investigates two key findings in psychology research to show the broad impact of anticipatory behavior mechanisms.

Predictive capabilities come into play on different levels and to different extensions. The very recent discovery of *mirror neurons* in neuroscience provides neurological evidence that at least "higher" animals, such as monkeys, form representations of their conspecifics [41, 15]. The findings show that there are neurons in monkeys that are active not only when performing a particular action, such as grasping an object, but also when watching another monkey or human performing the same action. This shows that predicting the action of other people is realized by the re-use of neuronal pathways that represent one's own actions. For now, it is unclear how the other agent's actions are linked to ones own action representation. Gallese [14] suggests that the link may be constituted by the embodiment of the intended goal, shared by the agent and the observer. Gallese [14] also argues that only due to mirror neurons it may be possible to become socially involved enabling understanding and prediction of other people's intentions by a *shared manifold* — the association of other peoples actions and feelings with ones own actions and feelings via mirror neurons. Arbib [1] proposed mirror neurons as a prerequisite for the evolution of language. He suggests that it may only be possible to comprehend other people's speech acts by simulating and predicting these acts with neurons identical to ones own speech acts.

In general, mirror neurons are strongly related to the *simulation theory* of mind reading which postulates that in simulating other person's minds ones own resources are used. Simulation and prediction of other people's mind states mediated by mirror systems in the brain causes anticipatory behavior due to resulting predispositions in the mind. Empathy, for example, can be seen as a special case of anticipatory behavior in which motivational and emotional resources become active due to predictions and simulation of other people's minds by the means of mirror systems [59].

Another clear benefit can be found in research on attention. Pashler [38] gives a great overview over the latest research knowledge on attention in humans. LaBerge [31] distinguishes between *selective* and *preparatory attention*. While he suggests that selective attention does not require any anticipatory mechanisms, preparatory attention does. Preparatory attention predicts the occurrence of a visual perception (spatial or object-oriented) and consequently biases the filtering mechanism. The prediction is done by the system's model of its environment and influences the state of the system by the means of the decision maker's actions that essentially manipulate attentional mechanisms in this case. Preparatory attention enables faster goal-directed processing but may also lead to *inattentional blindness* [34]. In inattentional blindness experiments it is revealed that attention can be directed spatially, temporally, and/or object-oriented. It is most strikingly shown in the famous "gorilla experiment" [44]. A tradeoff arises between faster processing and focusing capabilities due to preparatory, or anticipatory, attention and a possible loss of important information due to inattention. When the capability of faster goal-directed processing outweighs the possibility of blindness effects needs to be addressed in further detail.

The next section introduces a formal framework for the classification of anticipatory mechanisms in animats and proposes first important distinctions.

3   Anticipation in Adaptive Behavior

Adaptive behavior is interested in how so called *animats* (artificial animals) can intelligently interact and learn in an artificial environment [69]. Research in artificial intelligence moved away from the traditional predicate logic and planning approaches to intelligence without representation [7]. The main idea is that intelligent behavior can arise without any high-level cognition. Smart connections from sensors to actuators can cause diverse, seemingly intelligent, behaviors. A big part of intelligence becomes *embodied* in the animat. It is only useful in the environment the animat is *situated* in. Thus, a big part of intelligent behavior of the animat arises from the direct interaction of agent architecture and structure in the environment.

As suggested in the psychology literature outlined above, however, not all intelligent behavior can be accounted for by such mechanisms. Thus, hybrid behavioral architectures are necessary in which an embodied intelligent agent may be endowed with higher "cognitive" mechanisms including developmental mechanisms, learning, reasoning, or planning. The resulting animat does not only act intelligently in an environment but it is also able to adapt to changes in the environment, to handle unforeseen situations, or to become socially involved. Essentially, the agent is able to learn and draw inferences by the means of internal representations and mechanisms. Anticipatory mechanisms may be part of these processes.

The cognitive mechanisms employed in animats are broad and difficult to classify and compare. Some animats might apply direct reinforcement learning mechanisms, adapting behavior based on past experiences but choosing actions solely based on current sensory input. Others might be enhanced by making actual action decisions also dependent on past perceptions. Anticipatory behavior research is interested in those animats that base their action decisions also on future predictions. Behavior becomes anticipatory in that predictions and beliefs about the future influence current behavior.

In the remainder of this section we develop a framework for animat research allowing for a proper differentiation of various types of anticipatory behavioral mechanisms. For this purpose, first the environment is defined as a partially observable Markov decision process (POMDP). Next, a general animat framework is outlined that acts upon the POMDP. Finally, anticipatory mechanisms are distinguished within the framework.

3.1 Framework of Environment

Before looking at the structure of animats, it is necessary to provide a general definition of which environment the animat will face. States and possible sensations in states need to be defined, actions and resulting state transitions need to be provided, and finally, the goal or task of the animat needs to be specified. The POMDP framework provides a good means for a general definition of such environments.

We define a POMDP by the $< X, Y, U, T, O, R >$ tuple

- X, the state space of the environment;
- Y, the set of possible sensations in the environment;
- U, the set of possible actions in the environment;

- $T : X \times U \rightarrow \Pi(X)$ the state transition function, where $\Pi(X)$ is the set of all probability distributions over X;
- $O : X \rightarrow \Pi(Y)$ the observation function, where $\Pi(Y)$ is the set of all probability distributions over Y;
- $R : X \times U \times X \rightarrow \mathbb{R}^r$ the immediate payoff function, where r is the number of criteria;

A Markov decision process (MDP) is given when the Markov property holds: the effects of an action solely depend on current input. Thus, the POMDP defined above reduces to an MDP if each possible sensation in the current state uniquely identifies the current state. That is, each possible sensation in a state x (i.e., all $y \in Y$ for which $O(x)$ is greater than zero) is only possible in this state. If an observation does not uniquely identify the current state but rather provides an (implicit) probability distribution over possible states, the Markov property is violated and the environment turns into a *non-Markov problem*. In this case, optimal action choices do not necessarily depend only on current sensory input anymore but usually depend also on the history of perceptions, actions, and payoff.

3.2   Adaptive Agent Framework

Given the environmental properties, we sketch a general animat framework in this section. We define an animat by a 5-tuple $\mathcal{A} =< S, A, M^S, M^P, \Pi >$. This animat acts in the above defined POMDP environment.

At a certain time t, the animat perceives sensation $y(t) \in Y$ and reinforcement $P(t) \in \mathbb{R}$. The probability of perceiving $y(t)$ is determined by the probability vector $O(x(t))$ and similarly, the probability of $x(t)$ is determined by the probability vector $T(x(t-1), u(t-1))$ which depends on the previous environmental state and the executed action. The received reward depends on the executed action as well as the previous and current state, $P(t) = R(x(t-1), u(t-1), x(t))$.

Thus, in a behavioral act an animat $\mathcal{A}$ receives sensation $y(t)$ and reinforcement $P(t)$ and chooses to execute an action A. To be able to learn and reason about the environment, $\mathcal{A}$ has internal states denoted by S that can represent memory of previous interactions, current beliefs, motivations, intentions etc. Actions $A \subseteq U$ denote the action possibilities of the animat. For our purposes separated from the internal state, we define a state model M^S and a predictive model M^P. The state model M^S represents current environmental characteristics the agent believes in — an implicit probability distribution over all possible environmental states X. The predictive model M^P specifies how the state model changes, possibly dependent on actions. Thus, it describes an implicit and partially action-dependent probability distribution of future environmental states. Finally, Π denotes the behavioral policy of the animat, that is, how the animat decides on what to do, or which action to execute. The policy might depend on current sensory input, on predictions generated by the predictive model, on the state model, and on the internal state.

Learning can be incorporated in the animat by allowing the modification of the components over time. The change of its internal state could, for example, reflect the gathering of memory or the change of moods. The state model could be modified by generalizing over, for example, equally relevant sensory input. The predictive model could learn and adapt probabilities of possible state transitions as well as generalize over effects and conditions.

This rather informal agent framework suffices for our purposes of distinguishing between different classes of anticipatory behavior in animats.

3.3   Distinctions of Anticipatory Behavior

Within the animat framework above, we can infer that the predictive model M^P plays a major role in anticipatory animats. However, in the broader sense of anticipatory behavior also animats without such a model might be termed anticipatory in that their behavioral program is constructed in anticipation of possible environmental challenges. We term this first class of anticipations implicitly anticipatory. The other three classes utilize some kind of prediction to influence behavior. We distinguish between payoff anticipations, sensory anticipations, and state anticipations. All four types of anticipatory behavior are discussed in further detail below.

Implicitly Anticipatory Animats The first animat-type is the one in which no predictions whatsoever are made about the future that might influence the animat's behavioral decision making. Sensory input, possibly combined with internal state information, is directly mapped onto an action decision. The predictive model of the animat M^P is empty or does not influence behavioral decision making in any way. Moreover, there is no action comparison, estimation of action-benefit, or any other type of prediction that might influence the behavioral decision. However, implicit anticipations are included in the behavioral program of the animat. The basic structure of an implicit anticipatory mechanism is shown in Figure 1.

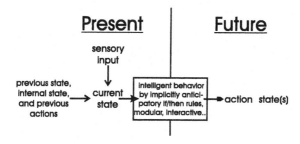

Fig. 1. Implicit anticipatory behavior does not rely on any explicit knowledge about possible future states. The behavior is anticipatory in that the behavioral architecture is predicted to be effective. For example, a genetic code is implicitly predicted (by evolution) to result in successful survival and reproduction.

In nature, even if a life-form behaves purely reactively, it has still implicit anticipatory information in its genetic code in that the behavioral programs in the code are (implicitly) anticipated to work in the offspring. Evolution is the implicit anticipatory learning mechanism that imprints implicit anticipations in the genes. Similarly, well-designed implicitly anticipatory animats, albeit without any prediction that might influence behavior, have implicit anticipatory information in the structure and interaction of algorithm, sensors, and actuators. The designer has included implicit anticipations of environmental challenges and behavioral consequences in the controller of the animat.

It is interesting to note that this rather broad understanding of the term "anticipation" basically classifies any form of life in this world as either implicitly anticipatory or more explicitly anticipatory. Moreover, any somewhat successful animat program can be classified as implicitly anticipatory since its programmed behavioral biases are successful in the addressed problems. Similarly, any meaningful learning mechanism works because it supposes that future experience will be somewhat similar to experience in the past and consequently biases its learning mechanisms on experience in the past. Thus, any meaningful learning and behavior is implicitly anticipatory in that it anticipates that past knowledge and experience will be useful in the future. It is necessary to understand the difference between such implicitly anticipatory animats and animats in which explicit future representations influence behavior.

Payoff Anticipations If an animat considers predictions of the possible payoff of different actions to decide on which action to execute, it may be termed payoff anticipatory. In these animats, predictions estimate the benefit of each possible action and bias action decision making accordingly. No state predictions influence action decision making. A payoff anticipatory mechanism is schematized in Figure 2.

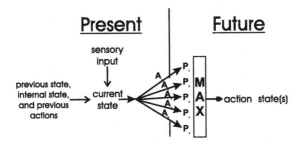

Fig. 2. Sensory anticipatory behavior influences sensory processing due to sensory predictions, expectations, or goal-dependent relevance measures.

A particular example for payoff anticipations is direct (or model-free) reinforcement learning (RL). Hereby, payoff is estimated with respect to the current behavioral strategy or in terms of possible actions. The evaluation of the estimate causes the alternation of behavior which again cause the alternation of the payoff estimates. It can be distinguished between on-policy RL algorithms, such as the SARSA algorithm [42, 52], and off-policy RL algorithms, such as Q-learning [65, 52] or recent learning classifier systems such as XCS [67].

Sensorial Anticipations While in payoff anticipations predictions are restricted to payoff, in sensory anticipations predictions are unrestricted. However, sensory anticipations do not influence the behavior of an animat directly but sensory processing is influenced. The prediction of future states and thus the prediction of future stimuli influences stimulus processing. To be able to form predictions, the animat must use a (not necessarily complete) predictive model M^P of its environment (see Section 3.2). Expected sensory input might be processed faster than unexpected input or unexpected input with certain properties (for example possible threat) might be reacted to faster. A sensory anticipatory mechanism is sketched in Figure 3.

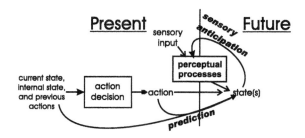

Fig. 3. Sensory anticipatory behavior influences, or predisposes, sensory processing due to future predictions, expectations, or intentions.

Sensory anticipations strongly relate to preparatory attention in psychology [31, 38] in which top-down processes such as task-related expectations influence sensory processing. Behavior is not directly influenced but sensory (pre-)processing is. In other words, sensory anticipatory behavior results in a predisposition of processing sensory input. For example, the agent may become more susceptible to specific sensory input and more ignorant to other sensory input. The biased sensory processing might then (indirectly) influence actual behavior. Also learning might be affected by such a bias as suggested in psychological studies on learning [22, 48].

State Anticipations Maybe the most interesting group of anticipations is the one in which animat behavior is influenced by explicit future state representations. As in sensory anticipations, a predictive model M^P must be available to the animat or it must be learned by the animat. In difference to sensory anticipations, however, state anticipations directly influence current behavioral decision making. Explicit anticipatory behavior is schematized in figure 4. The essential property is that prediction(s) about, or simply representations of, future state(s) influence actual action decision.

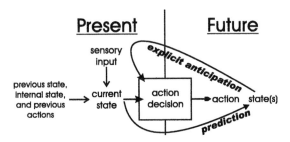

Fig. 4. Explicit anticipations influence actual action decision making due to future predictions, expectations, or intentions.

The simplest kind of explicit anticipatory animat would be an animat which is provided with an explicit predictive model of its environment. The model could be used directly to pursue actual goals by the means of explicit planning mechanisms such as diverse search methods or *dynamic programming* [5]. The most extreme cases of such high-level planning approaches can be found in early artificial intelligence work such as the general problem solver [36] or the STRIPS language [13]. Nowadays, somewhat related approaches try to focus on *local mechanisms* that extract only *relevant environmental information*.

In RL, for example, the dynamic programming idea was modified yielding indirect (or model-based) RL animats. These animats learn an explicit predictive model of the environment. Decisions are based on the predictions of all possible behavioral consequences and essentially the utility of the predicted results. Thus, explicit representations of future states determine behavior.

Further distinctions in state anticipatory animats are evident in the structure and completeness of the model representation, the learning and generalization mechanisms that may change the model over time, and the mechanisms that exploit the predictive model knowledge to adapt behavior. The structure of the predictive model can be represented by rules, by a probabilistic network, in the form of hierarchies and so forth. The model representation can be based on internal model states $M^S(t)$ or rather directly on current sensory input $y(t)$. State information in the sensory input can provide global state information or rather local state information dependent on the animat's current position in the environment. Learning and generalization mechanisms give rise to further crucial differences in the availability, the efficiency, and the utility of the predictive model. Finally, the bias of the behavioral component results in different anticipatory behavior mechanisms. For example, the number of steps that the animat can look into the future is a crucial measure as proposed in [45]. Moreover, anticipatory processes might only take place in the event of actual behavioral execution or the processes may be involved in adapting behavior offline. Proper distinctions between these different facets of state anticipatory behavior may be developed in future research.

With a proper definition of animats and four fundamental classes of anticipatory behavior in hand, we now provide a case study of typical existing anticipatory animats.

4   Payoff Anticipatory Animats

This section introduces several common payoff anticipatory animats. As defined above, these animats do not represent or learn a predictive model M^P of their environment but a knowledge base assigns values to actions based on which action decisions are made.

4.1   Model-Free Reinforcement Learning

The reinforcement learning framework [27, 52] considers adaptive agents involved in a sensory-motor loop acting upon a MDP as introduced above (extensions to POMDPs can be found for example in [9]). The task of the agents is to learn an optimal policy, i.e., how to act in every situation in order to maximize the cumulative reward over the long run.

In model-free RL, or *direct reinforcement learning*, the animat learns a behavioral policy without learning an explicit predictive model. The most common form of direct reinforcement learning is to learn utility values for all possible state-action combinations in the MDP. The most common approach in this respect is the Q-learning approach introduced in [65]. Q-learning has the additional advantage that it is policy independent. That is, as long as the behavioral policy assures that all possible state action transitions are visited infinitely often over the long run, Q-learning is guaranteed to generate an optimal policy.

Model-free RL agents are clearly payoff anticipatory animats. There is no explicit predictive model; however, the learned reinforcement values estimate action-payoff. Thus, although the animat does not explicitly learn a representation with which it knows the actual sensory consequences of an action, it can compare available action choices based on the payoff predictions and thus act payoff anticipatory.

Model-free RL in its purest form usually stores all possible state-action combinations in tabular form. Also, states are usually characterized by unique identifiers rather than by sensory inputs that allow the identification of states. This ungeneralized exhaustive state representation prevents RL to scale-up to larger problems. Several approaches exist that try to overcome the curse of dimensionality by function approximation techniques (cf. [52]), hierarchical approaches (cf. [54, 4]), or online generalization mechanisms. Approaches that generalize online over sensory inputs (for example in the form of a feature vector) are introduced in the following.

4.2 Learning Classifier Systems

Learning Classifier Systems (LCSs) have often been overlooked in the research area of RL due to the many interacting mechanisms in these systems. However, in their purest form, LCSs can be characterized as RL systems that generalize online over sensory input. This generalization mechanism leads to several additional problems especially with respect to a proper propagation of RL values over the whole state action space.

The first implementation of an LCS, called CS1, can be found in [25]. Holland's goal was to propose a model of a cognitive system that is able to learn using both reinforcement learning processes and genetic algorithms [23, 20]. The first systems, however, were rather complicated and lacked efficiency.

Reinforcement values in LCSs are stored in a set (the population) of condition-action rules (the classifiers). The conditions specify a subset of possible sensations in which the classifier is applicable thus giving rise to focusing mechanisms and attentional mechanisms often over-looked in RL. The learning mechanism of the population of classifiers and the classifier structure is usually accomplished by the means of a genetic algorithm (GA). Lanzi provides an insightful comparison between RL and learning classifier systems [33]. It appears from this perspective that a LCS is a rule-based reinforcement learning system endowed with the capability to generalize what it learns.

Thus, also LCSs can be classified as payoff-anticipatory animats. The generalization over the perceptions promises faster adaptation in dynamic environments. Moreover, the policy representation may be more compact especially in environments in which a lot of sensations are available but only a subset of the sensations is task relevant.

Recently, Wilson implemented several improvements in the LCS model. He modified the traditional Bucket Brigade algorithm [26] to resemble the Q-learning mechanism propagating Q-values over the population of classifiers [66, 67]. Moreover, Wilson drastically simplified the LCS model [66]. Then, he modified Holland's original strength-based criterion for learning — the more a rule receives reward (on average), the more fit it is [23, 24, 6] — by a new criterion relying on the accuracy of the reward prediction of each rule [67]. This last modification gave rise to the most commonly used LCS today, XCS.

5 Anticipations Based on Predictive Models

While the model-free reinforcement learning approach as well as LCSs do not have or use a predictive model representation, the agent architectures in this section all learn or have a predictive model M^P and use this model to yield anticipatory behavior. Due to the usage of an explicit predictive model of the environment, all systems can be classified as either sensory anticipatory or state anticipatory. Important differences of the systems are outlined below.

5.1   Model-Based Reinforcement Learning

The dynamical architecture *Dyna* [53] learns a model of its environment in addition to reinforcement values (state values or Q-values). Several anticipatory mechanisms can be applied such as biasing the decision maker toward the exploration of unknown/unseen regions or applying internal reinforcement updates. Dyna is one of the first state anticipatory animat implementations. It usually forms an ungeneralized representation of its environment in tabular form but it is not necessarily restricted to such a representation. Interesting enhancements of Dyna have been undertaken optimizing the internal model-based RL process [35, 40] or adopting the mechanism to a tile coding approach [30]. The introduction of Dyna was kept very general so that many of the subsequent mechanisms can be characterized as Dyna mechanisms as well. Differences can be found in the learning mechanism of the predictive model, the sensory input provided, and the behavioral policy learning.

5.2   Schema Mechanism

An architecture similar to the Dyna architecture was published in [11]. The implemented *schema mechanism* is loosely based on Piaget's proposed developmental stages. The model in the schema mechanism is represented by rules. It is learned bottom-up by generating more specialized rules where necessary. Although no generalization mechanism applies, the resulting predictive model is somewhat more general than a tabular model. The decision maker is — among other criteria — biased on the exploitation of the model to achieve desired items in the environment. Similar to Dyna, the schema mechanism represents an explicit anticipatory agent. However, the decision maker, the model learner, and the predictive model representation M^P have a different structure.

5.3   Expectancy Model SRS/E

Witkowski [70] approaches the same problem from a cognitive perspective giving rise to his *expectancy model SRS/E*. Similar to Dyna, the learned model is not generalized but represented by a set of rules. Generalization mechanisms are suggested but not tested. SRS/E includes an additional sign list that stores all states encountered so far. In contrast to Dyna, reinforcement is not propagated online but is only propagated once a desired state is generated by a behavioral module. The propagation is accomplished using dynamic programming techniques applied to the learned predictive model and the sign list.

5.4   Anticipatory Learning Classifier Systems

Similar to the schema mechanism and SRS/E, anticipatory learning classifier systems (ALCSs) [50, 8, 19, 17] contain an explicit prediction component. The predictive model consists of a set of rules (classifiers) which are endowed with a so called "effect" part. The effect part predicts the next situation the agent will encounter if the action specified by the rules is executed. The second major characteristic of ALCSs is that they generalize over sensory input.

ACS An *anticipatory classifier system* (ACS) was developed by Stolzmann [49, 50] and was later extended to its current state of the art, ACS2 [8]. ACS2 learns a generalized model of its environment applying directed specialization as well as genetic generalization mechanisms. It has been experimentally shown that ACS2 reliably learns a complete, accurate, and compact predictive model of several typical MDP environments. Reinforcement is propagated directly inside the predictive model resulting in a possible model aliasing problem [8]. It was shown that ACS2 mimics the psychological results of latent learning experiments as well as outcome devaluation experiments mentioned above by implementing additional anticipatory mechanisms into the decision maker [50, 51, 8].

YACS *Yet Another Classifier System* (YACS) is another anticipatory learning classifier system that forms a similar generalized model applying directed specialization as well as generalization mechanisms [17, 18]. Similar to SRS/E, YACS keeps a list of all states encountered so far. Unlike SRS/E, reinforcement updates in the state list are done while interacting with the environment making use of the current predictive model. Thus, YACS is similar to SRS/E but it evolves a more generalized predictive model and updates the state list online.

MACS A more recent approach by [16] learns a different rule-based representation in which rules are learned separately for the prediction of each sensory attribute. Similar to YACS, MACS keeps a state list of all so far encountered states and updates reinforcement learning in those states. The different model representation is shown to allow further generalizations in maze problems.

5.5   Artificial Neural Network Models of Anticipation

Also Artificial Neural Networks (ANN) can be used to learn the controller of an agent. In accordance with the POMDP framework, the controller is provided with some inputs from the sensors of the agent and must send some outputs to the actuators of the agent. Learning to control the agent consists in learning to associate the good set of outputs to any set of inputs that the agent may experience.

The most common way to perform such learning with an ANN consists in using the back-propagation algorithm. This algorithm consists in computing for each set of inputs the errors on the outputs of the controller. With respect to the computed error, the weights of the connections in the network are modified so that the error will be smaller the next time the same inputs are encountered.

The main drawback of this algorithm is that one must be able to decide for any input what the correct output should be so as to compute an error. The learning agent must be provided with a supervisor which tells at each time step what the agent should have done. Back-propagation is a supervised learning method. The problem with such a method is that in most control problems, the correct behavior is not known in advance. As a consequence, it is difficult to build a supervisor.

The solution to this problem consists in relying on anticipation [55, 57]. If the role of an ANN is to predict what the next input will be rather than to provide an output, then the error signal is available: it consists in the difference between what the ANN predicted and what has actually happened. As a consequence, learning to predict thanks to a back-propagation algorithm is straight-forward.

Baluja's Attention Mechanism Baluja and Pomerleau provide an interesting anticipatory implementation of visual attention in the form of a neural network with one hidden layer [2, 3]. The mechanism is based on the ideas of visual attention modeling in [28]. The system is for example able to learn to follow a line by the means of the network. Performance of the net is improved by adding another output layer, connected to the hidden layer, which learns to predict successive sensory input. Since this output layer is not used to update the weights in the hidden layer, Baluja argues that consequently the predictive output layer can only learn task-relevant predictions. The predictions of the output layer are used to modify the successive input in that the strong differences between prediction and real input are decreased assuming strong differences to be task irrelevant noise. Baluja shows that the neural net is able to utilize this image flattening to improve performance and essentially ignore spurious line markings and other distracting noise. It is furthermore suggested that the architecture could also be used to detect unexpected sensations faster possibly usable for anomaly detection tasks.

Baluja's system is a payoff anticipatory system. The system learns a predictive model which is based on pre-processed information in the hidden units. The predictive model is action-independent. Sensory anticipations are realized in that the sensory input is modified according to the difference between predicted and actual input.

Tani's Recurrent Neural Networks Tani published a recurrent neural network (RNN) approach implementing model-based learning and planning in the network [55]. The system learns a predictive model using the sensory information of the next situation as the supervision. *Context units* are added that feed back the values of the current hidden units to additional input units. This recurrence allows a certain internal representation of time [12]. In order to use the emerging predictive model successfully, it is necessary that the RNN becomes situated in the environment — the RNN needs to identify its current situation in the environment by adjusting its recurrent inputs. Once the model is learned, a navigation phase is initiated in which the network is used to plan a path to a provided goal.

The most appealing result of this work is that the RNN is actually implemented in a real mobile robot. The implementation is shown to handle noisy, online discretized environments. Anticipatory behavior is implemented by a lookahead planning mechanism. The system is a state anticipatory system in which the predictive model is represented in a RNN. In contrast to the approaches above, the RNN also evolves an implicit state model M^S represented and updated by the recurrent neural network inputs. This is the reason why the network has to become situated before planning is applicable. Tani shows that predicting the next inputs correctly helps stabilizing the behavior of its agents and, more generally, that using anticipations results in a bi-polarization of the behavior into two extreme modes: a very stable mode when everything is predicted correctly, and a chaotic mode when the predictions get wrong.

In a further publication [56], Tani uses a constructivist approach in which several neural networks are combined. The approach implements an attentional mechanism that switches between wall following and object recognition. Similar to the winner-takes-all algorithm proposed in [28], Tani uses a winner-takes-all algorithm to implement a visual attention mechanism. The algorithm combines sensory information with model prediction, thus pre-processing sensory information due to predictions. The resulting categorical output influences the decision maker that controls robot movement. Thus, the constructed animat comprises sensory anticipatory mechanisms that influence attentional mechanisms similar to Baluja's visual attention mechanism but embedded in a bigger modular structure.

In [57], a first approach of a hierarchical structured neural network suitable as a predictive model is published. While the lower level in the hierarchy learns the basic sensory-motor flow, the higher level learns to predict the switching of the network in the lower level and thus a more higher level representation of the encountered environment. Anticipatory behavior was not shown within the system.

5.6   Anticipations in a Multi-agent Problem

A first approach that combines low level reactive behavior with high-level deliberation can be found in [10]. The animats in this framework are endowed with a predictive model that predicts behavior of the other, similar animats. Although the system does not apply any learning methods, it is a first approach of state anticipations in a multi-agent environment. It is shown that by anticipating the behavior of the other agents, behavior can be optimized achieving cooperative behavior. Davidsson's agent is a simple anticipatory agent that uses the (restricted) predictive model of other agents to modify the otherwise reactive decision maker. Since the decision maker is influenced by the predictive model the agents can be classified as non-learning state-anticipatory animats.

6   Discussion

As can be seen in the above study of anticipatory systems, a lot of research is still needed to clearly understand the utility of anticipations. This section further discusses different aspects in anticipatory approaches.

6.1   Anticipating with or without a Model

One main advantage of model building animats with respect to model-free ones is that their model endows them with a planning capability. Having an internal predictive model which specifies which action leads from what state to what other state permits the agent to plan its behavior "in its head". But planning does not necessarily mean that the agent actually searches in its model a complete path from its current situation to its current goal. Indeed, that strategy suffers from a combinatorial explosion problem. It may rather mean that the agent updates the values of different state model states ($x \in M^S$) without having to actually move in its environment. This is essentially done in dynamic programming [5] and it is adapted to the RL framework in the Dyna architecture [53, 52]. The internal updates allow a faster convergence of the learning algorithms due to the general acceleration of value updates.

These ideas have been re-used in most anticipatory rule-based learning systems described above. Applying the same idea in the context of ANN, with the model being implemented in the weights of recurrent connections in the network, would consist in letting the weights of the recurrent connections evolve faster than the sensory-motor dynamics of the network. To our knowledge, though, this way to proceed has not been used in any anticipatory ANN animat, yet.

Pros and Cons of Anticipatory Learning Classifier Systems Having an explicit predictive part in the rules of ALCSs permits a more directed use of more information from the agent's experience to improve the rules with respect to classical LCSs. Supervised learning methods can be applied. Thus, there is a tendency in ALCSs to use heuristic search methods rather than blind genetic algorithms to improve the rules.

This use of heuristic search methods results then in a much faster convergence of anticipatory systems on problems where classical LCSs are quite slow, but it also results in more complicated systems, more difficult to program, and also in less general systems.

For example, XCS-like systems can be applied both to single-step problems such as Data Mining Problems [68] where the agent has to make only one decision independent from its previous decisions and to multi-step problems where the agent must run a sequence of actions to reach its goal [32]. In contrast, ALCSs are explicitly devoted to multi-step problems, since there must be a "next" situation after each action decision from the agent.

6.2   A Parallel Between Learning Thanks to Prediction in ANN and in ALCS

The second matter of discussion emerging from this overview is the parallel that can be made in the way ANN and rule-based systems combine predictions and learning to build and generalize a model of the problem.

We have seen that in Tani's system, the errors on predictions are back-propagated through the RNN so as to update the weights of the connections.

This learning process results in an improved ability to predict, thus in a better predictive model.

The learning algorithms in the presented ALCSs rely on the same idea. The prediction errors are represented by the fact that the predictions of a classifier are sometimes good and sometimes bad, in which case the classifier oscillates (or is called not reliable). In this case, more specific classifiers are generated by the particular specialization process. Thus, the oscillation of classifiers is at the heart of the model improvement process.

Specializing a classifier when it oscillates is a way to use the error of the prediction so as to improve the model, exactly as it is done in the context of ANN.

This way of learning is justified by the fact that both systems include a capacity of generalization in their models. Otherwise, it would be simpler just to include any new experience in the anticipatory model without having to encompass a prediction and correction process. The point is that the prediction can be general and the correction preserves this generality as much as it can. Interestingly, however, generalization is not exactly of the same nature in ANN and in ALCSs.

As a conclusion, both classes of systems exhibit a synergy between learning, prediction, and generalization, learning being used to improve general predictions, but also predictions being at the heart of learning general features of the environment.

6.3   Model Builders and Non-Markov Problems

As explained in section 3.1, a non-Markov problem is a problem in which the current sensations of the animat are not always sufficient to choose the best action. In such problems, the animat must incorporate an internal state model representation M^S providing a further source of information for choosing the best action. The information in question generally comes from the more or less immediate past of the animat. An animat which does not incorporate such an internal state model is said to be "reactive". Reactive animats cannot behave optimally in non-Markov problems.

In order to prevent misinterpretations, we must warn the reader about the fact that an internal state model differs from an internal predictive model. In fact, an internal predictive model alone does not enable the animat to behave optimally in a non-Markov problem. Rather than information about the immediate past of the animat, predictive models only provide information about the "atemporal" structure of the problem (that is, information about the possible future). In particular, if the animat has no means to disambiguate aliased perceptions, it will build an aliased model. Thus an animat can be both reactive, that is, unable to behave optimally in non-Markov environments, and explicitly anticipatory, that is, able to build a predictive model of this environment and bias its action decisions on future predictions, without solving the non-Markov problem.

7   Conclusion

This overview of internal models and anticipatory behavior showed that a lot of future research is needed to understand exactly when which anticipations are useful or sometimes even mandatory in an environment to yield competent adaptive behavior. Although psychological research proves that anticipatory behavior takes place in at least higher animals, a clear understanding of the *how*, the *when*, and the *which* is not available. Thus, one essential direction of future research is to identify environmental characteristics in which distinct anticipatory mechanisms are helpful or necessary.

Several more concrete research directions can be suggested. (1) It seems important to quantify when anticipatory behavior can be adapted faster than stimulus-response behavior. For example, in a dynamic environment some predictive knowledge may be assumed to be stable so that behavior can be adapted by the means of this knowledge. (2) It appears interesting to investigate how to balance reactive and anticipatory mechanisms and how to allow a proper interaction. A proper architecture of motivations and emotions might play an important role in this respect. (3) Adaptive mechanisms that are initially anticipatory and then become short circuited reactive demand further research effort. For example, initial hard practice of playing an instrument becomes more and more automatic and is eventually only guided by a correct feeling of its functioning. Can we create a similar adaptive motor-control mechanism? (4) The functioning of attentional processes influenced by sensory anticipations needs to be investigated further. When are such attentional mechanisms beneficial, when does the drawback due to inattentional blindness effects overshadow the benefits? (5) The benefit of simulating intentions and behavior of other animats requires further research effort. Which processes are necessary to create beneficial social relationships? Which mechanisms can result in mutual benefit, which mechanisms can cause unilateral benefit?

This small but broad list shows that future work in anticipatory learning systems promises fruitful research projects and new exciting insights in the field of adaptive behavior. We hope that our overview of current insights in anticipatory mechanisms and the available systems provide a basis for future research efforts. Moreover, we want to encourage the development of the distinctions between anticipatory behavior mechanisms. While implicit and payoff anticipatory mechanisms appear to be rather clear cut, sensory and state anticipatory behavior comprise many different forms and mechanisms. Future research will show which characteristics should be used to distinguish the different mechanisms further.

Acknowledgments

The authors would like to thank Stewart Wilson, Joanna Bryson, and Mark Witkowski for useful comments on an earlier draft of this introduction.

This work was funded by the German Research Foundation (DFG) under grant HO1301/4-3.

References

[1] Arbib, M.: The mirror system, imitation, and the evolution of language. In Dautenhahn, K., Nehaniv, C.L., eds.: Imitation in animals and artifacts. MIT Press, Cambridge, MA (2002)

[2] Baluja, S., Pomerleau, D.A.: Using the representation in a neural network's hidden layer for task-specific focus on attention. Proceedings of the Fourteenth International Joint Conference on Artificial Intelligence (1995) 133–141

[3] Baluja, S., Pomerleau, D.A.: Expectation-based selective attention for visual monitoring and control of a robot vehicle. Robotics and Autonomous Systems **22** (1997) 329–344

[4] Barto, A.G., Mahadevan, S.: Recent advances in hierarchical reinforcement learning. Discrete event systems (2003, to appear)

[5] Bellman, R.E.: Dynamic programming. Princeton University Press, Princeton, NJ (1957)

[6] Booker, L., Goldberg, D.E., Holland, J.H.: Classifier systems and genetic algorithms. Artificial Intelligence **40** (1989) 235–282

[7] Brooks, R.A.: Intelligence without reason. Proceedings of the 12th International Joint Conference on Artificial Intelligence (1991) 569–595

[8] Butz, M.V.: Anticipatory learning classifier systems. Kluwer Academic Publishers, Boston, MA (2002)

[9] Cassandra, A.R., Kaelbling, L.P., Littman, M.L.: Acting optimally in partially observable stochastic domains. Proceedings of the Twelfth National Conference on AI (1994) 1023–1028

[10] Davidsson, P.: Learning by linear anticipation in multi-agent systems. In Weiss, G., ed.: Distributed artificial intelligence meets machine learning, Berlin Heidelberg, Springer-Verlag (1997) 62–72

[11] Drescher, G.L.: Made-up minds, a constructivist approach to artificial intelligence. MIT Press, Cambridge, MA (1991)

[12] Elman, J.L.: Finding structure in time. Cognitive Science **14** (1990) 179–211

[13] Fikes, R.E., Nilsson, N.J.: STRIPS: A new approach to the application of theorem proving to problem solving. Artificial Intelligence **2** (1971) 189–208

[14] Gallese, V.: The 'shared manifold' hypothesis: From mirror neurons to empathy. Journal of Consciousness Studies: Between Ourselves - Second-Person Issues in the Study of Consciousness **8** (2001) 33–50

[15] Gallese, V., Goldman, A.: Mirror neurons and the simulation theory of mind-reading. Trends in Cognitive Sciences **2** (1998) 493–501

[16] Gérard, P., Meyer, J.A., Sigaud, O.: Combining latent learning and dynamic programming in MACS. European Journal of Operational Research (submitted, 2003)

[17] Gérard, P., Stolzmann, W., Sigaud, O.: YACS: A new learning classifier system with anticipation. Soft Computing **6** (2002) 216–228

[18] Gérard, P., Sigaud, O.: Adding a generalization mechanism to YACS. Proceedings of the Genetic and Evolutionary Computation Conference (GECCO-2001) (2001) 951–957

[19] Gérard, P., Sigaud, O.: YACS: Combining dynamic programming with generalization in classifier systems. In Lanzi, P.L., Stolzmann, W., Wilson, S.W., eds.: Advances in learning classifier systems: Third international workshop, IWLCS 2000. Springer-Verlag, Berlin Heidelberg (2001) 52–69

[20] Goldberg, D.E.: Genetic algorithms in search, optimization and machine learning. Addison-Wesley, Reading, MA (1989)

[21] Herbart, J.: Psychologie als Wissenschaft neu gegründet auf Erfahrung, Metaphysik und Mathematik. Zweiter, analytischer Teil. August Wilhem Unzer, Königsberg, Germany (1825)

[22] Hoffmann, J., Sebald, A., Stöcker, C.: Irrelevant response effects improve serial learning in serial reaction time tasks. Journal of Experimental Psychology: Learning, Memory, and Cognition 27 (2001) 470–482

[23] Holland, J.H.: Adaptation in natural and artificial systems. The University of Michigan Press (1975)

[24] Holland, J.H., Holyoak, K.J., Nisbett, R.E., Thagard, P.R.: Induction. MIT Press (1986)

[25] Holland, J.H., Reitman, J.S.: Cognitive systems based on adaptive algorithms. Pattern Directed Inference Systems 7 (1978) 125–149

[26] Holland, J.H.: Properties of the bucket brigade algorithm. Proceedings of an International Conference on Genetic Algorithms and their Applications (1985) 1–7

[27] Kaelbing, L.P., Littman, M.L., Moore, A.W.: Reinforcement learning: A survey. Journal of Artificial Intelligence Research 4 (1996) 237–285

[28] Koch, C., Ullmann, S.: Shifts in selective attention: Towards the underlying neural circuitry. Human Neurobiology 4 (1985) 219–227

[29] Kunde, W.: Response-effect compatibility in manual choice reaction tasks. Journal of Experimental Psychology: Human Perception and Performance 27 (2001) 387–394

[30] Kuvayev, L., Sutton, R.S.: Model-based reinforcement learning with an approximate, learned model. In: Proceedings of the ninth yale workshop on adaptive and learning systems, New Haven, CT (1996) 101–105

[31] LaBerge, D.: Attentional processing, the brain's art of mindfulness. Harvard University Press, Cambridge, MA (1995)

[32] Lanzi, P.L.: An analysis of generalization in the XCS classifier system. Evolutionary Computation 7 (1999) 125–149

[33] Lanzi, P.L.: Learning classifier systems from a reinforcement learning perspective. Soft Computing 6 (2002) 162–170

[34] Mack, A., Rock, I.: Inattentinal blindness. MIT Press (Cambridge, MA)

[35] Moore, A.W., Atkeson, C.: Prioritized sweeping: Reinforcement learning with less data and less real time. Machine Learning 13 (1993) 103–130

[36] Newell, A., Simon, H.A., Shaw, J.C.: Elements of a theory of human problem solving. Psychological Review 65 (1958) 151–166

[37] Pashler, H., Johnston, J.C., Ruthruff, E.: Attention and performance. Annual Review of Psychology 52 (2001) 629–651

[38] Pashler, H.E.: The psychology of attention. MIT Press, Cambridge, MA (1998)

[39] Pavlov, I.P.: Conditioned reflexes. London: Oxford (1927)

[40] Peng, J., Williams, R.J.: Efficient learning and planning within the dyna framework. Adaptive Behavior 1 (1993) 437–454

[41] Rizzolatti, G., Fadiga, L., Gallese, V., Fogassi, L.: Premotor cortex and the recognition of motor actions. Cognitive Brain Research 3 (1996) 131–141

[42] Rummery, G.A., Niranjan, M.: On-line Q-learning using connectionist systems. Technical Report CUED/F-INFENG/TR 166, Engineering Department, Cambridge University (1994)

[43] Schubotz, R.I., von Cramon, D.Y.: Functional organization of the lateral premotor cortex. fMRI reveals different regions activated by anticipation of object properties, location and speed. Cognitive Brain Research **11** (2001) 97–112

[44] Simons, D.J., Chabris, C.F.: Gorillas in our midst: Sustained inattentional blindness for dynamic events. Perception **28** (1999) 1059–1074

[45] Sjölander, S.: Some cognitive break-throughs in the evolution of cognition and consciousness, and their impact on the biology language. Evolution and Cognition **1** (1995) 3–11

[46] Skinner, B.F.: The behavior of organisms. Appleton-Century Crofts, Inc., New-York (1938)

[47] Skinner, B.F.: Beyond freedom and dignity. Bantam/Vintage, New York (1971)

[48] Stock, A., Hoffmann, J.: Intentional fixation of behavioral learning or how R-E learning blocks S-R learning. European Journal of Cognitive Psychology (2002) in press.

[49] Stolzmann, W.: Antizipative Classifier Systems [Anticipatory classifier systems]. Shaker Verlag, Aachen, Germany (1997)

[50] Stolzmann, W.: Anticipatory classifier systems. Genetic Programming 1998: Proceedings of the Third Annual Conference (1998) 658–664

[51] Stolzmann, W., Butz, M.V., Hoffmann, J., Goldberg, D.E.: First cognitive capabilities in the anticipatory classifier system. From Animals to Animats 6: Proceedings of the Sixth International Conference on Simulation of Adaptive Behavior (2000) 287–296

[52] Sutton, R.S., Barto, A.G.: Reinforcement learning: An introduction. MIT Press (1998)

[53] Sutton, R.: Reinforcement learning architectures for animats. From animals to animats: Proceedings of the First International Conference on Simulation of Adaptative Behavior (1991)

[54] Sutton, R., Precup, D., Singh, S.: Between MDPs and semi-MDPs: A framework for temporal abstraction in reinforcement learning. Artificial Intelligence **112** (1999) 181–211

[55] Tani, J.: Model-based learning for mobile robot navigation from the dynamical system perspective. IEEE Transactions on System, Man and Cybernetics **26** (1996) 421–436

[56] Tani, J.: An interpretation of the "self" from the dynamical systems perspective: A constructivist approach. Journal of Consciousness Studies 5 (1998) 516–542

[57] Tani, J.: Learning to perceive the world as articulated: An approach for hierarchical learning in sensory-motor systems. Neural Networks **12** (1999) 1131–1141

[58] Thistlethwaite, D.: A critical review of latent learning and related experiments. Psychological Bulletin **48** (1951) 97–129

[59] Thompson, E.: Empathy and consciousness. Journal of Consciousness Studies: Between Ourselves - Second-Person Issues in the Study of Consciousness 8 (2001) 1–32

[60] Thorndike, E.L.: Animal intelligence: Experimental studies. Macmillan, New York (1911)

[61] Tolman, E.C.: Purposive behavior in animals and men. Appletown, New York (1932)

[62] Tolman, E.C.: The determiners of behavior at a choice point. Psychological Review **45** (1938) 1–41

[63] Tolman, E.C.: Cognitive maps in rats and men. Psychological Review **55** (1948) 189–208

[64] Tolman, E.C.: Principles of purposive behavior. In Koch, S., ed.: Psychology: A study of science, New York, McGraw-Hill (1959) 92–157

[65] Watkins, C.J.: Learning with delayed rewards. PhD thesis, Psychology Department, University of Cambridge, England (1989)

[66] Wilson, S.W.: ZCS, a zeroth level classifier system. Evolutionary Computation **2** (1994) 1–18

[67] Wilson, S.W.: Classifier fitness based on accuracy. Evolutionary Computation **3** (1995) 149–175

[68] Wilson, S.W.: Mining oblique data with XCS. In Lanzi, P.L., Stolzmann, W., Wilson, S.W., eds.: Advances in learning classifier systems: Third international workshop, IWLCS 2000, Berlin Heidelberg, Springer-Verlag (2001)

[69] Wilson, S.W.: Knowledge growth in an artificial animal. In Grefenstette, J.J., ed.: Proceedings of an international conference on genetic algorithms and their applications, Carnegie-Mellon University, Pittsburgh, PA (1985) 16–23

[70] Witkowski, C.M.: Schemes for learning and behaviour: A new expectancy model. PhD thesis, Department of Computer Science, University of London, England (1997)

[71] Wolpert, D.H.: The lack of a priori distinctions between learning algorithms. Neural Computation **8** (1995) 1341–1390

Mathematical Foundations of Discrete and Functional Systems with Strong and Weak Anticipations

Daniel M. Dubois

Centre for Hyperincursion and Anticipation in Ordered Systems,
CHAOS asbl, Institute of Mathematics, B37, University of Liège,
Grande Traverse 12, B- 4000 Liège, Belgium
`Daniel.Dubois@ulg.ac.be`
`http://www.ulg.ac.be/mathgen/CHAOS`

Abstract. This paper deals with some mathematical developments to model anticipatory capabilities in discrete and continuous systems. The paper defines weak anticipation and strong anticipation and introduces the concepts of incursive and hyperincursive discrete processes as an extension to recursion. Functional systems represented by differential difference equations with anticipation and/or delay seem to be a very useful tool for describing strong anticipation. Anticipation and delay play a complementary role and synchronization mechanisms seem to be a powerful way to anticipate the evolution of systems with delay. This paper shows finally that the modelling of anticipation in predictive control is the basic mechanism for enhancing the control of the trajectory of systems toward a target.

1 Introduction

The purpose of this paper is to present some mathematical developments for formalizing systems with anticipative capabilities. The concept of anticipation is tentatively formalized in a general framework. The anticipative properties of systems depend on their structure and functionality. Indeed, the description of anticipation embedded in a natural biological system has some common features with the construction of anticipation in a control system.

In the field of bio-mathematics, Robert Rosen [20] was a precursor in defining an anticipatory systems as: "An anticipatory system is a system containing a predictive model of itself and/or of its environment, which allows it to change state at an instant in accord with the model's predictions pertaining to a latter instant". Robert Rosen related the anticipation to the concept of final causation of Aristotle. Moreover, Robert Rosen thought that anticipation is a key difference between natural living systems and non-living systems like physical ones. But, contrary to what Robert Rosen thought, anticipatory properties also exist in physical systems as shown for electromagnetism [12] and relativity transformations [11].

There are many applications dealing with predictive models. Let us give three examples. Firstly, internal models for prediction seems to play a role in neuroscience. During self-generated movement, it is postulated [2] that an efference copy of the descending motor command, in conjunction with an internal model of both the motor

M. Butz et al. (Eds.): Anticipatory Behavior ..., LNAI 2684, pp. 110-132, 2003.

system and environment, enables to predict the consequences of our own actions. Secondly, the application of model predictive control of active tracking of 3D moving targets [1]. Visual tracking is presented as a regulation control problem. Delays and system latencies substantially affect the performance of visually guided systems. Model predictive control strategies are proposed to compensate for the mechanical latency in visual control of motion. Thirdly, the application of generalized predictive control to discrete-time chaotic systems [17]. Both control performance and system sensitivity to initial conditions are compared with the conventional model-referenced adaptive control. Simulation results show that this controller yields faster setting time, more accurate tracking, and less initial sensitivity.

Let us point out that predictive models were developed over the past two or three decades in the field of control, what is referred as the Model-based Predictive Control (MPC). These include Model Predictive Heuristic Control (MPHC) [18], Dynamic Matrix Control (DMC) [7], Internal Model Control (IMC) [19] and Generalized Predictive Control (GPC) [6].

This paper is organized as follows. The section 2 introduces the concept of weak and strong anticipation and the definitions of recursive, incursive and hyperincursive discrete processes with a few examples. The section 3 gives a few mathematical methods for analysing and computing some functional systems represented by differential difference equations with delay and/or anticipation, with examples and numerical simulations. The last section 4 is a short survey of the role of anticipation in model predictive control with an example of control of chaos compared to an incursive control, with simulations.

2   Weak Anticipation and Strong Anticipation with Incursion and Hyperincursion

A tentative definition of anticipation could be: an anticipatory system is a system for which the present behaviour is based on past and/or present events but also on future events built from these past, present and future events. Any anticipatory system can obey, as any physical systems, the Maupertuis Least Action Principle.

In view of explicitly defining mathematically systems with anticipation, I introduced the concept of *incursion*, an inclusive or implicit recursion. An incursive system is a recursive system that takes into account future states for evolving. Some nonlinear incursive systems show several potential future states, what I called *hyperincursion*. An hyperincursive anticipatory system generates multiple potential states at each time step and corresponds to one-to-many relations. A selection parameter must be defined to select a particular state amongst these multiple potential states. These multiple potential states collapse to one state (among these states) which becomes the actual state.

The anticipation of a system can be based on a model of its environment. In this case, I introduced the notion of *exo-anticipation*, with the following definition: an exo-anticipation is an anticipation made by a system about external systems. In this case, anticipation is more related to predictions or expectations. This defines a *weak anticipation*.

The anticipation of a system can be based on itself, rather than its environment. In this case, I introduced the notion of *endo-anticipation*, with the following definition: an endo-anticipation is an anticipation built by a system or embedded in a system about its own behaviour. This is no more a predictive anticipation but a built anticipation. In this case, this is a *strong anticipation*.

Let us give more precise definitions of recursive, incursive and hyperincursive systems.

2.1   Recursive Systems

A recursive discrete system is a system, which computes its successive time states as a function of its past and present states as

$$x(t+1) = R(..., x(t-2), x(t-1), x(t); p) \qquad (1)$$

where $x(t)$ are the vector states at time t, R the recursive vector function and p a set of parameters to be adjusted. In knowing the function R, the values of the parameters p and the initial conditions ..., $x(-2)$, $x(-1)$, $x(0)$ at time t = 0, the successive states $x(t+1)$, $x(t+2)$, $x(t+3)$, ..., where the interval of time $\Delta t = 1$ is a duration, can be recursively computed.

2.2   Incursive Discrete Weak Anticipatory Systems

Definition of an incursive discrete weak anticipatory system: a weak incursive system is a system which computes its current state at time t, as a function of its states at past times, ..., t−3, t−2, t−1, present time, t, and even its predicted states at future times t+1, t+2, t+3, ...

$$x(t+1) = A(..., x(t-2), x(t-1), x(t), x^*(t+1), x^*(t+2), ...; p) \qquad (2)$$

where the variable x^* at future times t+1, t+2, ... are computed in using a predictive model of the system. Such weak anticipatory systems may be related to the Robert Rosen [20] definition: "An anticipatory system is a system containing a predictive model of itself and/or of its environment, which allows it to change state at an instant in accord with the model's predictions pertaining to a latter instant".

Paul Davidsson, Eric Astor and Bertil Ekdahl [8] gave the following equation, $s_{n+1} = f(s_1, s_2, ..., s_n, s_{n+1}, ..., s_k)$, k >n, representing an anticipatory system in formal terms, after the R. Rosen definition, which would be a function of past and future states whereas a causal system only depends on past states $s_{n+1} = f(s_1, s_2, ..., s_n)$. They continue in saying that a system "cannot normally have true knowledge of future states" (a system "in a closed world and having a perfect world model would be able to have true knowledge of future states"), "it is, of course, not possible to implement an anticipatory system in this strict sense". "The best we can do is to approximate such a system by using *predictions* of future states. Thus we have $s_{n+1} = f(s_1, s_2, ..., s_n, s^*_{n,1}, ..., s^*_{n,k-n})$, k >n, where $s^*_{n,i}$ is the predicted value of s_{n+i}.".

Contrary to what Paul Davidsson, Eric Astor and Bertil Ekdahl [8] and others believe, strong anticipatory systems can be mathematically formalized, as shown in the next sections.

2.3   Incursive Discrete Strong Anticipatory Systems

Definition of an incursive discrete strong anticipatory system: an incursive discrete system is a system which computes its current state at time t, as a function of its states at past times, ..., t–3, t–2, t–1, present time, t, and even its states at future times t+1, t+2, t+3, ...

$$\mathbf{x}(t+1) = \mathbf{A}(..., \mathbf{x}(t-2), \mathbf{x}(t-1), \mathbf{x}(t), \mathbf{x}(t+1), \mathbf{x}(t+2), ...; \mathbf{p}) \qquad (3)$$

where the variable $\mathbf{x}$ at future times t+1, t+2, ... is computed in using the equation itself. Such an incursive system is self-referential because it computes its future states from itself and not from a model-based prediction. Let us remark that the function $\mathbf{A}()$ must satisfy some properties out of the scope of this paper.

Example: Anticipation and Chaos
The recursive Pearl-Verhulst map

$$x(t+1) = 4\mu x(t)(1 - x(t)) \qquad (4)$$

shows bifurcations and chaos. A completely stable system for any value of μ can be obtained in considering the following self-referential strong anticipatory system [12]:

$$x(t+1) = 4\mu x(t)(1 - x(t+1)) \qquad (5)$$

in just changing $x(t)$ by $x(t + 1)$ in the saturation factor of eq. 4.
This is an incursive equation for which the value of $x(t+1)$ is computed from itself as

$$x(t+1) = 4\mu x(t)(1 - 4\mu x(t)(1 - 4\mu x(t)(1 - 4\mu x(t)(1 - 4\mu x(t)(1 - ...))))) \qquad (5a)$$

which is an infinite recursive equation. For this system, the alternate series

$$1 - 4\mu x(t)(1 - 4\mu x(t)(1 - 4\mu x(t)(1 - 4\mu x(t)(1 - ...)))) = 1 - [4\mu x(t)] + [4\mu x(t)]^2 - [4\mu x(t)]^3 ... = 1/[1 + 4\mu x(t)]$$

converges, so

$$x(t+1) = 4\mu x(t)(1 - x(t+1)) = 4\mu x(t)/[1 + 4\mu x(t)] \qquad (5b)$$

which can be called a Michaelis-Menten-Monod equation, which will no more show chaos. Monod showed that bacteria growth obeys such an equation rather than the Pearl-Verhulst one. So such an anticipatory equation can describe a biological growth process. This does not demonstrate that the natural growth process is anticipatory.
The fixed point in the recursive map is given by $x_0 = 4\mu x_0(1-x_0)$ which is unstable for $\mu > 3/4$. In the incursive map 5b, a second variable fixed point is given by $x(t+1)$.
This second fixed point converges to the first one $x_0 = 4\mu x_0(1-x_0)$. What is then the difference between the original Pearl-Verhulst map and the incursive one? For the original Pearl-Verhulst map, the fixed point is unstable: this means that the recursive

process never tends to a fixed value of x_0, but has a chaotic behaviour. The incursive map gives a stable fixed point. Thus the anticipation has the effect to stabilize an otherwise unstable solution. Such incursive anticipatory systems can give rise to another stability property than recursive one. So the objective of such anticipation is to change the stability property of a system and the implicit goal is that the system reaches an otherwise unstable state of the system. The unstable state becomes then an attractor with anticipation.

Interesting properties and extension of the incursive map can be found in [3, 4].

Let us remark that the above transformation of a recursive map 4 to an incursive map 5 can be obtained from the recursive map to be controlled

$$x(t+1) = 4\mu x(t)(1 - x(t)) + u(t) \qquad (4')$$

with an incursive control $u(t)$

$$u(t) = 4\mu x(t)[x(t) - x(t + 1)] \qquad (4'')$$

which is a non-linear anticipative control, for which $x(t + 1)$ is given by the eq. 5b

$$x(t+1) = 4\mu x(t)/[1 + 4\mu x(t)] \qquad (5b')$$

Such an incursive control, given by eq. 4", was successfully applied to the stabilization of the chaotic movement of a robot arm [14].

Look at the last section 4 for another method to stabilize the chaos of this map with a model predictive control and a comparison with this incursive control.

2.4   Hyperincursive Discrete Anticipatory System

Definition of a hyperincursive discrete anticipatory systems: a hyperincursive discrete anticipatory system is an incursive discrete anticipatory system generating multiple iterates at each time step (after Dubois and Resconi [15]).

Example: Hyperincursive Unpredictable Anticipatory System
The following equation

$$x(t) = a.x(t+1).(1-x(t+1)) \qquad (6)$$

defines an hyperincursive anticipatory system [12]. Hyperincursion is an incursion with multiple solutions.
With $a = 4$, mathematically $x(t+1)$ can be defined as a function of $x(t)$

$$x(t+1) = 1/2 \pm 1/2 \ \sqrt{(1-x(t))} \qquad (7)$$

where each iterate $x(t)$ generates at each time step two different iterates $x(t+1)$ depending of the plus minus sign. The number of future values of $x(t)$ increases as a power of 2. This system is unpredictable in the sense that it is not possible to compute its future states in knowing the initial conditions. It is necessary to define successive final conditions at each time step. As the system can only take one value at each time step, something new must be added for resolving the problem. Thus, the following decision function $u(t)$ can be added for making a choice at each time step:

$$u(t) = 2.d(t) - 1 \tag{8}$$

where $u = +1$ for the decision $d = 1$ (true) and $u = -1$ for the decision $d = 0$ (false). In introducing eq. 8 into eq. 7, the following equation is obtained:

$$x(t+1) = 1/2 + (d(t)-1/2). \sqrt{(1-x(t))} \tag{9}$$

The decision process could be explicitly related to objectives to be reached by the state variable x of this system. It is important to point out that the decisions $d(t)$ do not influence the dynamics of $x(t)$ but only guide the system which creates itself the potential futures. A similar hyperincursive anticipatory system was proposed as a model of a stack memory in neural networks [13]. Some interesting properties of hyperincursive anticipatory systems relate to a flip-flop binary memory and catastrophe theory (can be found in [12]).

3   Functional Anticipatory Systems

The anticipatory systems defined until now were represented by discrete equations. Most models of systems are described by differential equations for which analytical solutions can sometimes be found. But in practice, most differential equations are simulated on computer and must, thus, be transformed to discrete equations.

We can define weak and strong anticipatory systems in using differential difference equations, also called functional differential equations, with retarded and anticipated time shifts. See [16] for an early work in functional differential equations.

Definition of a functional strong and weak anticipatory system: a functional strong anticipatory system is a system that computes its states at time t as a function of the current state at the time t, of the past state at time $t - \tau$ and of the future states at time $t + \tau$, where τ is a time shift:

$$d\mathbf{x}(t)/dt = \mathbf{A}(\mathbf{x}(t - \tau), \mathbf{x}(t), \mathbf{x}(t + \tau); \mathbf{p}) \tag{10}$$

Other time shifts can be added for generalization.

In such a strong anticipation, $\mathbf{x}(t + \tau)$ is computed from the system itself, as well as the delayed variable $\mathbf{x}(t - \tau)$. When $\mathbf{x}(t + \tau)$ is not computed from the system, a functional weak anticipatory system is defined as:

$$d\mathbf{x}(t)/dt = \mathbf{A}(\mathbf{x}(t - \tau), \mathbf{x}(t), \mathbf{x}^*(t + \tau); \mathbf{p}) \tag{10a}$$

where $\mathbf{x}^*(t + \tau)$ is then computed from a model of the system.

Notice that the retarded factor $\mathbf{x}(t - \tau)$ could be also approximated by a model.

A closed form solution of such functional systems can be found for strong anticipation when retardation is taken into account. In general, retardation and anticipation are conjugated as shown in the following section.

3.1   Conjugate Retardation and Anticipation Variables

Let us consider the two differential difference equations

$$dx(t)/dt = F(y(t + \tau)) - ax(t) \qquad (11a)$$

$$dy(t)/dt = G(x(t - \tau)) - by(t) \qquad (11b)$$

which are mixed advanced-retarded differential equations by the time shift τ. The first equation of $x(t)$ depends on $x(t)$ and a function F of the anticipatory $y(t + \tau)$ and the second equation of $y(t)$ depends on $y(t)$ and a function G of the past $x(t - \tau)$. Thus, this means that the system depends on the future of y, the present of x and y and the past of x.

Let us show that it is possible to transform such a delayed-advanced system to a system defined at the current time t [9]. For that, we will use the method I described in the paper [10], where a slave system can be synchronized to a master system for anticipating the future (past) states of the master system.

Following this synchronizing method, let us consider the system 11a-b as the master equation system, and let us define two new variables $x_1(t)$ and $y_1(t)$ defined at the current time t, the slave equation system of which being given by

$$dx_1(t)/dt = F(y(t)) - ax_1(t) \qquad (12a)$$

$$dy_1(t)/dt = G(x(t)) - by_1(t) \qquad (12b)$$

These new variables $x_1(t)$ and $y_1(t)$, defined at the current time t, will synchronize, respectively, with the delayed variable $x(t - \tau)$ and the advanced variable $y(t + \tau)$.
Indeed, in defining the variables $v(t)$ and $w(t)$ as the difference between the two couples of variables as

$$v(t) = x_1(t) - x(t - \tau) \ \text{ and } \ w(t) = y_1(t) - y(t + \tau)$$

one obtains

$$dv(t)/dt = dx_1(t)/dt - dx(t - \tau)/dt = F(y(t)) - ax_1(t) - [F(y(t)) - ax(t - \tau)]$$
$$= - a[x_1(t) - x(t - \tau)] = - av(t)$$

the solution of which is $v(t) = v(0).\exp(-at)$, so $v(t)$ will tend to zero (for $a > 0$) and $x_1(t)$ will tend to $x(t - \tau)$. And, similarly,

$$dw(t)/dt = dy_1(t)/dt - dy(t + \tau)/dt =$$
$$G(x(t)) - by_1(t) - [G(x(t)) - by(t + \tau)] = - b\,[y_1(t) - y(t + \tau)] = - bw(t)$$

the solution of which is $w(t) = w(0).\exp(-bt)$, so $w(t)$ will tend to zero (for $b > 0$) and $y_1(t)$ will tend to $y(t + \tau)$. In taking the following constraints on the initial conditions $x_1(0) = x(-\tau)$ and $y_1(0) = y(+\tau)$, $v(t)$ and $w(t)$ will remain equal to zero during all the evolution of the system. If perturbations occur, the fact to have positive values, for a and b, assures the convergence of the system. We may consider the slave equation of $x_1(t)$ as the mechanism to build the past value $x(t - \tau)$, at the current time t, and the slave equation of $y_1(t)$ as the mechanism to build the future value of $y(t + \tau)$, at the current time t. In other words, this means that the past and future states of the delayed and advanced variables can be built from additional slave equation systems.

So, $x(t - \tau)$ and $y(t + \tau)$ can be replaced by $x_1(t)$ and $y_1(t)$, respectively, in the original master equation system, and one thus obtains four coupled equations

$$dx(t)/dt = F(y_1(t)) - ax(t) \tag{13a}$$

$$dy(t)/dt = G(x_1(t)) - by(t) \tag{13b}$$

$$dx_1(t)/dt = F(y(t)) - ax_1(t) \tag{13c}$$

$$dy_1(t)/dt = G(x(t)) - by_1(t) \tag{13d}$$

which are now defined at the current time t.

For explicitly pointing out that these new variables represent past and future values of the master system, let us introduce the notation, $x_1(t) = x(t \mid t - \tau)$, where the notation $x(t \mid t - \tau)$ means the value of x at the time $t - \tau$, defined at the current time t. Similarly, the same notation is used for the future state as $y_1(t) = y(t \mid t + \tau)$, where the notation $y(t \mid t + \tau)$ means the value of y at the time $t + \tau$, defined at the current time t. In using the notations for the delayed and advanced variables at the current time, this system can be rewritten as

$$dx(t)/dt = F(y(t \mid t + \tau)) - ax(t) \tag{14a}$$

$$dy(t \mid t + \tau)/dt = G(x(t)) - by(t \mid t + \tau) \tag{14b}$$

$$dx(t \mid t - \tau)/dt = F(y(t)) - ax(t \mid t - \tau) \tag{14c}$$

$$dy(t)/dt = G(x(t \mid t - \tau)) - by(t) \tag{14d}$$

where we observe that, in fact, there are two similar quasi-independent coupled equation systems. If $\tau = 0$, $x(t \mid t - \tau) = x(t)$ and $y(t \mid t + \tau) = y(t)$, then the two coupled equation systems become identical. In fact, the two coupled systems are linked by the initial conditions. Indeed, the initial conditions are given by $x(0)$ and $y_1(0) = y(0 \mid 0 + \tau) = y(+\tau)$, and, $y(0)$ and $x_1(0) = x(0 \mid 0 - \tau) = x(-\tau)$.

What is innovative in this method, is the fact that such an anticipatory system is tractable because the delayed equation computes the advanced variable and the advanced equation computes the delayed variable. The two retardation and anticipation variables are time shift conjugates of each other. This is a necessary condition for transforming any delayed-advanced systems to a current time system.

Let us now show some fundamental properties of systems with retardation and then with anticipation. It must be pointed out that the properties and computation of a system with anticipation is not trivially similar to a system with retardation in simply changing the sign of the time shift τ. This will be clear in looking at the algorithm to compute the retarded system and the anticipated system.

The next section will explain some properties of a few functional equations with retardation. This is a key point to understand the complementary role of retardation and anticipation in systems.

3.2   Systems with Retardation

Let us first consider the following simple process with a retardation

$$dx(t)/dt = a.x(t - \tau) - b.x(t) \tag{15}$$

where the state of a system is represented by the variable $x(t)$, as a function of the current time t. The rate of change of its state, $dx(t)/dt$, is equal to the difference between its value, $x(t - \tau)$, at the past time, $t - \tau$, where τ is a constant time shift of the delay, and its value, $x(t)$, at the current time t. Two parameters, a and b, define the rate at which these processes work. The fact that the system depends on its past state is called a delayed system or a system with a retardation. A system with retardation represented by such an equation is also called a delayed equation. The parameters "a" and "b" can have positive or negative values. We also must understand $x(t - \tau)$ as the value of x at the past time $t - \tau$, known at the current time t, so we can write explicitly, $x(t - \tau) \equiv x(t|t - \tau)$, so the equation becomes

$$dx(t)/dt = a.x(t|t - \tau) - b.x(t) \tag{15a}$$

The initial condition at time $t = 0$ is given by $x(0)$, and the initial values of $x(t|t - \tau)$ must be known for the past time t from $-\tau$ to 0. There exists a hidden sub-system which keeps the value of x during this time shift as a memory.

The numerical simulation of such an eq. 15a with delay can be performed easily by Euler's discretization:

$$x(t + \Delta t) = x(t) + \Delta t.[a.x(t - n.\Delta t) - b.x(t)] \tag{15b}$$

where Δt is the discrete interval of time and the delay given by $n.\Delta t = \tau$. Let us give a few methods for finding the evolution of the solution of such an equation.

3.2.1 First Order Retardation System

Firstly, when the retardation τ is small, we can extrapolate $x(t|t - \tau)$, until the first order as

$$x(t|t - \tau) = x(t) - \tau.dx(t)/dt \tag{16}$$

With this expression 16, the delayed eq. 15a is then represented by the following equation defined at the current time

$$dx(t)/dt = a.[x(t) - \tau.dx(t)/dt] - bx(t)$$

or, with a simple mathematical transformation,

$$dx(t)/dt = (a - b).x(t)/(1 + a.\tau) \tag{17}$$

the solution being

$$x(t) = x(0) \exp((a - b).t/(1 + a.\tau)) \tag{18}$$

Discussion for $(1 + a.\tau) > 0$: When "a" is positive, the effect of the delay is to slowdown the evolution process. Indeed, without retardation, $\tau = 0$, the growth is given by $x(t) = x(0) \exp((a - b).t)$. The effect of the retardation, τ, is to divide the difference between a and b by $1 + a.\tau$. When "a" is negative, the effect is to speedup the evolution process. Moreover, the evolution process will show a growth or a decay depending on the fact that a > b or a < b. When a = b, the process shows a constant value.

3.2.2 Second Order Retardation System
Secondly, it is also possible to extrapolate until any order: the greater the retardation the more orders are necessary. For example, with the second order

$$x(t|t - \tau) = x(t) - \tau.dx(t)/dt + (\tau^2/2).d^2x(t)/dt^2 \qquad (19)$$

so, putting this expression 19 into eq. 15a, one obtains

$$dx(t)/dt = a.[x(t) - \tau.dx(t)/dt + (\tau^2/2).d^2x(t)/dt^2] - bx(t)$$

or

$$d^2x(t)/dt^2 - [2.(1 + a\tau)/a\tau^2].dx(t)/dt = - 2.(a - b).x(t)/a.\tau^2 \qquad (20)$$

which is a second order differential equation that gives rise to an oscillatory process when $(1 - b/a) > 0$. When "a" is positive, the amplitude of the oscillations increases with time. When "a" is negative, with $(1 + a\tau) > 0$, the amplitude of the oscillations decreases with time, and in the particular case, $(1 + a\tau) = 0$, the amplitude of the oscillations is constant.

3.2.3 Anticipated Synchronization of System with Retardation
It is possible to anticipate the states of the delayed equation by synchronizing it to a slave equation y (with the method described in [10]), as

$$dx(t)/dt = a.x(t|t - \tau) - b.x(t) \qquad (21a)$$

$$dy(t)/dt = a.x(t) - b.y(t) \qquad (21b)$$

Indeed, in defining a difference variable $z(t) = x(t|t + \tau) - y(t)$, we obtain

$$dz(t)/dt = a.x(t) -b.x(t|t + \tau) - a.x(t) + b.y(t) = - b.z(t) \qquad (21c)$$

so z(t) will tend to zero (for b > 0) and y(t) will synchronize to $x(t|t + \tau)$.

Let us point out that $x(t|t + \tau)$ means the potential value of x at the future time $t + \tau$, known at the current time t. This is actually a true strong anticipatory process.

So the values of the state of a system with retardation can be anticipated by an interval of time equal to $t + \tau$, where τ is the retardation time.

A narrow link thus exists between retardation and anticipation. Without retardation, such an anticipation cannot be made. Other methods exist (e.g. [10]) if this method fails (when b < 0, for example).

Contrary to systems with retardation, systems with anticipation were not so largely studied in the scientific literature. However, the concept of anticipation is very important in many practical domains. A similar model as for systems with retardation is now considered for pointing out the key properties of both.

3.3 Systems with Anticipation

Instead of considering a system with retardation, let us now consider a system with anticipation. We will start with the same equation as in the preceding section, but where the retardation is replaced by an anticipation

$$dx(t)/dt = a.x(t|t + \tau) - b.x(t) \tag{22}$$

This means that now the system will depend no more of its state at a past value, but will depend on its state at a future value given by an anticipation time τ. A system with anticipation represented by an equation is also called an advanced equation. The anticipated state written as $x(t|t + \tau)$ means the state of the system at the anticipated time $t+\tau$, known at the current time t.

The initial condition at time $t = 0$ is given by $x(0)$, and the initial values of $x(t|t + \tau)$ must be known for the time t from τ to 2τ.

These initial conditions are not the initial conditions of the retarded system 15a in simply changing the sign of the time shift τ !

As for the delay case, one can try to use the same Euler's discretization for the numerical simulation of eq. 22 :

$$x(t + \Delta t) = x(t) + \Delta t.[a.x(t + n.\Delta t) - b.x(t)] \tag{22a}$$

where Δt is the discrete interval of time and the anticipation given by $n.\Delta t = \tau$, the numerical computation can be made for the $n-1$ first steps, from the initial conditions $x(0)$ and $x(n.\Delta t + i.\Delta t)$, $i = 0$ to $n-2$. For the following time steps, this algorithm fails.

A new algorithm is obtained as follows. From eq. 22a, $x(t + n.\Delta t)$ can be written as

$$x(t + n.\Delta t) = [x(t + \Delta t) - x(t) + \Delta t.b.x(t)]/(\Delta t.a) \tag{23a}$$

Then $x(t + \Delta t)$ is obtained by a translation of $(1 - n).\Delta t$ time steps in eq. 23a as

$$x(t+n.\Delta t+(1-n).\Delta t)= \tag{23b}$$
$$[x(t+\Delta t+(1-n).\Delta t)-x(t+(1-n).\Delta t)+\Delta t.b.x(t+(1-n).\Delta t)]/(\Delta t.a)$$

or, after simplification

$$x(t + \Delta t) = [x(t+(2-n).\Delta t))-x(t+(1-n).\Delta t))+\Delta t.b.x(t+(1-n).\Delta t))]/(\Delta t.a) \tag{23c}$$

The relation 23c for $x(t + \Delta t)$ is put into eq. 23a, and the new algorithm is given by the following two eqs. 24-25:

$$x(t + n.\Delta t) = [[x(t+(2-n).\Delta t))-x(t+(1-n).\Delta t))+\Delta t.b.x(t+(1-n).\Delta t))]/(\Delta t.a) \tag{24}$$
$$- x(t) + \Delta t.b.x(t)]/(\Delta t.a)$$

$$x(t + \Delta t) = x(t) + \Delta t[a.x(t + n\Delta t) - b.x(t)] \tag{25}$$

which can be computed at all the following successive steps n, n+1, n+2, ..., in starting with the initial condition $x(0)$ and the $n-1$ values $x(1)$, ..., $x(n-1)$ computed with the first algorithm 22a.

This is a drastic difference in comparison with the delayed equation.

So, it is wrong to naively think that a system with a delay $-\tau$ can be simply transformed to a system with an anticipation $+\tau$, in simply changing the sign of the time shift τ !

Let us now consider a useful way to obtain an approximated solution of eq. 22, that represents a weak anticipation, in extrapolating the anticipative variable as follows.

3.3.1 First Order Anticipation System

Firstly, when τ is very small, we can extrapolate $x(t|t + \tau)$, until the first order, as

$$x(t|t + \tau) = x(t) + \tau.dx(t)/dt \qquad (26)$$

so eq. 22 becomes

$$dx(t)/dt = a.[x(t) + \tau.dx(t)/dt] - bx(t)$$

or, with a simple mathematical transformation

$$dx(t)/dt = (a - b).x(t)/(1 - a.\tau) \qquad (27)$$

the solution being

$$x(t) = x(0) \exp((a - b).t/(1 - a.\tau)) \qquad (28)$$

Discussion for $(1 - a.\tau) > 0$:

When "a" is positive, the effect of the anticipation is to speedup the evolution process. When "a" is negative, the effect of the anticipation is to slowdown the evolution process.

This is exactly the opposite effect in comparison with the system with retardation. When $(a - b) > 0$, the evolution is a growth, and when $(a - b) < 0$, the evolution is a decay.

3.3.2 Second Order Anticipation System

Secondly, it is also possible to extrapolate $x(t|t + \tau)$ until any order, for example, with the second order

$$x(t|t + \tau) = x(t) + \tau.dx(t)/dt + (\tau^2/2).d^2x(t)/dt^2 \qquad (29)$$

so eq. 22 becomes

$$dx(t)/dt = a.[x(t) + \tau.dx(t)/dt + (\tau^2/2).d^2x(t)/dt^2] - bx(t)$$

or

$$d^2x(t)/dt^2 - [2.(1 - a\tau)/a\tau^2].dx(t)/dt = - 2.(a - b).x(t)/a.\tau^2 \qquad (30)$$

which becomes a second order equation that exhibits oscillations for $(1 - b/a) > 0$.

When "a" is positive, with $(1 - a\tau) > 0$, the system exhibits an oscillatory behaviour with a decreasing amplitude, and in the particular case, $(1 - a\tau) = 0$, the amplitude of the oscillations is constant.

When "a" is negative, the amplitude of the oscillations increases.

3.3.3 Retarded Synchronization of System with Anticipation

It is possible to compute the past states of the advanced equation 22 by synchronizing it to a slave equation y (with the method described in [10]) as

$$dx(t)/dt = a.x(t|t + \tau) - b.x(t) \tag{31}$$

$$dy(t)/dt = a.x(t) - b.y(t) \tag{31a}$$

Indeed, in defining a difference variable $z(t) = x(t|t - \tau) - y(t)$, we obtain

$$dz(t)/dt = a.x(t) - b.x(t|t - \tau) - a.x(t) + b.y(t) = -b.z(t) \tag{31b}$$

so z(t) will tend to zero (when b > 0) and y(t) will synchronize to $x(t|t - \tau)$, where $x(t|t - \tau)$ means the value of x at the past time $t - \tau$, known at the current time t.

So the past values of the state of a system with anticipation can be known by an interval of time equal to $t - \tau$. A narrow link thus exists between anticipation and retardation. Without anticipation, such an history of the system cannot be made.

Let us now consider an anticipatory oscillator.

3.4   Properties of an Oscillatory System with Anticipation

Let us consider the anticipatory differential difference equation, in using the following notations : $x''(t) = d^2x(t)/dt^2$ and $x'(t) = dx(t)/dt$,

$$x''(t) + cx'(t) + ax(t) - bx(t + \tau) = 0 \tag{32}$$

where a, b and c are parameters and τ the time shift. This is an oscillator with damping and anticipation factors.

Let us remark that the term with anticipation can be considered as an advanced control of a damped oscillator

$$x''(t) + cx'(t) + ax(t) = u(t) \tag{32a}$$

with

$$u(t) = bx(t + \tau) \tag{32b}$$

This is a classical feedback delayed control for $\tau < 0$.

Two mathematical methods will be considered [9] for describing the properties of eq. 32.

3.4.1 Mathematical Characteristic Equations for Strong Anticipation Solution

Let us take a solution of the form

$$x(t) = \exp(\phi t), \phi = \sigma + i\omega \tag{33}$$

In replacing 33 in 32, one obtains

$$\phi^2.\exp(\phi t) + c.\phi.\exp(\phi t) + a - b.\exp(\phi t).\exp(\phi \tau) = 0$$

or, after division by $\exp(\phi t)$,

$$\phi^2 + c\phi + a - b.\exp(\phi \tau) = 0 \tag{34}$$

or, in taking the relation 33 into account,

$$(\sigma + i\omega)^2 + c(\sigma + i\omega) + a - b.\exp((\sigma + i\omega)\tau) = 0$$

or

$$\sigma^2 + 2i\sigma\omega - \omega^2 + c\sigma + ic\omega + a - b.\exp(\sigma\tau).(\cos(\omega\tau) + i.\sin(\omega\tau)) = 0 \qquad (35)$$

Let us separate the real and imaginary parts of eq. 35 as

$$\sigma^2 - \omega^2 + c\sigma + a - b.\exp(\sigma\tau).\cos(\omega\tau) = 0 \qquad (36a)$$

$$2\sigma\omega + c\omega - b.\exp(\sigma\tau).\sin(\omega\tau) = 0 \qquad (36b)$$

For obtaining an oscillatory behaviour, σ is to be taken equal to zero. Relations 36ab become

$$-\omega^2 + a - b.\cos(\omega\tau) = 0 \qquad (37a)$$

$$c\omega - b.\sin(\omega\tau) = 0 \qquad (37b)$$

From the second relation 37b one obtains

$$\omega = (b/c).\sin(\omega\tau)$$

which can be replaced in the first relation 37a as

$$-((b/c).\sin(\omega\tau))^2 + a - b.\cos(\omega\tau) = 0$$

or

$$-(b/c)^2.(1 - \cos^2(\omega\tau)) + a - b.\cos(\omega\tau) = 0$$

In view of obtaining an easy solution, let us choose $(b/c)^2 = a$, so

$$(a/b)\cos^2(\omega\tau) = \cos(\omega\tau)$$

A first solution is $\cos(\omega\tau) = 0$,
so $\omega\tau = \pi/2 + n\pi$, $n = 0,1,2,...$, and $\omega = \pm b/c$, $\tau = \pm (c/b)\pi/2$.
A second solution is given by $\cos(\omega\tau) = b/a$.
For example, for $b/a = 1/\sqrt{2}$, $\cos(\omega\tau)) = 1/\sqrt{2}$, so $\omega\tau = \pm\pi/4 + 2n\pi$, and so, for $a = 2$, $b = \sqrt{2}$ and $c = 1$, one obtains $\omega = 1$ and $\tau = \pi/4$.
With these values, the following equation with anticipation

$$x''(t) + x'(t) + 2.x(t) - \sqrt{2}.x(t + \pi/4) = 0 \qquad (38)$$

exhibits oscillations without damping, so anticipation can regulate amplitudes of oscillations.

3.4.2 Model by Extrapolation for Weak Anticipatory Solution
In taking the same anticipatory differential difference equation 32, as in the previous section, we can find models in view of computing the anticipatory factor $x(t + \tau)$.

For a small anticipatory time shift τ, one can extrapolate $x(t + \tau)$ by a development in Taylor's series as

$$x(t + \tau) = x(t) + \tau.x'(t) + (\tau^2/2).x''(t) + ... \qquad (39)$$

If we restrict to the second order, the anticipatory differential difference equation 32, with the relation 39, becomes

$$x''(t) + cx'(t) + ax(t) - bx(t) - b.\tau.x'(t) - b.(\tau^2/2).x''(t) = 0 \qquad (40)$$

or

$$x''(t) + [(c - b.\tau)/(1 - b\tau^2/2)].x'(t) + [(a - b)/(1 - b\tau^2/2)].x(t) = 0 \qquad (40a)$$

which is an oscillator defined at the current time t. The condition for obtaining an oscillation without damping is $c = b.\tau$, and in taking $a = 2$, $b = \sqrt{2}$, $c = 1$, as for eq. 38, one obtains $\tau = 0.71$ and $\omega = 0.95$, a good approximation of $\tau = \pi/4$ and $\omega = 1$.

We defined, in a preceding section, two classes of anticipatory systems: strong and weak anticipatory systems related to anticipation built by the system itself and anticipation based on a model of the system. With such an extrapolation of the anticipatory factor, the resulting system can be called a weak anticipatory system.

Now let us consider two coupled oscillators with retardation and anticipation.

3.5   Two Coupled Oscillators with Anticipatory Behaviours

More complex oscillatory systems can be built such as

$$dx(t)/dt = a_1.y(t + \tau) + a_2.x(t) \qquad (41a)$$

$$dy(t)/dt = - b_1.x(t - \tau) - b_2.y(t) - \alpha.u(t) \qquad (41b)$$

$$du(t)/dt = - c_1.v(t - \tau) - c_2.u(t) \qquad (41c)$$

$$dv(t)/dt = d_1.u(t + \tau) + d_2.v(t) + \beta.x(t) \qquad (41d)$$

These are two coupled oscillators with retardation and anticipation. The retardation and anticipation in this system are conjugated. In the framework of control theory, $\alpha.u(t)$ can be considered as the control function of the first oscillator 41ab and, $\beta.x(t)$ as the control function of the second oscillator 41cd.

As already shown at the preceding section, the anticipatory variables $y(t + \tau)$ and $u(t + \tau)$ can be computed by synchronization as

$$dy_1(t)/dt = - b_1.x(t) - b_2.y_1(t) - \alpha.u(t + \tau) \qquad (42a)$$

where $y_1(t) = y(t + \tau)$, which depends on the anticipated $u(t + \tau)$ that will be computed by $u_1(t)$ as

$$du_1(t)/dt = - c_1.v(t) - c_2.u_1(t) \qquad (42b)$$

where $u_1(t) = u(t + \tau)$ is an anticipative control computed at the current time.

Numerical simulation of eqs. 41abcd, with eqs. 42ab, are given in Figures 1abc. The parameters are:
$a_1 = 1$, $a_2 = 0.5$, $b_1 = 1$, $b_2 = 0.5$, $\alpha = 0.5$, $c_1 = 1$, $c_2 = 0.5$, $d_1 = 1$, $d_2 = 0.5$, $\beta = 0.5$, and the time anticipation $\tau = 100$, for $t = 0$ to 600 with a time step $\Delta t = 0.05$.
Initial conditions are given at time $t = 100$ for y, u, y_1, u_1, and the initial delayed functions x and v are given for $t = 0$ to 100.

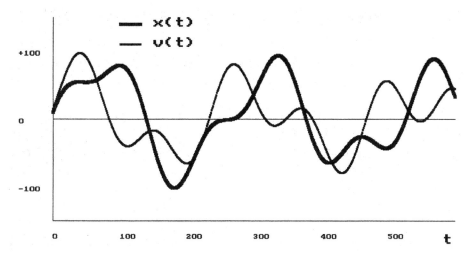

Fig. 1a. Simulation of x(t) and v(t) as a function of the time $t = 100$ to 600. The delayed function $x(t - 100)$ and $v(t - 100)$ are given between $t = 0$ and 100.

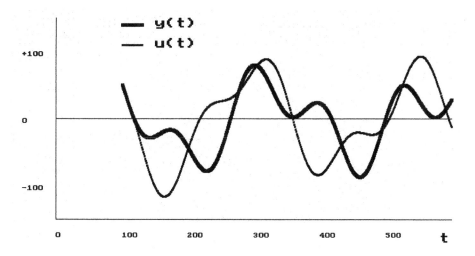

Fig. 1b. Simulation of y(t) and u(t) as a function of the time $t = 100$ to 600.

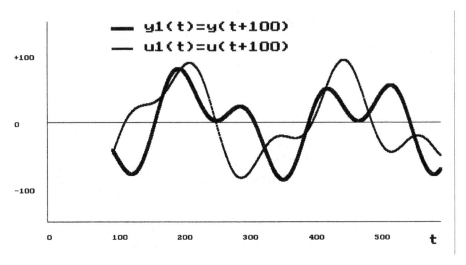

Fig. 1c. Simulation of $y_1(t)$ and $u_1(t)$ as a function of the time t = 100 to 600. It is well seen that $y_1(t) = y(t + 100)$ and $u_1(t) = u(t + 100)$ are the anticipations of $y(t)$ and $u(t)$, in comparing with Fig. 1b.

As shown with this simple example of two coupled oscillators with retardation and anticipatory behaviours, the computation of the anticipatory behaviour is based on the synchronisation mechanism between the two coupled oscillators.

Synchronisation is a key mechanism in many natural and artificial systems, and in particular for adaptive learning systems within neurons in the brain. Natural and artificial neural networks are similar to coupled oscillators for which the adaptive learning process is based on synchronisation.

What is very important to point out is the fact that such anticipatory synchronization permits to define new additional variables which give at the current time t the value of anticipatory variables depending on a future time t + τ.

This is drastically different from other types of methods like, for example, predictive methods, as shown in the following section.

4   The Role of Anticipation in Model Predictive Control

The key difference between a conventional feedback control and a predictive control is that the control error e = y − r , which is the difference between the process output y and the setpoint r (the desired output), used by the predictive controller is based on future and/or predicted values of the setpoint r(t + τ), and also on future and/or predicted values of the process output y(t + τ), rather than their current values.

See [5] for an overview of MPC.

A very basic explanation of predictive control is as follows.

Firstly, one-step future time horizon predictive control based on e(t + Δt) is required for discrete control systems because the output y(t) of the practical process does not respond instantly to changes in the input u(t). More specifically, the effect of a change in the process at the t-th control interval is not observed, measured or predicted, until the $(t + \Delta t)$th control interval. Therefore, for perfect control the controller must compute u(t) such that $y(t + \Delta t)$ will equal $r(t + \Delta t)$.

Secondly, multi-step future time horizon predictive control is used for systems with time delays since by definition the effect of a change in the process input does not affect the output until after the delay. For discrete controllers, this means that the best possible output control is to compute u(t) so that $y(t + \tau) = r(t + \tau)$, where τ is the delay related to the n-step future time horizon $\tau = n\Delta t$, n being an integer.

Thirdly, multi-step future time horizon predictive control is also used to give more practical control. Consider a step change in setpoint at time t on a process without delay operating at steady state. The best possible output control would be achieved with one-step future time horizon control. However this type of control can be impractical. For example, the required control action u(t) could be greater than available in exceeding the upper limit on the input. So a much smaller control change u(t) would be required to achieve a given setpoint change in m-step future time horizon control intervals rather than in a single control interval and the output will change more slowly.

Fourthly, more advanced controllers can be formulated that make it possible to specify the path or trajectory that the output follows as it moves from the current value y(t) to the desired value $y(t + \tau)$. Similarly, it is possible to formulate optimal controllers that compute a series of Nu control future moves $\{u(t), u(t + \Delta t), ..., u(t + Nu\Delta t)\}$. For that, a series of N output future values are to be computed $\{y(t + \Delta t), ..., y(t + N\Delta t)\}$. The error to be minimized is then related to the difference between the series of future outputs $y(t + n\Delta t)$ and the series of desired future setpoints $r(t +n\Delta t)$, n = 1 to N, representing the desired path or trajectory. N is called the future time horizon of the outputs and Nu the future time horizon of the control. These horizons are not necessarily equal. Only the control move u(t), resulting from the minimization of the error on the path or trajectory, is applied and the series control moves and the series outputs are updated at each time step interval Δt.

So, the principle of predictive control consists in minimizing a cost function given, for example, by a weighted least squares criterion for estimation

$$J = E\{ \sum_{i=1}^{N} [y(t + i.\Delta t) - r(t + i.\Delta t)]^2 + \sum_{i=1}^{Nu} w_i.u(t - \Delta t + i.\Delta t)^2 \} \qquad (43)$$

where $\{r(t + i.\Delta t)\}$ is the setpoint sequence (target tracking), $\{w_i\}$ is the weight sequence, and N and Nu are fixed integers representing the time horizons of the predicted outputs and control sequence.

How are the anticipative outputs $y(t + i.\Delta t)$, i = 1 to N, defined ?

There are many possibilities depending on the knowledge of the process to be controlled.

The general predictive control (GPC) method does not require a precise mathematical model of the system to be controlled. For example, Park et al [17] combine an ARMAX (AutoRegressive Moving Average model with eXogenous noise input) modelling with the GPC technique for developing a strategy for adaptive control of uncertain discrete-time chaotic systems. They used a cost function like eq.

43 and give explicitly the setpoint r equal to the unstable equilibrium of the chaotic systems.

The model predictive control (MPC) uses a model of the system to be controlled.

Very often, a discrete approximation of the dynamic equations (e.g. Euler's step) is used.

To show explicitly an example of the model predictive control, let us consider the map given in eq. 4. This map 4, to be controlled by the control action $u(t)$, in the chaos regime with $\mu = 1$, is written as

$$x(t+1) = 4x(t)(1 - x(t)) + u(t) \tag{44}$$

with a cost function given by

$$J = [x(t+1) - x_0]^2 + w_1 u(t)^2 \tag{45}$$

corresponding to the cost function 43, with $N = Nu = 1$, and with a constant setpoint $r(t+1) = x_0 = 3/4$, that is the unstable equilibrium of the map.

The objective of the control is to stabilize the map at its unstable equilibrium given by $x_0 = 3/4$, obtained from $x_0 = 4x_0(1 - x_0)$, corresponding to the equilibrium condition $x(t+1) = x(t) = x_0$ in eq. 44 (without the control $u(t)$).

In applying the classical criteria of stability given by

$$| dx(t+1)/dx(t) | < 1 \tag{46}$$

to this map 44 at the equilibrium $x_0 = 3/4$, one obtains

$$| dx(t+1)/dx(t) | = | 4 - 8x_0 | = 2 \tag{46a}$$

which is greater then 1, so this equilibrium state x_0 is unstable.

The eq. 44 is used for prediction of $x(t+1)$ in the cost function 45.

So, in putting eq. 44 into eq. 45, one obtains

$$J = [4x(t)(1 - x(t)) + u(t) - 3/4]^2 + w_1 u(t)^2 \tag{45a}$$

The minimum of J is obtained with the condition $dJ/du(t) = 0$, so

$$u(t) = [3/4 - 4x(t)(1 - x(t))]/[1 + w_1] \tag{47}$$

Now, let us put eq. 47 into eq. 44, so that

$$x(t+1) = 4x(t)(1 - x(t)) + [3/4 - 4x(t)(1 - x(t))]/[1 + w_1] \tag{48}$$

After elementary mathematical transformations, eq. 48 becomes

$$x(t+1) = [3/4 + 4w_1 x(t)(1 - x(t))]/[1 + w_1] \tag{49}$$

The equilibrium condition $x(t+1) = x(t) = x_0$ of eq. 49 gives $x_0 = 3/4$, as desired.

In applying the stability criteria 46 to this eq. 49, one obtains

$$| dx(t+1)/dx(t) | = | [4w_1 - 8w_1 x_0]/[1 + w_1] | = | -2w_1/[1 + w_1] | < 1 \tag{46a}$$

and the chaotic map is stabilized to $x = x_0 = 3/4$ for a weight in the range $0 < w_1 < 1$.

Let us show the differences between the incursive control and the model predictive control in considering the control of the chaos map 44,

$$x(t+1) = 4x(t)(1 - x(t)) + u(t)$$

with the three following control functions:

a) The incursive control (eq. 4" with 5b' in the chaos regime with $\mu = 1$)

$$u(t) = 4x(t)[x(t) - 4x(t)(1 - x(t))]/[1 + 4x(t)] \qquad (50a)$$

b) The model predictive control (eq. 47), with $w_1 = 1/2$

$$u(t) = [3/4 - 4x(t)(1 - x(t))]/[1 + w_1] \qquad (50b)$$

c) The incursive model predictive control, with $w_1 = 1/2$

$$u(t) = [x(t) - 4x(t)(1 - x(t))]/[1 + w_1] \qquad (50c)$$

This third control is based on the incursive control applied to the model predictive control in replacing the setpoint x_0 by $x(t)$ in the cost function 45 as follows

$$J = [x(t + 1) - x(t)]^2 + w_1 u(t)^2 \qquad (51)$$

This new controller 50c is obtained in minimizing this cost function 51. The control 50a could be obtained from this control 50c in taking a variable weight $w_1 = 1/4x(t)$.

Figures 2a-b-c show the simulations of the chaos map 44, with the three controllers 50a-b-c.

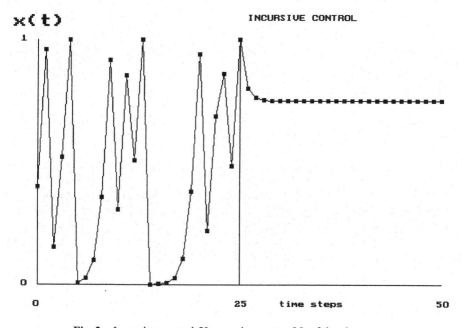

Fig. 2a. Incursive control 50a, starting at step 25, of the chaos map.

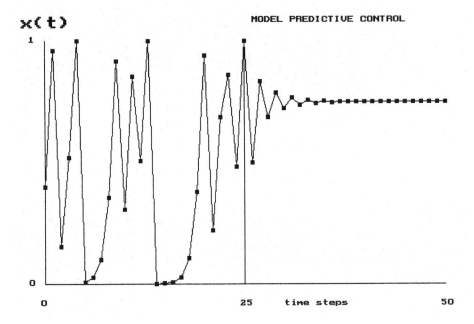

Fig. 2b. Model predictive Control 50b, starting at step 25, of the chaos map

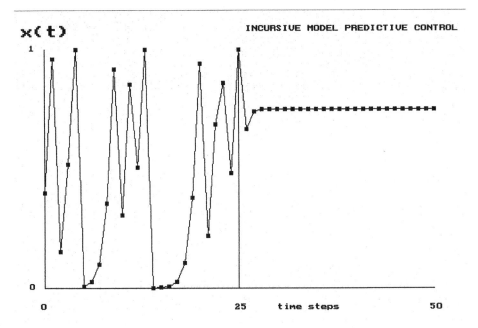

Fig. 2c. Incursive Model Predictive Control 50c, starting at step 25, of the chaos map.

From the simulations, the best control is the incursive control, in figure 2a, where it is well seen that the control is optimal. In figure 2b, the model predictive control shows damped oscillations around the setpoint $x_0 = 3/4$. Let us notice that the damping effect depends on the value of the weight w_1. In Figure 2c, the incursive model predictive control shows a better control than the model predictive control, with the same weight w_1. The incursive control, as well as the incursive model predictive control, finds by itself the setpoint which is the unstable equilibrium of the chaos map. Indeed, the incursive control does not use an explicit setpoint, but an implicit setpoint given by the unstable equilibrium state which is stabilized. The control minimizes the distance between $x(t + 1)$ and $x(t)$, which corresponds to an adaptive behaviour for obtaining a fixed point, $x(t + 1) = x(t)$, that could be assimilated to an anticipative learning. Indeed, it is well known that most algorithms of learning are based on similar processes consisting in minimizing a cost function for obtaining a system that converges to a stable state.

In conclusion of this section, such an incursive model predictive control could be used to model adaptive behaviour in anticipative learning systems.

5   Conclusion

The purpose of this paper deals with some mathematical methods to analyze and compute systems with anticipatory capabilities. One important tool for describing such anticipatory systems is the mathematical modelling. In the scientific literature, there is a lot of work dealing with delay systems and less with anticipatory systems. An interesting result shown in this paper is the fact that the anticipation and the delay play complementary roles. In predictive control the modelling is also an important aspect and anticipation is used for enhancing the control of systems in view of compensating some structural delays. A lot of work is still to be done in this field and this paper is hoping to be a contribution to the understanding and development of mathematical tools for anticipatory systems that could be applied to several fields including adaptive and learning systems with anticipatory behaviour.

Acknowledgments

The author would like to thank the three anonymous referees for their valuable comments for enhancing the comprehension of this paper. Dr Martin Butz is also thanked for having suggested to relate the incursive processes to model predictive control.

References

1. Barreto João P., Jorge Batista, Helder Araújo, "Model Predictive Control to Improve Visual Control of Motion: Applications in Active Tracking of Moving Targets", *ICPR'00 - 15th Int. Conference on Pattern Recognition*, Barcelona, Spain, September 3-8, 2000.

2. Blakemore Sarah J., Susan J. Goodbody, and Daniel M. Wolpert, "Predicting the Consequences of Our Own Actions: The Role of Sensorimotor Context Estimation", *The Journal of Neuroscience*, September 15, *18*(18):7511-7518, 1998.
3. Burke Mark E., "Properties of Derived Scalar Anticipatory Systems", *Computing Anticipatory Systems : CASYS 2001¯ Fifth International Conference*, edited by D. M. Dubois, Published by the American Institute of Physics, AIP CP 627, 2002, pp. 49-58.
4. Burke Mark E., Niall D. Fitzgerald, "Scalar Weak Anticipatory Systems", *Computing Anticipatory Systems : CASYS 2001¯ Fifth International Conference*, edited by D. M. Dubois, Published by the American Institute of Physics, AIP CP 627, 2002, pp. 85-94.
5. Camacho, E. F. and Bordons, C., *Model Predictive Control*, Springer Verlag, 1999.
6. Clarke, D. W., Mohtadi, C., " Generalized Predictive Control: Part I - The Basic Algorithm; Part II - Extensions and Interpretations", *Automatica*, Vol. 23, pp. 137-160, 1987.
7. Cutler, C. R., Ramaker, B. L., "Dynamic Matrix Control - A Computer Control Algorithm", *Proceedings of the 1980 Joint Automatic Control Conference*, pp. WP5-B, San Francisco, 1980.
8. Davidsson Paul, Astor Eric and Ekdahl Bertil, "A framework for autonomous agents based on the concept of anticipatory systems". *CYBERNETICS AND SYSTEMS'94*, editor : Robert Trappl, World Scientific, volume II, 1994, pp. 1427-1434.
9. Dubois Daniel M., ""Theory of Computing Anticipatory Systems Based on Differential Delayed-Advanced Difference Equations". *Computing Anticipatory Systems : CASYS 2001¯ Fifth International Conference*, edited by D. M. Dubois, Published by the American Institute of Physics, AIP CP 627, 2002, pp. 3-16.
10. Dubois Daniel M., "Theory of Incursive Synchronization and Application to the Anticipation of a Chaotic Epidemic". *International Journal of Computing Anticipatory Systems*, volume 10, 2001, pp. 3-18.
11. Dubois Daniel M., "Anticipatory Effects in Relativistic Electromagnetism". *Computing Anticipatory Systems: CASYS 2000 - Fourth International Conference*. Published by The American Institute of Physics, AIP Conference Proceedings 573, pp. 3-19, 2001.
12. Dubois, Daniel M., "Review of Incursive, Hyperincursive and Anticipatory Systems - Foundation of Anticipation in Electromagnetism". *Computing Anticipatory Systems: CASYS'99 - Third International Conference*. Edited by Daniel M. Dubois, Published by The American Institute of Physics, AIP Conference Proceedings 517, 2000, pp. 3-30.
13. Dubois, D. M., "Boolean Soft Computing by Non-linear Neural Networks with Hyperincursive Stack Memory", Invited Paper, in *Computational Intelligence: Soft Computing and Fuzzy-Neuro Integration with Applications*, Edited by Okyay Kaynak, Lotfi A. Zadeh, Burhan Türksen, Imre J. Rudas, NATO ASI Series, Series F: Computer and Systems Sciences, vol. 162, Springer-Verlag, (1998), pp. 333-351.
14. Dubois, D. M., "Feed-In-Time Control by Incursion", in *Robotics and Manufacturing: Recent Trends in Research and Applications*, volume 6, edited by M. Jamshidi, F. Pin, P. Dauchez, published by the American Society of Mechanical Engineers, ASME Press, New York, pp. 165-170, 1996.
15. Dubois D., and G. Resconi, *Hyperincursivity: a new mathematical theory*, Presses Universitaires de Liège, 260 p., 1992.
16. Hale J., *Functional Differential Equations*. Applied Mathematical Sciences, 3, Springer-Verlag, New York (1971).
17. Park, K.-S., Park, J.-B., Choi, Y.-H., Yoon, T.-S., Chen, G., "Generalized Predictive Control of Discrete-Time Chaotic Systems", *International Journal of Bifurcation and Chaos*, Vol. 8, No. 7, pp. 1591-1587, 1998.
18. Richalet, J., Rault, A., Testud, J. L., Papon, J., "Model Predictive Heuristic Control: Application to Industrial Processes", *Automatica*, Vol. 14, pp. 413-417, 1978.
19. Rivera, D. E., Morari, M., "Internal Model Control: PID Controller Design", *Ind. Eng. Chem. Process Des. Dev.*, Vol. 25, pp. 252-265, 1986.
20. Rosen, R., *Anticipatory Systems*. Pergamon Press: New-York (1985).

Anticipation Driven Artificial Personality: Building on Lewin and Loehlin

Stevo Bozinovski

Mathematics and Computer Science Department
South Carolina State University
sbozinovski@sets.scsu.edu

Abstract. This paper addresses the issue of personality of an animat in terms of anticipation, motivation and emotion. It also discusses some relevant models and theories of personality, and their relation to the consequence driven systems theory. The main result of this work is a fundamental mathematical equation between the emotion, motivation, and behavior. In essence the result can be stated that what motivates an animat behavior is the value of the anticipated emotional consequence of that behavior. Experimental research with an artificial personality architecture is provided, supporting the obtained result.

1   Introduction: Problem Statement

This work is our further investigation in the problem of building and/or understanding the relation between concepts like personality, consequence, and anticipatory behavior in the adaptive learning systems. The work is continuation of our previous effort in developing a sound consequence driven system theory. It includes a search for relationships between the consequence driven systems theory as a personality theory on one side, and other theories and approaches for understanding personality on the other side.

In particular, we found interesting relationships to the model of personality proposed by Loehlin [16], as well as to the research concept of force field proposed by Lewin [12-14]. To the best of our knowledge those relevant works have not been, to date, investigated by researchers in AI or cognitive science.

In the sequel, after giving a definition of the concept of personality, we will give a short review of the above-mentioned works. In the second part, we will introduce a *fundamental equation of anticipatory behavior*, and then through the concept of motivational graph we will present an artificial personality architecture that exhibits anticipatory learning behavior.

2   Artificial Personality

Personality can be understood as *expectancy of a consistent pattern of attitudes and behavior*. All people exhibit recognizable individual actions that serve to identify them. Classical problems in personality include: Where do those individual characteristics come from? Are they truly unique or just particular combination of

M. Butz et al. (Eds.): Anticipatory Behavior ..., LNAI 2684, pp. 133-150, 2003.

characteristics all people possess? Are they learned, inherited, or both? Can personality be altered, and if so, how? [15]

For many years Artificial Intelligence research attention was primarily given to the *concept of intelligence,* not necessarily an embodied one. In recent years, growing interest has been in building agents that will interact with humans, and will exhibit features of personality. The effort can be described as understanding the *concept of personality* and building an *artificial personality,* with its features like emotions and motivations, among others.

The problems that arise in the study of human personality are even less addressed for animat personality. There are also specific problems that should be addressed, examples being how can we build a control architecture for an agent that will inherit and/or learn and/or exhibit personality; what is the role of anticipation in exhibiting personality; and how can we design an experiment that will show personality features of agents? In the sequel we address some of the challenges of artificial personality research.

3   Building on Previous Works

Searching through the literature we found relevant works that can be considered as grounds on which our work is a continuation. In particular we found a highly relevant connection with the work of Lewin [12-14] and Loehlin [16]. In this chapter we will describe those works along with the basic concepts of our consequence driven systems theory [6, 7].

3.1   Lewin's Metaphor: Force Field

Among many theories describing personality and behavior, Lewin [12-14] proposed what is probably the most abstract one. In his research he applied a fundamental approach: *using metaphors* from a well-known system (e.g. physics) in order to understand/explain a rather unknown system (e.g. psychology). In explaining phenomena around the concept of behavior he used the metaphor of a "force field" (Table I).

Table I. Basic constructs of Lewin's force field theory

Term	Operational definition	Conceptual properties
Psychological Tension (t)	empirical syndrome indicating a "need"	tendency of spreading to neighboring systems
Psychological force (f)	psychological locomotion	vector
Fluidity (fl)		factor determining the velocity of equalization of tension with neighboring systems

As Table I shows, the principal constructs of his force field theory are *psychological tension, psychological force,* and *fluidity.* In his theory, it is understood that an *intention* is setting up a tension. It is assumed that a tension needs a *satisfaction,* since satisfaction is a release of tension. In symbolical terms if S^G is a person having a goal G in his system S, then for such a system it is true that $t(S^G) > 0$, i.e. *a system having a goal is always in tension, and that is a basis of a behavior.* If the system is in state P it will tend to behave such that $P \subset G$ in order to reach $t(S^G) = 0$. So a behavior is always a tension-derived phenomenon. *Behavior is an observable manifestation of a rather unobservable phenomenon, tension.* Behavior is a manifestation of a psychological force, since it produces movement. The forces can be at the level of thinking and/or at the level of acting.

An important concept in relation with behavior is the concept of a valence of a state. *While tension is the push part, the valence is a pull part of a behavior.* Another concept relevant to the theory is the concept of recall. *The recall of a state G is a function of satisfaction felt being in that state.* Other concepts in his theory are motivation, and emotion, as unavoidable concepts in a discussion about a behavior. However, those are explanatory ones, used in describing the crucial metaphors from physics, tension and force.

3.2   Loehlin's Metaphor: Programming Flowchart

Another relevant approach toward personality is the one proposed by Loehlin [16]. In modeling a personality, he actually suggested that what is to be modeled is a *basic behavioral cycle,* rather than just some personality properties. A basic behavioral cycle is modeled using a metaphor from computer programming, a *flowchart.* He proposed an artificial personality, a character named Aldous, whose basic behavioral cycle is shown in Figure 1.

In addition to proposing a need for a basic behavioral routine, as we can see from Figure 1, Loehlin [16] proposes the basic concepts of his theory to be recognition, emotion, action, consequence, and learning. He emphasizes the importance of emotion and consequence in the basic routine of a personality, inseparable from processes such as situation recognition, action, and learning.

We recognize that Aldous is a *consequence driven system,* since the consequence is a crucial concept in any discussion involving learning and behavior. The work of Loehlin [16] is a remarkable example of a consequence driven systems research, both in psychology and Artificial Intelligence. Using the flowchart description rather than verbose one, Loehlin [16] clearly describes the most relevant issues about *agents that learn from emotional reactions to the consequences of their actions.*

3.3   Consequence Driven Systems Theory

Consequence Driven Systems theory [6, 7] is an attempt to understand and build an *agent (animat) personality.* It tries to *find an architecture* that will ground the notions such as motivation, emotion, learning, disposition, anticipation, curiosity, confidence, and behavior, among other notions usually present in a discussion about an agent personality.

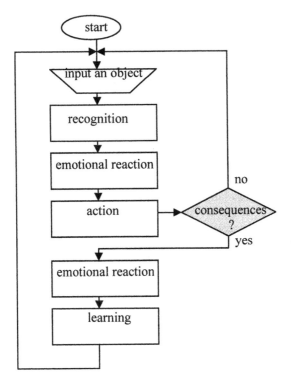

Fig. 1. Basic behavioral cycle of a personality proposed by Loehlin [16]

Among several related efforts in contemporary research on emotion-based architectures, we distinguish the work of Gadanho [9-11] by its basic concepts, architecture engineering, and realization aspects that resulted in developed an emotion learning architecture that has been implemented successfully in simulated as well as real robots. Starting from concepts such as feelings and hormonal system, the approach has similarity with our approach in issues like having innate emotions that will define goals and learning emotional associations between states and actions that will determine the agent's future decisions.

In the sequel we will briefly describe the origin of the Consequence Driven Systems theory and some of its features.

3.3.1 Origin: The Challenge of Secondary Reinforcement

Consequence driven systems theory originated in an early *Reinforcement Learning* (e.g. [3]) research effort to solve the *assignment of credit problem* using a neural network. The problem is related the problem of engineering a mechanism of a backpropagated, secondary reinforcement. Such an effort was undertaken in 1981 within the Adaptive Networks (ANW) Group at the Computer and Information Science (COINS) Department of the University of Massachusetts at Amherst. Two instances of the assignment of credit problem were considered: the maze learning problem and the pole balancing learning problem. *Two anticipatory learning architectures were proposed*: The Actor/Critic (A/C) architecture and the Crossbar Adaptive Array (CAA) architecture (Figure 2).

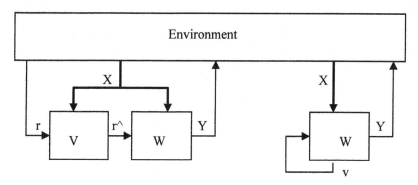

Fig. 2a. A/C architecture   Fig. 2b. CAA architecture

Fig. 2. Examples of two anticipatory learning systems

As Figure 2 shows, the obvious difference is that A/C architecture needs two identical memory structures, *V* and *W*, to compute the internal reinforcement r^, and the action, while CAA architecture uses only one memory structure, *W*, the same size as one of the A/C memory structures.

The most important difference, however, is the *design philosophy*: In contrast to A/C architecture, *CAA architecture does not use any external reinforcement r*; it only uses the current situation *X* as input. A state value is used as secondary reinforcement concept, while genetically predefined state value, rather than an immediate reinforcement, is used as primary reinforcement. The CAA architecture *introduces the concept of state evaluation* and connects it to *the concept of feeling*.

Although both architectures were designed to solve the same basic problem, the maze learning problem, the challenges of the design were different. The A/C architecture effort was challenged by the mazes from animal learning experiments, where there are many states and there is only one rewarding state (food). The CAA architecture effort from the start was challenged by the mazes defined in the computer game of Dungeons and Dragons, where there is one goal state, but many reward and punishment states along the way. So, from the very beginning CAA effort adopted the concept of dealing with *pleasant and unpleasant states*, feelings and emotions.

Interestingly enough, the CAA approach proved more efficient in solving the maze learning instance of the credit assignment problem, and was the only architecture that presented the solution in front of the ANW group as well as the COINS department in 1981. The AC architecture had problems in solving the maze learning problem for a longer time [17].

The original CAA idea of having *one memory structure for crossbar computation of both state evaluations and action evaluations* was later also implemented in reinforcement learning architectures such as Q-learning system [2, 18] and its successor architectures.

Interested readers may find more details in the first published reports on the subject [6, 1] as well as later reviews [8]. It is also interesting that while [1] stays at the basic issue of assignment of credit, [6] extends toward the issue of delayed reinforcement learning, later accepted by Watkins [18] as the issue of delayed reward learning.

The CAA approach introduced neural learning systems that can *learn without external reinforcement.* It has now come to our attention that other authors came to the same conclusion, one remarkable example being Loehlin [16]. Consequence driven systems approach actually proposes a *paradigm shift* in learning theories, from the concept of reinforcement (reward, punishment, payoff, ...) to the concept of *state as a consequence* and *emotion as a state evaluation.*

3.3.2 Main Concepts of the Theory

After the exposition of origins of the Consequence Driven Systems theory, we will present its grounding concepts. These are:

Three environments. The theory assumes that an agent should always be considered as a three-environment system. The agent expresses itself in its *behavioral environment,* where it behaves. It has its own *internal environment* where it synthesizes its behavior. It also has access to a *genetic environment* where from it receives its initial conditions for existing in its behavioral environment. The genetic and the behavioral environment are related: The initial knowledge transferred through the imported *species (or personality) genome* properly reflects, *in the value system of the species,* the dangerous situations for the species in the behavioral environment. It is assumed that all the agents import some genomes at the time of their creation.

Three tenses. The theory emphasizes that *an agent should be able to understand the temporal concepts,* like past, present, *and the future,* in order to be able to self-organize in an environment. The past tense is associated with the evaluation of its previous performance, the present tense is associated with the concept of emotion that the agent computes toward the current situation (or the current state), and the *future* with the *moral (self-advice)* about a behavior that in future should be associated with a situation.

A *Generic Architecture for Learning Agents* (GALA architecture) is proposed within the theory (Figure 3).

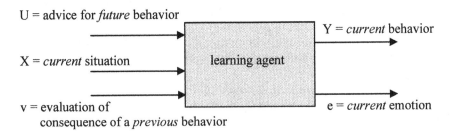

Fig 3. The GALA architecture

Note that GALA is a genuine generic, black-box-only architecture: only the inputs and outputs are specified. Yet it is very specific the way inputs and outputs are defined. The GALA architecture is a reconfigurable architecture: The derivation of various types of learning architectures from the GALA architecture is described elsewhere [6, 7].

Personality architecture. The Consequence Driven System theory introduced an architecture of personality of an agent capable of learning using secondary reinforcement principle. The CAA architecture (Figure 4) is derived as an emotion learning agent from the GALA architecture.

ENVIRONMENT

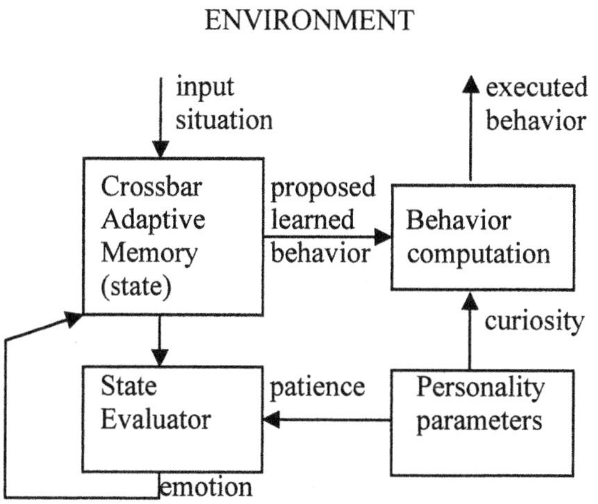

Fig. 4. Crossbar Adaptive Array architecture

In a crossbar fashion, the CAA architecture computes both state evaluations (emotions) and behavior evaluations. It contains three basic modules: crossbar learning memory, state evaluator, and behavior selector. In its basic behavioral cycle, the CAA architecture firstly computes the emotion of being in the current state (like previously done by Loehlin [16]). Then, using a feedback loop, it computes the possibility of choosing again, in a next time, the behavior to which the current situation is the consequence. The state evaluation module computes the global emotional state of the agent and broadcasts it (e.g. by way of a neuro-hormonal signal) to the crossbar learning memory. The behavior computation module using some kind of behavior-algebra initially performs a *curiosity driven*, default behavior, but gradually that behavior is replaced by a learned behavior.

The forth module defines specific *personality parameters* of a particular agent, such as curiosity, level (threshold) of tolerance (patience), etc. It also provides some physiological parameters, such as urge signals. Urge signals produce urge behaviors in the behavior algebra of the agent.

Emotions. The subjective emotional graph [6] is the basic concept of the *mental representations* of the agents in the consequence driven systems theory. The environment situations are represented as emotionally colored nodes and they represent an emotion space through which an agent can search for a behavior trajectory. The emotional value can be represented in different ways, for example using numerical values, but stylizations of *facial expressions* are preferred whenever possible. Transitions between states are behaviors. The term "behavior" covers both the simple actions, and also possible complex network of actions. Sometimes there are states that are not reachable by the behavior repertoire of the agent. The emotional value is given to a state either by the genetic mechanism, or by learning, after the agent visits that state. A state can be a neutral one and can change its emotional value after being visited.

There are different approaches of what an agent learns in an environment. It can learn the whole graph, like a *cognitive map*, or it can learn only a *policy* (the set of states and behaviors associated to those states). In case of policy learning the actual map (interconnection network between the states) is provided by the environment and is not stored within the agent. An experiment of policy learning is described further in the text.

Motivations. The *motivation is anticipated behavior evaluation*, before actual execution of that behavior. The motivations are computed from the values of their (anticipated) consequence states. The Consequence Driven System theory introduced the *principle of learning and remembering only motivations (behavior evaluations), not the state evaluations.* There is no need to store the state values, since they can be computed from the (current) behavior values. The approach of storing action values rather than state values was also emphasized by Watkins [18], who discovered the relationship between Reinforcement Learning and Dynamic Programming [4].

Agent-environment interaction: Parallel programming metaphor. In Consequence Driven Systems theory parallel programming was understood as *a way of thinking about the agent-environment interaction.* Since the introduction of CAA, parallel programming was used in carrying out the experiments of pole balancing learning [5], with the CAA controller running on one VAX/VMS terminal, while the environment, the pole balancing dynamics, (programmed by Charles Anderson) running as a separate process on another terminal.

4   The Main Result: Emotion Potential, Motivation Field, and Anticipatory Behavior

In this chapter we will state our main result. Extending from the force field metaphor [14], we will use the motivational hieroglyphs to denote concepts like *emotion, motivation, and behavior*, among other terms in describing anticipatory systems.

4.1   A Fundamental Equation of Anticipatory Systems

The main idea about the motivational field is given in Figure 5. As Figure 5 shows, it is assumed that each state j of a system (e.g. animat) is assigned an *emotion potential* $\mathcal{E}(j)$. The emotion potential space defines by itself a vector field denoted as *motivational field*, $\heartsuit^{\rightarrow}$. In such a way each point in the space is assigned a *motivation gradient* toward the highest emotional value at that region of the space.

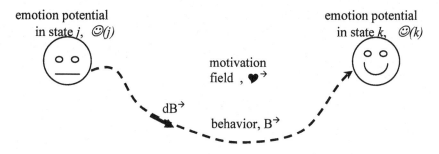

Fig. 5. Motivation field metaphor

In such a setup, a behavior $\vec{B}$ that is motivated by the motivation field $\vec{\heartsuit}$ induced by the emotional potential $\copyright$ obeys a basic ("Maxwell like") *field equation*:

$$\copyright = p \int_{j}^{k} \vec{\heartsuit} \, d\vec{B} \qquad . \qquad (1)$$

We propose this equation as a *fundamental equation of anticipatory systems*. It connects the emotion potential in a current state j, an *anticipated emotion potential* in a next state k, the motivation of moving toward state k, and behavior toward the state k. Here $d\vec{B}$ is a differential segment of a behavior, an action, that may sometimes not be in the same direction as the motivational field, for example due to some obstacles (resistances) along the way. The constant p allows for some personality features of a particular agent. The equation (1) explicitly states that *a satisfaction (positive emotion) is to be earned* by executing a proper behavior that follows a proper motivation (anticipation).

From the rather abstract mathematical *metaphor of field* we can step down to the *metaphor of circuit*. Figure 6 shows a branch of a motivation circuit.

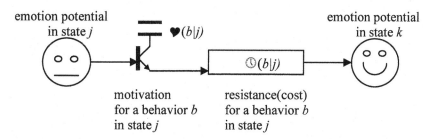

Fig. 6. An example of a motivational circuit

As figure 6 shows, we use a (transistor controlled) *switched capacitor* metaphor to represent a motivation $\heartsuit(b|j)$ for behavior b in state j. A *resistor metaphor* is used for representation of a cost $\copyright(b|j)$ (or any other obstacle) for performing b in j. The circuit in Figure 6 shows a predefined value of a motivation in a capacitor. To represent an adaptive behavior we need a circuit that will enable modulation of a motivation according to some emotional experience. The circuit in Figure 7 completes a circuit metaphor of an anticipatory behavior. As Figure 7 shows, the switched capacitor is able to update its motivation value due to experienced emotion after visiting state k.

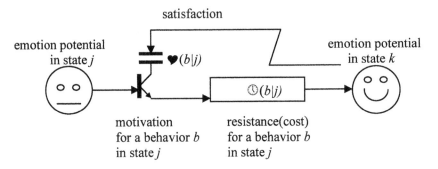

satisfaction

emotion potential in state j

$\heartsuit(b|j)$

emotion potential in state k

$\circledcirc(b|j)$

motivation for a behavior b in state j

resistance(cost) for a behavior b in state j

Fig. 7. Learning in a switched capacitor network

For the sake of simplicity further in the text we will not draw the transistors and capacitors. Instead, we will simply show the motivational value in a box.

4.2   Motivational Graphs: Goal-Seeking and Self-maintaining Behaviors

An agent performing an *anticipatory behavior* is often forced to deviate from such a behavior in order to execute some *self-maintaining behavior* (thirst, hunger, repair, etc). We distinguish such *interrupt (urge) behaviors* with respect to the main, purposeful, goal seeking, anticipatory behavior.

Figure 8 shows a segment of a *motivational graph* including both anticipatory and self-maintaining behaviors.

In a motivational graph each behavior is assigned a *motivation* for that behavior. The notation $\heartsuit(B2|S1)$ means motivation for *B2 given S1*. It is assumed that *motivations are functions of anticipations, which are functions of emotional potentials (attractors or repellers)*.

Each state is assumed to have the possibility of executing a behavior which we denote as *urge behavior* (e.g. [16]). Under that term we understand an urgent, interrupt behavior, which is used to *service some need* of the agent. In contrast to the motivated behavior, we do not assume the urge behavior as an anticipatory cognitive one, but rather as result of energy supplying and metabolism balance needs (hunger, thirst etc). It is usually assumed that the urge produces a circular path in a motivational graph, (see Figure 8), however the models allows other paths produced by urges. As special types of motivations, urges are usually represented by motivation monomials like ⬛, 🕿, ⬧, ◼, ◢. Usually a monomial such as ◓ would not be considered urges, but the model includes that possibility too.

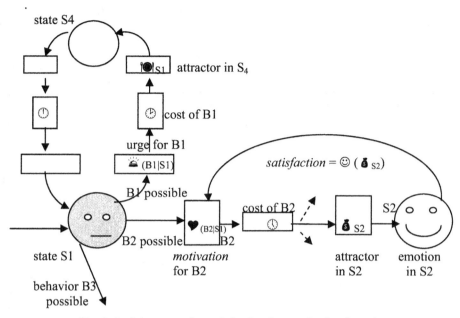

Fig. 8. Anticipatory and urge behaviors in a motivational graph

The motivational graph concept includes the possibility that the environment is non-deterministic, or stochastic, in a sense that (see Figure 8) after executing the behavior, say B2, it is not guaranteed that the next state will be the anticipated state S2. It is possible that after executing B2 the environment will present a state different from S2.

4.3   Learning Equation

Using circuit metaphor (e.g. Figure 7) we are able to define a *motivation learning function* for anticipation learning agents as

$$\heartsuit(b|s) \; += \; (\mathcal{O}(s\,') - \mathbb{O}(b|s)) \tag{2}$$

where += is the update operator we borrowed from object oriented languages for use in the simulation experiment described further in this text. The equation (2) simply states that *the motivation update is proportional to the satisfaction achieved.* In more detail, it states that the motivation $\heartsuit(b|s)$ of performing behavior b in s depends on the personally anticipated emotion $\mathcal{O}(s\,')$ in the *consequent situation s'* but also on the personally anticipated effort $\mathbb{O}(b|s)$ needed for b in s.

Urges are not considered elements of the learning system. However, *urge functions compete with motivation functions* for executing their behaviors. If an urge function reaches a greater value than the motivation functions in a state, then the behavior selection module would execute that urge behavior. So motivations define a priority scheme for executing behaviors.

5   Simulation Experiment: Two Personalities in the Same Environment

In this chapter we will show a simulation experiment in order to illustrate the presented theory. The experiment presents two personalities in the same environment exhibiting different behaviors due to their internal genetic features of personality. We will describe the genetic environment, the agent architecture, the behavioral environment, the result of the experiment in the behavioral environment, and the result of the experiment as learning process in the agent inner environment.

5.1   Genetic Environment

We will consider two characters, R1 and R2, having different genotypes represented by their imported (one-chromosome) genomes. Let their input chromosomes be

input_chromosome (R1) = (OOOOOOOOO⊗OOOO☺OOOO⊗)
input_chromosome (R2) = (OOOOOOO⊗O⊗OOOO☺OOOO⊗)   (3)

It means that the sensory systems of both personalities can distinguish 20 different situations in the related behavioral environment. For both the personalities, the situation 15 is considered desirable, while 10 and 20 are considered undesirable ones. The personality R1 will consider the situation 8 as neutral, while personality R2 will consider it dangerous. The behaviors are executed with a same cost, such that the cost does not play a significant role in the simulation experiments. As *experiment hypothesis*, it is expected that both personalities will develop different motivations in their memories, and consequently different learned behaviors in the behavioral environment.

5.2   Agent Personality Architecture

Figure 9 shows the CAA architecture used as the control architecture of both the personalities in the experiment. The initial, genetic information about the environment situations is received through the imported genome. The learning process is carried out through interaction with the behavioral environment. The learning routine is: Suppose in a situation s, a behavior b is executed and as a consequence a new situation is obtained from the environment. The new situation is evaluated on desirability, depending on either genetic information, or learning information in the matrix column addressed by the received situation. Satisfaction is computed and is sent back as a signal to the whole memory structure where only one element, indeed ♥$(b|s)$, is updated by that satisfaction signal.

The completion of the learning process is decided, according to some criterion, by the personality module which is a kind of operating system for the whole architecture. In case of learning a policy as to how to behave in an environment, as is the case in this experiment, the end of the learning process is decided *when a policy is assigned to the starting situation*, since the policy learning goes backwards. At the end of the learning process (or any time before), the personality module might decide to export a genome about what has been learned so far.

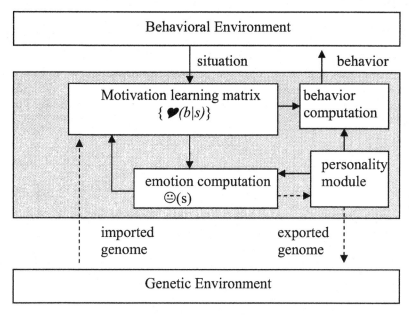

Fig. 9. Personality architecture used in the experimental setup

5.3 The Agent -Environment Interaction

Figure 10 shows the parallel programming version of the agent-environment interaction used in the experiment. Here *(*/*)* means matrix and *(*/s)* means column vector.

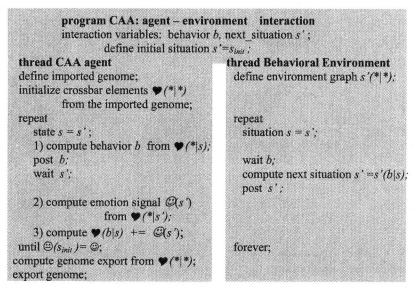

Fig. 10. The parallel programming model of inter-process communication between a CAA agent personality and the behavioral environment

5.4   Result of the Experiment: Behavioral Environment

Figure 11 shows the result of the experiment *in a behavioral environment that corresponds to the given genome from the genetic environment.* As hypothesized, Figure 11 shows that as a result of different developed plans in their mental representations, the two personalities indeed exhibit different behaviors in the same behavioral environment, as hypothesized. The learned behavior toward the goal state for personality R1 is 6-1-2-3-8-9-14-15, while for personality R2 it is 6-7-12-13-18-19-14-15.

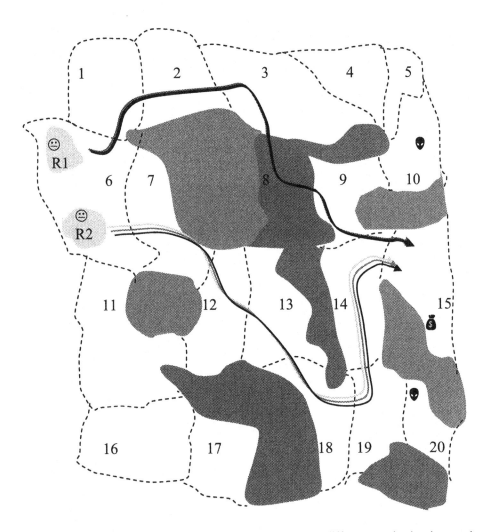

Fig. 11. Learned behaviors of agents R1 and R2 due to different motivational mental representations of the environment

5.5   Result of the Experiment: Inner Agent Environment

Figure 12 shows the result of the learning experiment in the inner mental representation of the personality R2. It is a crossbar adaptive array representation, a memory array with motivation elements ♥ *(b|s)*. The values are taken from the ordered set {□, ≈, ♥, 2♥, 3♥,…}, where □ denotes negative, ≈ denotes neutral, and ♥ denotes a positive motivation of performing behavior *b* in situation *s*.

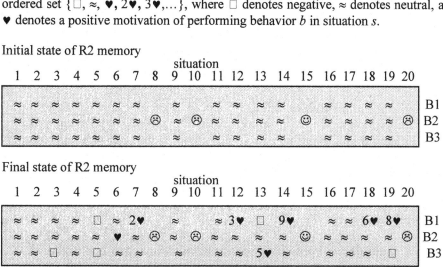

Fig. 12. Inner environment of the personality R2 after learning

As we can see from Figure 12, the initial state of the memory of the personality R2 as expanded from the imported genome (3) has one attractive situation, situation 15, and three repellent situations. All behaviors can be executed with equal probability, including the behaviors B1, B2, B3, as well as some default behavior B0. After learning, motivations about performing behaviors changed, and in some situations the agent has developed a policy as how to behave. For example, in situation 3 it will not tend to execute behavior B3 again, while in situation 18 it will tend to execute the behavior B1 again. In some situations, like the situations 1, 2, 4, 9, 11, 16, and 17, it will tend to execute the default behavior, which can be a random walk. Learning backwards, the learning process converged when the initial situation 6 became desirable, ☺(S_6) = ☺, since ♥($B_2|S_6$) = ♥. In other words the experiment ends when the agent being in the starting situation 6 gains confidence as how to execute a behavior that will possibly lead toward a goal state.

Figure 13 emphasizes that a gradient has been built between the situations 6 and 14, with motivational values ♥ and 9♥ respectively.

As Figure 13 shows, the agent will simply climb the gradient field in its inner environment, and thus will exhibit a particular learned behavior in behavioral environment (Figure 11). The gradient defines a learned plan of how to behave in this particular environment. The plan has been developed by learning a *partial policy for behaving* in this environment. It is represented as *a field gradient in a motivational field defined by the emotion potential* in state 15. The *path-following behavior* in the behavioral environment is *expressed* by the *hill-climbing behavior* in the inner environment field gradient. Note that the learner is not aware of actual situation transitions; that knowledge is local to the behavioral environment.

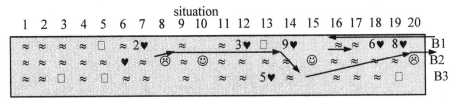

Fig. 13. Developed gradient in the potential field due to learning

The experiment described above was carried out in 1981 and subsequently reported [6]. The experiment is described here as illustration of the theoretical concepts proposed in the paper.

6   Conclusion

The work addresses the question of a fundamental relationship that would contain crucial grounding concepts of an anticipatory behavior. The result is presented in a form of an integral field equation, relating anticipatory behavior with a motivation for that behavior and the anticipated emotion after executing that behavior. The proposed equation suggests analogies to phenomena in other potential fields (e.g. electric field in physics).

From the potential field level of presentation we step down to a circuit level, and we introduce a concept of a motivational circuit. We use the concept of switched capacitor, which to our knowledge has not been used so far in cognitive science modeling.

The paper advances toward understanding the concept of artificial personality. It emphasizes that among current efforts of building autonomous robots that will communicate with humans, the issue of artificial personality is important and includes concepts like intelligence, emotion, and motivation, among others.

Within a theory of personality, the paper makes a relation between the concept of emotion and motivation. In a verbal form, a motivation for anticipative behavior in an adaptive system is its anticipation of a future consequence of that behavior; the emotion system is an evaluation system for computing the value of that consequence. A temporal distinction between emotions and motivations is proposed: Emotions are about current states; motivations are about anticipated emotions in future states. The motivational system can be understood as a priority system for executing behavioral processes.

Although the paper is basically theoretical, an early experimental work is described in terms of the new concepts, to illustrate the learning process in terms of concepts discussed in the paper.

A special effort in search of relevant works that can be considered as early efforts in anticipatory behavior and consequence driven systems research is made in this paper. The paper points out the attention of the anticipatory systems research community to some early works of Lewin and Loehlin as valuable sources of further inspiration.

The paper is believed to contribute toward computational theories of personality as well as toward the search for basic principles of the anticipatory behavior in adaptive learning systems.

Acknowledgement

The author wishes to express gratitude to the Editors and the anonymous reviewers for valuable comments on the earlier draft of this paper. The author also wishes to express gratitude to the 1981 ANW Group (Harry Klopf, Nico Spinelli, Michael Arbib, Andrew Barto, Richard Sutton, Charles Anderson, Ted Selker, and Jack Porterfield) and to his Cognitive Psychology teacher James Chumbley as well as to his Animal Learning teacher John Moore, for the knowledge and inspiration gained at that time that lead to the results presented in this paper.

References

1. Barto A., Sutton R., Anderson C.: Neuronlike Elements that can Solve Difficult Learning Control Problems. IEEE Transactions on Systems, Man, and Cybernetics 13: (1983) 834-846
2. Barto A., Sutton R., Watkins C.: Learning with Sequential Decision Making. In: M. Gabriel, J. Moore (eds.): Learning and Computational Neuroscience: Fundamentals of Adaptive Networks. MIT Press (1990) 539-602
3. Barto A.: Reinforcement learning. In: O. Omidvar and D. Elliot (eds.): Neural Systems for Control, Academic Press. (1997)
4. Bellman R.: Dynamic Programming. Princeton University Press. (1957)
5. Bozinovski S.: Inverted pendulum learning control. ANW Memo. COINS, University of Massachusetts, Amherst (1981)
6. Bozinovski S.: A Self-learning System Using Secondary Reinforcement. In: R. Trappl (ed.) Cybernetics and Systems Research. North-Holland (1982) 397-402,
7. Bozinovski S.: Consequence Driven Systems. Gocmar Press (1995)
8. Bozinovski S.: Crossbar Adaptive Array: The First Connectionist Network that Solved the Delayed Reinforcement Learning Problem. In: A. Dobnikar, N. Steele, D. Pearson, R. Alberts (eds.) Artificial Neural Networks and Genetic Algorithms Springer Verlag (1999) 320-325
9. Gadanho S.: Reinforcement Learning in Autonomous Robots: An Empirical Investigation of the Role of Emotions. PhD Thesis. University of Edinburgh (1999)
10. Gadanho S., Hallam J.: Robot Learning Driven by Emotion. Adaptive Behavior 9(1) (2001) 42-62
11. Gadanho S.: Emotional and Cognitive Adaptation in Real Environments. Proc Symp ACE'02 of the 16th European Meeting on Cybernetics and Systems Research, Vienna, Austria (2002)
12. Lewin K.: Dynamic Theory of Personality. McGraw Hill (1935)
13. Lewin K.: Principles of Topological Psychology. McGraw Hill (1936)
14. Lewin K.: Field Theory in Social Science. Harper and Row (1951)
15. Liebert R., Spiegler M.: Personality. The Dorsey Press (1974)
16. Loehlin J. Computer Models of Personality. Random House (1968)

17. Sutton R.: Temporal credit assignment in reinforcement learning. Ph.D. Thesis. Computer and Information Sciences (COINS) Department, University of Massachusetts at Amherst (1984)
18. Watkins C.: Learning from Delayed Rewards. Ph.D. Thesis. King's College, Cambridge. (1989).

A Framework for Preventive State Anticipation

Paul Davidsson

Department of Software Engineering and Computer Science
Blekinge Institute of Technology
Soft Center, 372 25 Ronneby, Sweden
Paul.Davidsson@bth.se

Abstract. A special kind of anticipation is when an anticipated undesired situation makes an agent adapt its behavior in order to prevent that this situation will occur. In this chapter an approach is presented that combines low level reactive and high level deliberative reasoning in order to achieve this type of anticipatory behavior. A description of a general framework for preventive state anticipation is followed by a discussion of different possible instantiations. We focus on one such instantiation, linear anticipation, which is evaluated in a number of empirical experiments in both single- and multi-agent contexts.

1   Introduction

In this book many different types of anticipatory behavior are described. A special kind is when an anticipated undesired situation makes an agent adapt its behavior in order to *prevent* that this situation will ever occur. For example, assume that you are going out for a walk and that the sky is full of dark clouds. Using your internal weather model and your knowledge about the current weather situation, you anticipate that it will probably begin to rain during the walk. This makes you foresee that your clothes will get wet which, in turn, might cause you to catch a cold, something you consider a highly undesirable state. So, in order to avoid catching a cold you will adapt your behavior and bring an umbrella when going for the walk.

We will here present an approach that combines low level reactive and high level deliberative reasoning in order to achieve this type of anticipatory behavior. This approach is based on the concept of *anticipatory systems* as described by Rosen [9]. We will begin with a brief description of anticipatory systems followed by a presentation of a general framework for *anticipatory autonomous agents* [2]. We then discuss different possible instantiations of this general framework and focus on one such instantiation, *linearly anticipatory autonomous agents* [3]. To evaluate this approach, some empirical results from a number of experiments in both single- and multi-agent contexts are presented. Basically, this chapter is a compilation of previous work, together with some refinements of the framework and some reflections and lessons learnt. Some pointers to future work conclude the chapter.

M. Butz et al. (Eds.): Anticipatory Behavior ..., LNAI 2684, pp. 151-166, 2003.
© Springer-Verlag Berlin Heidelberg 2003

2   Anticipatory Systems

According to Rosen [9], an anticipatory system is "...a system containing a predictive model of itself and/or of its environment, which allows it to change state at an instant in accord with the model's predictions pertaining to a latter instant." (p. 339) Thus, such a system uses the knowledge concerning future states to decide which actions to take in the present. The next state of an *ideal* anticipatory system would be a function of past and future states, whereas a causal system depends only on past states. However, since an agent acting in the real world normally does not have true knowledge of future states, it is not possible to implement an anticipatory system in this strict sense. (An agent acting in a closed world and having a perfect world model, on the other hand, would be able to have true knowledge of future states.) The best we can do is to approximate such a system by using *predictions* of future states.

Let us describe a simple class of anticipatory systems suggested by Rosen [8]. It contains an ordinary causal (i.e., non-anticipatory) dynamic system, S. With S he associates another dynamical system, M, which is a model of S. It is required that the sequence of states of M is parameterized by a time variable that goes faster than real time. That is, if M and S are started out at some time t_0, then after an arbitrary time interval Δt, M's sequence of states will have proceeded $t_0 + \Delta t$. In this way, the behavior of M predicts the behavior of S: by looking at the state of M at time t, we get an estimate about the state that S will be in at some later time than t. In addition, M is equipped with a set E of effectors which allows it to operate either on S itself, or on the environmental inputs to S, in such a way as to change the dynamical properties of S. If S is modified the effectors must also update M.

Then, how should the predictions about the behavior of S be used to modify the properties of S? Rosen [8] argues that this could be done in many ways, but suggests that the following is the simplest:

> "Let us imagine the state space of S (and hence of M) to be partitioned into regions corresponding to "desirable" and "undesirable" states. As long as the trajectory in M remains in a "desirable" region, no action is taken by M through the effectors E. As soon as the M-trajectory moves into an "undesirable" region (and hence, by inference, we may expect the S-trajectory to move into the corresponding region at some later time, calculable from a knowledge of how the M- and S-trajectories are parameterized) the effector system is activated to change the dynamics of S in such a way as to keep the S-trajectory out of the "undesirable" region." (p. 248)

3   Preventive State Anticipatory Agents

How could we transfer this type of systems into an agent framework? There are some additions and changes to the simple system suggested by Rosen that we have found necessary. First, we need a meta-level component that runs and monitors the model, and evaluates the predictions and decides how to change S or the input to S. (We will in the following regard both these types of changes as changes to S.) Thus, we also include the effectors, E, in this meta-level component that we here will call the *Anticipator*. Second, in order to predict future environmental inputs to S we need to extend the model M to include also the environment. This inclusion is in line with later work of Rosen (cf. [9]).

3.1   An Agent Architecture for Preventive State Anticipation

In the suggested framework, an anticipatory agent consists mainly of three entities: an object system (S), a world model (M), and a meta-level component (Anticipator). The object system is an ordinary (i.e., non-anticipatory) dynamic system. M is a description of the environment *including* S, but excluding the Anticipator. (The importance of having an internal model that includes both the agent as a part of the environment and (a large portion of) its abilities has been stressed by, for instance, Zeigler [11].) The Anticipator should be able to make predictions using M and to use these predictions to change the dynamic properties of S. Although the different parts of an anticipatory agent certainly are causal systems, the agent taken as a whole will nevertheless *behave* in an anticipatory fashion.

When implementing an anticipatory agent, what should the three different components correspond to, and what demands should be made upon these components? To begin with, it seems natural that S should correspond to some kind of reactive system similar to the ones mentioned above. We will therefore refer to this component as the *Reactor*. It must be a fast system in the sense that it should be able to handle routine tasks on a reactive basis and, moreover, it should have an architecture that is both easy to model and to change. The Anticipator would then correspond to a more deliberative meta-level component that is able to "run" the world model faster than real time. When doing this it must be able to reason about the current situation compared to the predicted situations and its goals in order to decide whether (and how) to change the Reactor. The resulting architecture is illustrated in Figure 1.

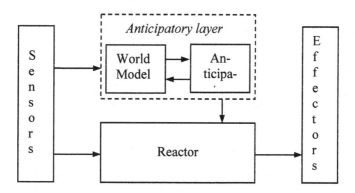

Figure 1. The basic architecture of an anticipatory agent.

To summarize: The sensors receive input from the environment. This data is then used in two different ways: (1) to update the World Model and (2) to serve as stimuli for the Reactor. The Reactor reacts to these stimuli and provides a response that is forwarded to the effectors, which then carry out the desired action(s) in the environment. Moreover, the Anticipator uses the World Model to make predictions and on the basis of these predictions the Anticipator decides if, and what, changes of the dynamical properties of the Reactor are necessary. Every time the Reactor is modified, the Anticipator should, of course, also update the part of the World Model describing the agent accordingly. Thus, the working of an anticipatory agent can be viewed as two concurrent processes, one reactive at the object-level and one more deliberative at the meta-level.

If we adopt Rosen's suggestion of the simplest way of deciding when (and how) to change the Reactor, i.e., by dividing the state space into desired and undesired regions and assume that the Reactor consists of situation-action rules, an anticipatory agent can be specified as a 4-tuple, $< W, R, U, X >$, where

- W is the description of the environment (the world model).
- R is the set of situation-action rules defining the Reactor.
- U is the set of undesired states.
- X is the set of rules describing how to modify R.

The Anticipator is defined by U and X. For each element in U there should be a corresponding rule in X, which should be applied when an undesired state of this kind is anticipated. Thus, we need in fact also a function, $f: U \rightarrow X$, which determines which rule for modifying the Reactor should be applied given a particular type of undesired state. However, as this function typically is obvious from the specification of U and X, it will not be described explicitly. Moreover, in all simulations described below, W will consist of the positions of all obstacles, targets, and agents present in the environment. Sometimes it is useful to make a distinction between *static* and *dynamic* rules in R. A static rule usually does not change during the lifetime of an agent, whereas a dynamic rule is typically valid only until the current goal is fulfilled.

3.2   Different Types of Anticipatory Agents

It is possible to make a distinction between different types of anticipatory agents based on the way the predictions are made by the Anticipator. How the predictions are made is partially determined by (i) the properties of the environment and (ii) the quality of the world model. If the environment is deterministic and the world model complete, the Anticipator is able to always predict exactly what will happen. Thus, for each state of the world it is possible to foresee what the next event in the world will be and what the consequences of this event will be, i.e., the next state of the world. This type prediction will result in a *sequence* of future states (see Figure 2) and we will therefore refer to this as *linear anticipation*.

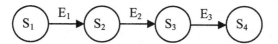

Figure 2. The result of a linear anticipation. S is the state of the world (model), including e.g. the state and position of all objects and agents in the world. E_i is the event that occurs when the world is in state S_i, e.g., an action of one of the agents.

In many situations, however, an agent only has a partial model of the environment. This would lead to that the Anticipator does not know what the next event will be, e.g., which action another agent will perform. Consequently, it has to consider a number of different possible events at each point in time. (A sophisticated Anticipator may assign probabilities to each of the possible events.) Thus, the predictions will form a tree structure of future states (see Figure 3). In addition, the environment may not be deterministic. This will lead to an even larger tree structure since each event may result in a number of world states.

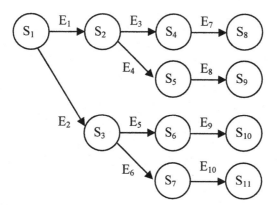

Figure 3. The result of a tree anticipation. At each state of the world (model) more than one event is considered.

The reminder of this chapter will focus on agents using linear anticipation. As mentioned, this type of preventive state anticipation makes strong assumptions concerning the quality of the world model and the behavior of the environment. However, we believe that there exist non-trivial applications where it is possible to have a *sufficiently* good world model and where the environment is *sufficiently* deterministic so that linear anticipation can be used successfully.

4   Linearly Anticipatory Agents

Agents using preventive linear state anticipation, which we will refer to as *linearly anticipatory agents*, is a specialization of the general architecture described above. In this section this type of agent will be described and in the next section results from a number of experiments with this type of agent will be presented.

In the experiments below, the Reactor and the Anticipator are run as two separate processes where the Reactor is given a higher priority than the Anticipator, which runs whenever the Reactor is waiting, e.g., for an action to be performed. Since the Reactor is able to preempt the Anticipator at any time, reactivity is always guaranteed.

The Reactor carries out a never-ending cycle of: perception of the environment, action selection by situation-action rules, and performance of action. Rather than having explicit goals, the goals of the Reactor are only implicitly represented in its collection of situation-action rules. The basic algorithm of the Reactor is:

procedure REACTOR;
while true **do**
   Percepts ← Perceive;
   Action ← SelectAction(Percepts);
   Perform(Action);

The Anticipator, on the other hand, carries out a never-ending cycle of anticipation sequences. Each such sequence begins with making a copy of the agent's world model, i.e., a description of the environment containing the agent as a physical entity in the environment, and a copy of the agent's current set of situation-action rules. These are then used to make a sequence of one-step predictions. After each prediction step, it is checked whether the simulated agent has reached an undesired state, or whether it has achieved the goal. If it has reached an undesired state, the Reactor will be modified in order to avoid reaching this state. Thus, this functioning corresponds to that of the simplest kind of anticipatory system suggested by Rosen. The basic algorithm of the Anticipator is:

procedure ANTICIPATOR;
while true **do**
  WorldModelCopy ← WorldModel;
  ReactorCopy ← Reactor;
  UndesiredState ← false;
  while not UndesiredState **and**
     not GoalAchieved(WorldModelCopy) **do**
   Percepts ← WorldModelCopy.Perceive;
   Action ← ReactorCopy.SelectAction(Percepts);
   WorldModelCopy.Perform(Action);
   UndesiredState ← Evaluate(WorldModelCopy);}
  if UndesiredState **then**
   Manipulate(Reactor);

The function Evaluate can be described as checking whether the current anticipated state belongs to U, and Manipulate as first applying f on the anticipated undesired state and then using the resulting rule from X to modify R.

It is important to note that since the behavior of the Reactor in each situation is determined by situation-action rules, the Anticipator always "knows" which action the Reactor would have performed in that situation. Also the environment, including all other agents present in the environment, is treated as being purely reactive. Thus, since everything is governed by deterministic rules, the anticipation mechanism requires no search, or in other words, the anticipation is *linear*. It should also be noted that the agent is not limited to have only a singular goal. In a multi-goal scenario, some of the modifications of the Reactor should only hold for a limited interval of time, e.g., until the current goal has been achieved. Otherwise, there is a danger that these changes might prevent the agent to achieve other goals. Furthermore, it seems straightforward to generalize the Anticipator algorithm to handling goals of maintenance rather than of achievement (e.g., by removing the GoalAchieved condition).

5   Evaluation

We will present three sets of experiments where linearly anticipatory agents have been used to solve different tasks, starting by a very simple task and then increasing the complexity. An application concerning guidance of the plot of a story using a similar approach is described by Laaksolahti and Boman [7].

5.1   Linearly Anticipatory Agents versus Reactive Agents for Path Finding in Simple Worlds

This is a summary of the experiment described in [3]. The environment is a small two-dimensional grid (10x10 units) in which a number of unit-sized square obstacles form a maze. The goal of an agent is to pick up a number of targets by finding paths from its current position to the positions of the targets. The agent is able to move in four directions (north, south, east, and west), unless there is an obstacle that blocks the way. The agent is always able to perceive the direction to the target and whether there are any obstacles immediately north, south, east, or west of the agent.

5.1.1 Single Agent Worlds
We compared a simple reactive agent to a simple anticipatory agent using this reactive agent as its Reactor. The reactive agent has only one simple rule: *reduce the distance to the target if possible, else increase it as little as possible*. If there are two possible directions, it chooses the one yielding the greatest reduction in distance to the target. If there is an obstacle that blocks the way in this direction, it tries the other direction that reduces the distance to the target, and if this direction is also blocked the agent chooses the direction that increases the distance to the target the least. (Although this seems as a very dumb strategy, many primitive animals behave in this manner.) If the maze is not too complex this simple agent will in most situations be able to find a path to the target. However, sometimes it will "loop" between two or more positions and never reach the target.

Let us now consider a linearly anticipatory agent that has this reactive agent as its Reactor. That is, R = {reduce the distance to the target if possible, else increase it as little as possible}. We then define the undesired states as those in which the agent is trapped in a loop, i.e., U = {being in a loop}. One way of modifying R to prevent the agent to get trapped in the loop is to avoid entering one of the positions in the loop, e.g., X = {avoid the position in the loop closest to the target}. (Which position in the loop to avoid can, of course, be selected according to other principles.) Thus, the Anticipator will detect the loop (in which the reactive agent would get trapped) and make the Reactor avoid the position closest to the target. It should be noted that the rule is added as a dynamic rule to R, meaning that it will cease to exist as soon as the current goal is fulfilled, i.e., the agent has reached the target position. The reason is that which positions to avoid depend on the current position and the target position. Thus, in this case we make a distinction between *static* and *dynamic* rules, where the original rule in R is static.

We have compared the performance of these two agent types in a series of experiments. In all experiments there was one target in the environment whereas the number of obstacles varied between 0 and 35. In each experiment the positions of the agent, the target and the obstacles were randomly chosen. In order to avoid too many trivial scenarios there was also a requirement that the distance between the agent and the target should be at least five unit-lengths. Moreover, only scenarios in which there actually is a path to the target were selected. The results are presented in Figure 4. We see that the more complex the environment gets, the more useful is the anticipatory behavior. If there are no obstacles at all, even the reactive agent will, of course, always find the target.

Thus, this anticipatory agent is able to reach the target (when this is possible) in almost all kinds of mazes. However, there are some in which it will not succeed and they are typically of the kind where the Anticipator detects a loop and blocks a position which is the only possible way to the target, i.e., where a part of the loop belongs to the path to the goal. The problem is that it is too eager to block a position. The reason for this is that the Reactor is inclined to turn around 180° as soon as it is not decreasing the distance to the target. By augmenting R with the condition that it should only change its direction 180° when there are no other alternatives (i.e., if there are obstacles in the three other directions), we will solve this problem. This rule together with the U and X used in the last example will result in an anticipatory agent which seems always to reach the target (if the Anticipator is given enough time to anticipate, that is). This Reactor (i.e., R = {reduce the distance to the target if possible, else increase it as little as possible, but do not turn around if not forced to}) will be used in all experiments below.

5.1.2 Multi-agent Worlds

In this section some experiments are presented that point out the advantages of being anticipatory, both when competing and when cooperating with other agents. We used the same environment as above but with two agents and five targets (and 30 obstacles). The goal of both the agents is to pick up as many targets as possible. In addition to the basic algorithm described above, the Anticipator needs a model of the other agent which it uses to predict that agent's actions (in the same manner as it predicts its

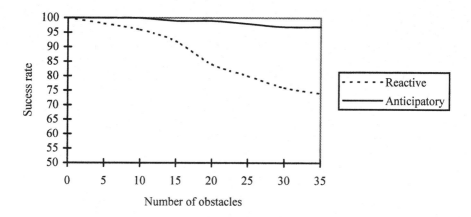

Figure 4. Comparison between a reactive and a linearly anticipatory agent. The number of successful experiments (averages over 200 runs) was measured for environments of different complexity. An experiment is considered successful if the agent finds a path to the target.

own actions). Although the experiments have been carried out with two competing or cooperating agents in order to make things as clear as possible, it would be trivial to extend the experiments to permit a larger number of agents.

Competing Agents
The main idea is that an anticipatory agent should be able to use its knowledge about the behavior of other agents in order to detect future situations in which the other agents' intentions interfere with the agent's own intentions. When such a situation is detected, the Anticipator should manipulate the Reactor in order to minimize the probability that this situation will ever occur. For example, when the Anticipator realizes that another agent will reach a target first, it notifies the Reactor that it should ignore this target. Thus, we have that: U = {being in a loop, pursuing targets that presumably will be picked up by another agent} and X = {avoid the position in the loop closest to the target, avoid the target that presumably will be picked up by another agent}.

Two experiments were performed, which each was repeated 1000 times. In the first both agents were reactive and in the second there was one anticipatory agent competing with one reactive agent. The result of the first experiment was that both agents picked up 2.5 targets each on average. This was not very surprising but confirmed that there were no bias towards either of the agents. In the second experiment the anticipatory agent pricked up on average 2.6 targets whereas the reactive agent picked up the remaining 2.4 targets. Even though this may seem as a small difference the performance of the anticipatory agent is almost optimal. (Given the behavior of the reactive agent, it was estimated that optimal behavior would result is 2.64 targets.)

Cooperating Agents

These experiments were done to show that linear anticipation could be used also for cooperation. The task for the two agents is to pick up all the targets in *shortest possible time*. It does not matter which agent picks up a particular target.

We did three experiments. In the first and second we used the same pair of agents that were used above. In the third we used a cooperative anticipatory agent and a cooperative reactive agent. The idea was that as soon as the anticipatory agent realized that it will arrive to one of the targets before the reactive agent it will tell this to the reactive agent, which then will ignore this target. Thus, we add {other agent pursuing target that presumably will be picked up by me} to U and {send message to other agent that it should avoid the target} to X.

The experimental results were that the two reactive agents needed in average 18.5 time units to pick up all five targets and that the competing anticipatory and reactive agents needed in average 17 time units to do this. The cooperating agents performed best and needed only 15 time units, which is close to the optimum that was estimated to circa 14 time units.

In this scenario, only one of the agents is anticipatory, whereas the other agent is an ordinary reactive agent. But, even if we also let the other agent be anticipatory with the same model of the world, we would not increase the performance. The reason for this is that both agents would have made the same predictions and therefore send messages to each other about things that both agents have concluded by themselves. However, in a more realistic setting where the agents do not have exactly the same information about the world, such an approach would probably be fruitful. In such a case things get quite complicated if one anticipatory agent simulates the anticipatory behavior of another anticipatory agent which in turn simulates first agent's anticipatory behavior. One solution to this problem is to simulate only the reactive component of other agents and when an agent modifies its reactive component, it should broadcast information about this modification to the other agents. In this way we are still able to make linear anticipations. This approach can be contrasted with the Recursive Modeling Method suggested by Gmytrasiewicz and Durfee [4] in which an agent modeling another agent includes that agent's models of other agents and so on, resulting in a recursive nesting of models.

5.2   Linearly Anticipatory Agents versus A*-based Agents for Path Finding in More Complex Worlds

In this experiment, which is described by Seger and Törnqvist [10], the task was still path finding but with increased complexity. The environment consisted of 10000 positions instead of 100 and the agents were able to move in 8 directions instead of 4. The maze built in the 100x100 grid was quite complex and 12 tasks were defined in this environment, i.e., pairs of starting and target positions. Most of these chosen to provide very challenging situations with which path finding agents are known to have difficulties to cope. The performance of an anticipatory agent (very similar to the one used above) was compared to an agent based on the A* algorithm [5, 6].

Traditionally path-finding problems have been solved using different types of search techniques. The simplest method is to search through the whole problem space, e.g., by using breath-first, and return the shortest path. However, as this is extremely time-consuming for non-trivial problems, heuristic methods are often used. The A* algorithm, which is one of the most well-known methods, is based on a function that estimates the cost to reach the goal from the current state. A* uses this cost function (and the cost to get to the current state from the initial state) to keep a prioritized queue of the possible moves, from which it always tries the move with the highest priority first. The A* algorithm is optimal in the sense that no other known algorithm visits fewer nodes given the same heuristics and finds the shortest path. In the experiments the straight-line distance was used to compute the cost function. The A* agent was implemented as a purely deliberative agent, i.e., it first computes the path and then moves to the target following this path.

Two measures that are interesting to compare are the number of steps that the agents need to reach the target and the number of states that are considered in the search/anticipation phase. As the A* agent always will find the shortest path, the anticipatory agent will never perform better. In the experiment, it was able to find this optimal path for one of the twelve tasks. For the other tasks, the anticipatory agent chose a path that was twice as long or more than the optimal path. However, if we consider the number of states considered during the search/anticipation, the anticipatory agent consider less states than the A* agent in 75% of the cases. Another important measure is the time it takes the agent to reach the target. As the anticipatory agent starts to move immediately, whereas the A* agent first does the search, the anticipatory agent will reach the target faster in simple tasks. (At least this holds when the anticipatory agent finds the optimal path, otherwise it depends on the relation between CPU performance and the time it takes to make the physical moves.) The result in more complex task depends on, e.g., the relation between the time it takes to consider one move and to actually performing that move. A possible conclusion, which needs further experimentation to be confirmed, is that anticipatory agents perform better than search-based agents when the state space is very large.

It should also be noted that the results in this experiment (and the following) are not a definitive measure how good linear anticipation is in this task. It is just a measure of how good this particular anticipatory agent (defined by the *WRUX*-tuple) is. There are other possible anticipatory agents that solve path-finding problems that may be interesting to evaluate. For instance, the major problem for the Reactor considered in the experiment is concave obstacles. These could be handled by a Reactor based on tracing rules, so that when the Reactor encounters an obstacle, it navigates around the obstacle by following the boundary of the obstacle until it is moving in the direction of the target again.

5.3   Linearly Anticipatory Agents versus Biter for Soccer Playing in the RoboCup Simulator

Due to the RoboCup (cf. www.robocup.org), soccer playing has become a very popular application domain for intelligent and adaptive systems. This domain has several properties that makes it a great challenge for such systems, e.g., the state space is enormous, the real-time requirements are hard, and all sensory information is uncertain as well as the outcome of the actions performed by the soccer-playing agents.

The experiment that will be described below concerns the simulated league which should be contrasted to the physical robot leagues. Although the problems of perception and action are harder in the robot leagues, the simulated league has all the properties listed above, e.g., each player only sees a small portion of the field, it has to estimate its position (and the other players' positions) in the field, communication with other agents is limited and distorted, the server which simulates the game only accept commands at certain time intervals (about each 100 ms). Thus, to cope with these complexities is a grand challenge for linear anticipation and will mercilessly point out any weaknesses. Just as the comparison with A* in the path finding domain, this experiment is described in detail by Seger and Törnqvist [10].

Small children usually play a very primitive type of soccer which can be classified as a purely reactive behavior. If they are in control of the ball they kick it towards the opponent's goal, else if they see the ball they run towards it, if they do not, they look around for it. The Reactor used in the experiment was based on this behavior but instead of always kicking towards the goal when possessing the ball, it only kick if it is close to the goal and there is nothing in the way. If there is a teammate closer to the goal, it passes the ball to that player, otherwise (i.e., if it is not close enough to the goal, or there is not clear shot at the goal, and no teammate is closer to the goal) it dribbles towards the opponents goal.

The Anticipator used in the experiment was very simple, having just one undesired state: *some player in the opponent team controls the ball*. If such a state is anticipated by a player currently controlling the ball, the anticipator will tell the Reactor to kick the ball in a different angle so that it will "trick" the opponent. During the anticipation, an opponent player is assumed to have control of ball if it is closer than one meter to the ball.

Typically we have very little information about how the opponent's players are implemented and therefore there are no perfect models of their behavior available for the Anticipator. In the experiments a very simple model was used, namely that each player just runs towards the ball. A more advanced possibility would be to learn their behavior from observation, but this seems as a very difficult task. The anticipation horizon (how far into the future the agent anticipates) used was one second, corresponding to 10 command intervals. Due to the quality of the world model (e.g., the uncertainties of the perception), longer anticipation horizon would probably have been useless.

The opponent in the experiment was a publicly available team called Biter [1]. First the reactive agents defined by the Reactor described above played 10 five-minute games against Biter and then the anticipatory agents did the same. The results were that the reactive agents lost to Biter, the average score being 1,7 − 2,6, whereas the anticipatory agents played equally well as Biter with the average score 1,5 − 1,5.

6   Conclusions and Discussion

A general framework for preventive state anticipation has been presented. Linear anticipation, which is one class of instantiations of this framework and which seems promising due to its simplicity and efficiency, was selected for further study and evaluated in three sets of experiments. The first two sets of experiments indicated the viability of this approach in relatively simple path-finding tasks. In the first set of experiments it was shown that by adding an anticipatory layer to a reactive agent, it is possible to improve the performance of the agent. The anticipatory agent adapts to the environment by letting the Anticipator manipulate the Reactor component according to anticipated future undesired states.

In the second set of experiments an anticipatory agent was compared to an agent based on a traditional search-based AI method, the A* algorithm. Whereas the anticipatory agent just tries to predict what will happen if nothing unexpected happens and changes its behavior only to prevent undesired states to occur, the search-based agent continuously evaluates what will happen if all of the possible actions are performed. The results from the second set of experiments showed that traditional search-based methods find more optimal solutions, but consider more possible states in order to find that solution. Thus, the anticipatory approach reduces the amount of search needed but does not always find the optimal solution. Moreover, it requires only a small amount of heuristic domain knowledge (encoded in the rules of the Reactor) as it relies on the anticipation mechanism to achieve complex behaviors.

In the two first experiments the environment and the tasks were quite simple and the agents had complete knowledge of the environment available. In the third set of experiments, the task got more complex; a much larger number of possible states, only partial and uncertain knowledge available etc. Although the results showed that it was possible to improve performance of a reactive agent by adding the anticipatory layer, the complexity caused some new difficulties to the designers. What is the appropriate anticipation horizon? How should the behavior of the Reactor be changed in order to prevent an anticipated undesired state? When should this change be made?

6.1 Choosing Anticipation Horizon

It is necessary to balance the anticipation horizon according to the quality of the world model, W, and the real-time requirements of the application domain. If W is complete and correct, we can let the agent anticipate as many time steps into the future as possible as long as the real-time requirements are met. In most realistic scenarios however, W is not complete and correct and therefore the anticipation will probably fail now and then. There are a number of possible reasons for such failures, e.g.:

-   W does not accurately describe the current state of the environment.
-   W does not accurately describe the dynamical behavior of the environment.
-   The environment is simply not predictable.

The first type of failure typically is caused by low quality or partial sensor data (cf. the RoboCup domain). One way of dealing with this problem is to use only partial models of the world that describes the relevant parts of the environment (of which it is possible to get sensor data of sufficient quality).

The second type of failure is often caused by that the quality of the models of other agents' behavior is too low. Figure 5 illustrates what may happen when imperfect models of other agents are used. It is assumed that the quality of W degrades with a certain percentage (1, 5 and 10% respectively) for each prediction step and that the value of the quality at the start of the anticipation was 1 (corresponding for instance to a complete and correct of the current environment). For instance, we see that if the degradation is 10%, the quality of W is below 0,4 after just 9 prediction steps. This is indeed a very simple model which makes strong assumptions (e.g., different event may differently hard to predict), but it indicates the importance of the quality of the world model in order to make reliable predictions.

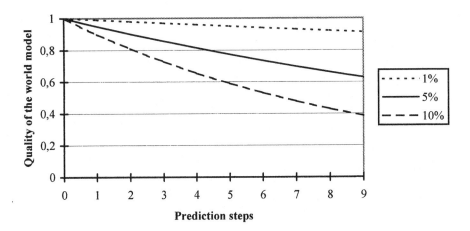

Figure 5. The quality of the world model degenerates as more anticipation steps are performed.

If there are competing agents in the environment (that do not want to share a description of their behavior) this is a difficult problem to deal with. As mentioned earlier, one approach would be to try to refine the model of those agents by learning from observation.

If no other solution is found to remove these two types of failures, the best thing to do is probably to shorten the anticipation length to keep the discrepancy between predicted and actual future states sufficiently low to make reliable predictions.

One way of dealing with environments that are not completely predictable (and also with the first two types of failures), is to increase the level of abstraction of W (decrease its granularity). Even though all details cannot be predicted accurately, it is often possible to find a higher level of description where predictions of a sufficient quality can be achieved.

6.2   Future Work

Although some promising work has been carried out regarding the foundations and the application of linear anticipation, there are a lot of interesting topics that requires further study. Some examples are given below.

- It seems that the more time the Anticipator gets to anticipate, the better will the anticipatory agent behave if the world model is sufficiently good, e.g., it will choose shorter paths. However, more (empirical) evidence is needed in order to get a clearer picture of the relation between the anticipation horizon, the quality of the world model, and the performance of the agent.

- Usually it is quite easy to define the undesired states, but to decide how to change the behavior in order to prevent this state from occurring is usually much more difficult. Especially, this is true in dynamic environments including other competing agents. Even if a preventive change of behavior is found, there still may be a timing problem. In some situations it is crucial to change the behavior at the exact right point in time, otherwise the result can be even worse than the anticipated undesired state. Also, there is a possibility that a high level looping behavior might occur where the Anticipator switch the rules of the Reactor back and forth in order to prevent two (or more) different undesired states. Thus, to learn more about how and when to change the reactive behavior is a very important area for future study.

- What are the appropriate application domains of linear anticipation?
  - If the environment is predictable and completely known, the approach obviously works very well. However, for this type of problems there may be other approaches that are more efficient, especially if the state space is small and the environment static. How large must the state space be in order to make linear anticipation a good choice? When can optimal behavior be achieved? What requirements should be put on predictability of the environment? The answers to these questions may be application dependent.
  - On the other extreme, if there is very little knowledge available to build a world model, linear anticipation does not work at all. How good must the world model be in order to make linear anticipation a good choice? Also to this question the answer may be application dependent.

Acknowledgements

I would like to thank Christian Seger and Björn Törnqvist for carrying out some of the experiments described in this chapter as their Master's thesis work, and the anonymous reviewers for constructive comments.

References

1. P. Buhler and J.M. Vidal. Biter: A Platform for the Teaching and Research of Multiagent Systems' Design Using Robocup. *RoboCup 2001: Robot Soccer World Cup V.* LNCS Vol. 2377. Springer-Verlag, 2002
2. P. Davidsson and E. Astor and B. Ekdahl. A Framework for Autonomous Agents Based on the Concept of Anticipatory Systems, *Cybernetics and Systems '94*, pp. 1427-1434, World Scientific, 1994.
3. P. Davidsson. Learning by Linear Anticipation in Multi-Agent Systems, *Distributed Artificial Intelligence Meets Machine Learning*, LNAI Vol. **1221**, pp. 62-72, Springer Verlag, 1997.

4. P.J. Gmytrasiewicz and E.H. Durfee. A rigorous, operational formalization of recursive modeling. *First International Conference on Multiagent Systems*, pages 125-132. AAAI Press, 1995.
5. P.E. Hart, N.J. Nilsson, and B. Raphael. A Formal Basis for the Heuristic Determination of Minimum Cost Paths, *IEEE Transactions on SSC*, Vol. 4, 1968.
6. P.E. Hart, N.J. Nilsson, and B. Raphael. Correction to "A Formal Basis for the Heuristic Determination of Minimum Cost Paths", *SIGART Newsletter*, Vol. 37, 1972.
7. J. Laaksolahti and M. Boman. Anticipatory Guidance of Plot. *In this volume*. 2003.
8. R. Rosen. Planning, Management, Policies and Strategies: Four Fuzzy Concepts, *International Journal of General Systems*, Vol. 1, pp. 245-252, 1974.
9. R. Rosen. *Anticipatory Systems - Philosophical, Mathematical and Methodological Foundations*, Pergamon Press, 1985.
10. C. Seger and B. Törnqvist. Linear Quasi Anticipation – An evaluation in real-time domains, Master Thesis, MSE-2002:10, Blekinge Institute of Technology, Ronneby Sweden, 2002.
11. B.P. Zeigler. *Object-Oriented Simulation with Hierarchical, Modular Models - Intelligent Agents and Endomorphic Systems*, Academic Press, 1990.

Symbols and Dynamics in Embodied Cognition: Revisiting a Robot Experiment

Jun Tani

Brain Science Institute, RIKEN
2-1 Hirosawa, Wako-shi, Saitama, 351-0198 Japan
tani@brain.riken.go.jp

Abstract. This paper introduces novel analyses that clarify why the dynamical systems approach is essential for studies of embodied cognition by revisiting author's prior robot experiment studies. Firstly, we argue that the symbol grounding problems as well as the "situatedness" problems should be the consequences of lacking a shared metric space for the interactions between the higher cognitive levels based on symbol systems and the lower sensory-motor levels based on analog dynamical systems. In our prior studies it was proposed to employ recurrent neural networks (RNNs) as adaptive dynamical systems for implementing the top-down cognitive processes by which it is expected that dense interactions can be made between the cognitive and the sensory-motor levels. Our mobile robot experiments in prior works showed that the acquired internal models embedded in the RNN is naturally situated to the physical environment by means of entrainment between the RNN and the environmental dynamics. In the current study, further analysis was conducted on the dynamical structures obtained in the experiments, which turned out to clarify the essential differences between the conventional symbol systems and its equivalence realized in the adaptive dynamical systems.

1 Introduction

We speculate that the problems of cognition commence with the robots' attempt to acquire internal models of the world in certain forms so that they can mentally simulate or plan their own behavior consequences. By this means, we may not consider purely reactive-type robots that base their actions on simple sensori-motor reflexes since those robots do not deal with any mental processes that employ internal models. When discussing internal models, it is important to consider how they can be grounded to the physical environments and how the mental processes manipulating them can be situated in the behavioral contexts. This question addresses one of the observation problems in cognition which asks us where the observer, dealing with the descriptions, is positioned. We examine these problems by revisiting our prior studies on robot navigation learning experiments [14]. We attempt to conduct further dynamical systems analysis for the results of this experiment. This analysis will clarify the essential differences

M. Butz et al. (Eds.): Anticipatory Behavior ..., LNAI 2684, pp. 167–178, 2003.
© Springer-Verlag Berlin Heidelberg 2003

between symbols in conventional symbol systems and those embedded in analog dynamical systems through learning processes.

In the traditional approach of the robot navigation problems, the robots are forced to acquire exact maps of the environment measured in the global coordinate systems. Such robots apparently use the external views to describe their environments, since the descriptions are made by assuming the global observation from the outside.

On the other hand, the recent approach based on landmark-based navigation [8, 9] does not assume any global observations of the environments. In this approach, the observer sits inside the robot and looks at the outside through the sensory device focusing on upcoming events or landmarks. The observer collects the sequences of landmark-types and tries to build chain representations of them in the form of finite state machines (FSM) as the topological map of the environment. Although it is true that this approach provides us with much more successful results in the navigation tasks compared to the global map strategies, the symbolic representation of the FSM can still cause the symbol grounding problems. The symbol grounding problem is a general problem, as discussed by Harnad [5]. The problem is that discrepancies occur between the objects in the physical environment and their symbolic representations in the system which cannot be resolved autonomously through the system's own operations.

Let us consider the situation where the robot navigates in a pre-learned environment by identifying the current position from trying to match the state transitions in the FSM. A problem can arise when the robot fails to recognize an oncoming landmark because of some noise. The robot will be lost because it has received an erroneous sensory input which is different from the one expected using the FSM. The FSM will simply halt upon receiving this illegal input[1].

Although some may argue that this problem can be resolved by further development of the categorization schemes for landmark recognition, we consider that this approach leaves the underlying problem unsolved. We believe that the underlying problem exists in the position of the observers who look over the symbolic representations and try to manipulate them. The observer here is external to the descriptions. As long as such external observers are allowed for the robots, the robots face the symbol grounding problems.

We have investigated this problem from the dynamical systems perspectives [1, 6]. We speculate that real number systems best represent the mental activities of robots. We expect that the nonlinear dynamics characterized by chaos or fractal structures may serve as a basis for the mental activities of robots, as the theories of symbolic dynamics [2, 12, 17] have shown that such nonlinear dynamics exhibit a certain linguistic complexity. When the internal dynamics, which describe the mental processes of the robot, and the environment dynam-

---

[1] When an illegal input is received, the current state cannot be identified correctly. There could be an extra algorithm by which the current state can be estimated by means of the maximum likelihood. Such an extra algorithm, however, could generate another symbol grounding problem. The author will argue that there are intrinsic mechanisms to avoid these problems in the dynamical systems approach

ics are coupled together through the sensory-motor loop, those two dynamics would share the same metric space. We consider that the mental processes of the robots can be naturally situated to the environments as the coherence is achieved between those two dynamics interacting with each other in the same phase space. An important objective here is to unify the two separate entities for "descriptions" and their "manipulations" in the systems into one entity within the framework of the time-development of dynamical systems. We speculate that the internal observer [10, 4] finally appears in the cognitive processes of robots if this objective is accomplished. The next section reviews our embodied work on mobile robot learning of cognitive maps based on the dynamical systems approach.

2 Formulation

Firstly we review our navigation scheme which is applied to the *YAMABICO* mobile robot (cf. Fig. 1). *YAMABICO* can obtain the range image by a laser range

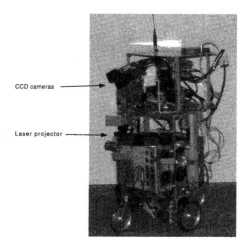

CCD cameras

Laser projector

Fig. 1. The *YAMABICO* mobile robot. It is equipped with a laser range sensor.

finder in real-time. In our formulation, maneuvering commands are generated as the output of a composite system consisting of two levels. The control level generates a collision-free, smooth trajectory using a variant of the potential field method i.e. the robot simply proceeds towards a particular potential hill in the range profile (direction toward an open space). The navigation level focuses on the topological changes in the range profile as the robot moves through a given workspace. The profile gradually changes until another local peak appears when the robot reaches a branching point. At this moment of branching the navigation level decides whether to transfer the focus to the new local peak or to stick with the current one. The navigation level functions only at branching points which appear in unconstructed environments. The importance here is that the navigation of the robot consists of the topological trajectories which are determined by the branching sequences. The control level is pre-programmed and the learning takes place only in the navigation level. Hereafter, our discussion focuses on how to learn and determine the branching sequences using neural learning schemes.

In the learning phase, the robot explores a given obstacle environment by randomly determining branching. Suppose that the robot comes to the nth branch point with receiving sensory input (range image vector) p_n and randomly determine branching (0 or 1) as x_n, then it moves to the next branch point n+1th (see Fig 2.) Through the entire exploratory travel, the robot acquires the sensory-

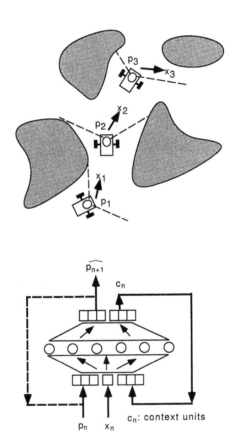

Fig. 2. Sensory-motor sequence in branching and RNN architecture.

motor sequence of (p_i, x_i). Using this sample of the sensory-motor sequence, a recurrent neural net (RNN) is trained so that it can predict the next sensory input p_{n+1} in terms of the current sensory input p_n and the branching motor command x_n (see Fig 2). We employ the idea of the context re-entry by Jordan [6] which effectively adds internal memory to the network. The current context input c_n (a vector) is a copy of the context output in the previous time: by this means the context units remember the previous internal state. The navigation problem is an example of a so-called "hidden state problem" (or non-Markov problem) where a given sensory input does not always represent a unique situation/position of the robot. Therefore, the current situation/position is identifiable, not by the current sensory input only, but additionally the memory of the sensory-motor sequence stored during travel is necessary. The memory structure is self-organized through the learning process. We expect that the RNN can learn certain "grammatical" structure hidden in the obstacle environment as embedded in its intrinsic dynamical structure by utilizing the context re-entry. (As many have shown the capability of RNNs for grammar learning.) We employ the back-propagation through time algorithm [13] for the RNN learning.

Once the RNN is trained, it can conduct the following two types of mental processes. (1) The RNN can conduct lookahead prediction of the sensory sequences for arbitrary given motor programs (branching sequences) by the closed-loop forward computation. In this computation the sensory prediction outputs in the current step is copied to the sensory inputs in the next step as shown by a dashed loop in the left hand side of the RNN in Fig 2. In this way, lookahead prediction of the future sensory sequence can be recursively computed with a given branching sequence. (2) The RNN can conduct goal-directed planning. It can generate the motor programs (branching sequences) for the robot to reach a goal specified by the corresponding distal sensory image. The inverse dynamics of the RNN with the minimum travel distance criteria can determine an optimal motor program. Details of goal-directed planning are not shown here, but in [14].

3   Experiment

Here, we review a part of our experiments of lookahead prediction. The robot explored a given workspace and the RNN was trained with 193 samples of the sensory-motor sequence. After this learning, the robot is started to travel from arbitrary positions. The robot maneuvers following an arbitrary motor program (branching sequence) and it tries to predict the coming sensory input of the next branch using the sensory input and a given motor command at each current branch. (This is one-step lookahead prediction.) Fig 3 shows an example of the results. The upper part of the figure shows the measured trajectory of the robot. The lower part shows the comparison between the actual sensory sequence and the predicted one. The figure shows the nine steps of the branching sequence, where five units in the most left are the sensory input, the next five units are its prediction, the next one unit is the motor command (0 or 1 of branching), and the most right four units are the context units. Initially the robot cannot predict correctly. It, however, becomes able to predict correctly

trajectory of robot travel

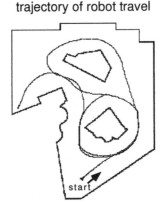

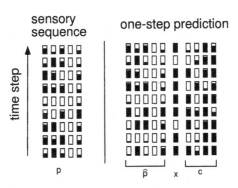

Fig. 3. One-step prediction.

after the 4th step. Since the context units are randomly set initially, the prediction fails at the very beginning. However as the robot continues to travel, the sequence of the sensory input "entrain" the context activations into the normal state transition sequence, thereafter the RNN becomes able to predict correctly. We repeated this experiment with various initial settings (positions and motor programs), which showed that the robot always starts to predict correctly within 10 steps. Furthermore we found that although the context is easily lost when perturbed by large sensory noise (e.g. when the robot fails to detect a branch or receiving totally different values of the sensory inputs from the ones expected for some branching steps), the prediction can be always recovered as long as the robot continues to travel. This auto-recovery of the cognitive process is made in consequence that a sort of coherence is organized between the internal and the environmental dynamics in their interactions.

Once the robot is "situated" in the environment (i.e. the robot becomes able to conduct one-step predictions correctly as the context is recovered after the travel), the robot can conduct multiple steps of lookahead predictions from a branching point. An example of the comparison between a lookahead prediction and its outcome of the actual sensory sequence during the travel is shown in Fig. 4. In (a) the arrow denotes the branching point where the robot conducted a lookahead prediction using a motor program given by 1100111. The robot, after conducting the lookahead prediction, traveled following the motor program, generating an "eight-figure" trajectory, as shown. In (b) the left-hand side shows the sensory input sequence, while the right-hand side shows the lookahead sequence, the motor program and the context sequence. This sequence consists of

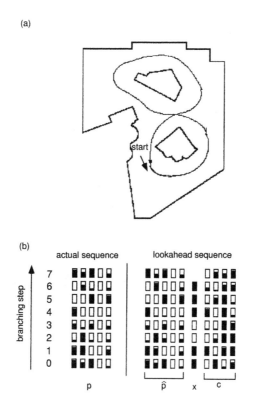

Fig. 4. The robot conducted lookahead prediction for a motor program from a branching point.

eight branching steps (from the 0th to the 7th step) including the initial one in the "start" point. It can be seen that the lookahead for the sensory input agrees very well with the actual values. It is also observed that the context as well as the prediction of sensory input at the 0th and the 7th steps are almost the same. This indicates that the robot predicted its return to the initial position at the 7th step in its "mental" simulation. The robot actually returned back to the "start" point at the 7th step in its test travel. We repeated this experiments of lookahead prediction for various branching sequences, and found that the robot can predict the sensory sequences correctly for arbitrary motor programs unless severe noise affects the branching sequence. From this result, it can be assumed that the robot successfully learned to extract grammatical structure hidden in the obstacle workspace.

4   Analysis

It is assumed that there exists an essential dynamical structure which can generate the coherence between the internal and the environment system, as we have discussed. We conducted the phase space analysis of the obtained RNN in order to see such structure. Phase plots show shapes and structures of attractors (invariant sets) of dynamical systems. For the purpose of drawing the phase plot of the RNN trained, the re-entry loop is connected from the sensory output nodes to the sensory input nodes so that the RNN can conduct lookahead predictions for arbitrary length of motor command sequences. Then the RNN was activated for 2000 steps with feeding randomly generated branching sequences of $x*$. (Here, the RNN conducts mental simulations for the random branching sequences.) The state space trajectory of the context units was plotted using the activation sequences of two context units (we took a 2-D projection of the entire state space) excluding the first 100 step points, which are regarded as transient. The result is a one-dimensional like invariant set as shown in Fig. 5 (a). Our mathematical analysis shows that this invariant set is topologically transitive[2].

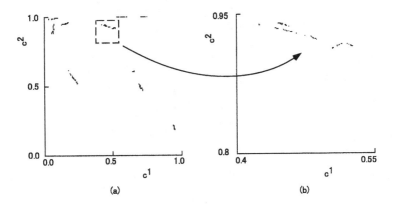

Fig. 5. (a)The global attractor appeared in the RNN dynamics and (b) enlargement of a part of the attractor.

We also found that the invariant set is a global attractor since the plotting always converges into the same figure independent of the initial setting of the

---

[2] This means that there are always finite step transition paths between any given two points in the invariant set

context values or the branching sequences used. More intuitively, the state starting from arbitrary points in the state space goes around the phase widely during the initial transient period. After convergence, the state transits only within the invariant set (among segments) shown in Fig. 5.

In this plot, one may interpret that the invariant set shows the boundary of rationality/cognition for the mental processes of the robot. When the RNN is perturbed by receiving noisy inputs, the state goes out of the invariant set where the rationality in terms of predictability is lost. However, as the RNN continues its dynamical iterations, the state always returns to the rational region, i.e. the invariant set, and the RNN is able to predict correctly again. This cognitive boundary is self-maintained solely from the system's own dynamical iterations, as stated by Maturana and Varela [11]. Here, we see that an inherently robust mechanism of dynamic cognition is achieved by self-organizing the global attractor.

Our further analysis of this invariant set revealed the fact that each line segment corresponds to each identical branching point. Each segment has two ways of transitions depending on the binary branching. And each segment is accessible from all other segments within finite steps of state transitions (Remember that the invariant set is topologically transitive.). It was also found that each segment is not just a one-dimensional collection of points but it is organized as a Cantor set [22] where the history of past branching sequences is represented by the current position in this Cantor set. Mathematically speaking, if two branching sequences share an infinite number of steps of same branching sequences in the past, the current states of the two sequences will correspond to two points that are epsilon-neighbors on the Cantor set of the same segment. On the other hand, two points will be distanced from each other if their recent past sequences are different. (This is due to the dynamic characteristics of the RNN as an iterated function system. See [7] for the details.) An interesting point is that the RNN naturally takes a context-dependent representation in which the history of the robot travel is encoded tacitly. This idea is also related to Tsuda's [21] Cantor set coding of the episodic memory in the hippocampus.

What we see in the phase plot is the so-called dynamical closure which some might interpret as equivalent to an FSM representation. However, a segment shown in the phase plot is not equivalent to a node in an FSM since it maintains more rich information of the context in terms of the Cantor set coding. The "symbols" appeared in the dynamical systems scheme are not arbitrarily shaped tokens in Harnad's terminology [5], but they maintain a certain metric structure in a tacit manner.

5   Summary and Discussion

It can be said that the representation and manipulation capabilities of symbols have been the most significant power of Artificial Intelligence. However, cognitive robotics researchers found that such computational symbols cannot be grounded easily in the physical environment. Then, they attempted to employ pattern categorizers as ideal interfaces between the real world analog dynamical systems and the computational symbolic systems. However, such trials could not produce much successful solutions to the problems. The failure is due to the fact that those two systems cannot share the same metric space where they can interact densely with each other.

Our studies have shown an alternative approach based on the dynamical systems view. We proposed that a RNN, as an adaptive dynamical system, could be an alternative to symbol systems which can naturally interact with the physical real world by sharing the same metric space of analog dynamical systems. Our experiments with a real robot have shown some interesting results. Firstly, it was shown that the internal system, once perturbed by possible accidental events or noise, is naturally re-situated to the environment system by means of entrainment between the two systems. This sort of entrainment between the internal and environmental systems becomes possible because those two systems share the same metric space of analog dynamical systems in our scheme.

Secondly, the phase space analysis indicated that the dynamical system's iterations by the RNN can be equivalent to the symbol manipulations processes of FSMs or language as has also been indicated by Pollack [12], Kolen [7] and Elman [3] [3]. We, however, found genuine differences between the symbol systems embedded in the proposed dynamical system and those of the computational FSMs in the way they are constituted. In the FSM formulation, first nodes are allocated as discrete states and then transitions among them are defined. In our dynamical systems scheme using a RNN, the dynamical flow, which is represented as a local vector in the state space, is organized through the sensory-motor learning processes. As a result, a mechanism equivalent to the FSM becomes visible in the phase space. The dynamical flow includes both the outer flow attracted towards the invariant set and the internal flow going only within the invariant set (see Fig. 6 for the illustration).

A crucial point is that an equivalent function of the FSM can be generated only in the form of an attractor and that the outer flow of attraction is indispensable for its existence. It is this flow that explains the auto-recovery mechanism of the system from its perturbed states. Important here is that the auto-recovery mechanism is intrinsic to the dynamical systems scheme since the internal flow and the outer attraction flow are generated as inseparable units in the process of self-organizing the attractor. This suggests that the transient dynamics might be more crucial in cognitive systems than believed previously since once conflicts or gaps arise between the mental images and its reality they are resolved during such transient periods. Symbol systems, which support neither notions of attraction nor transient dynamics, just halt if conflicts occur unless extra exception-handling type mechanisms are initiated.

In the proposed dynamical systems scheme, the system itself neither sees the descriptions of an FSM nor involves their direct manipulations. When the

---

[3] Their studies, however, could not articulate enough the advantage of RNNs as an alternative of symbol systems since their studies never addressed the embodied cognition of sensory-motor systems.

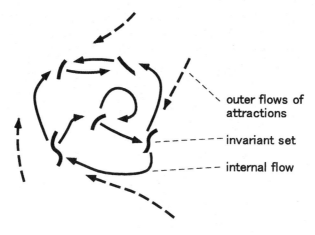

outer flows of
attractions

invariant set

internal flow

Fig. 6. The invariant set, the outer flow attracted to the invariant set, and the internal flow going inside the invariant set are illustrated.

internal system merely repeats its dynamical iterations, the emergent structure of the state transitions observed in the phase plots by an outside observer (like myself) may be perceived as if symbols actually existed and were manipulated internally. In fact, all what exists is the dynamical structure and the resultant dynamical flow in the system. The descriptions and manipulations appear to be an inseparable entity in the dynamical system. Since there are no observers dealing with the descriptions, we finally find the internal observer [10, 4] in our robot. Consequently, there are no descriptions or symbols to cause the symbol grounding problem from the view of this internal observer.

In the end, I would like to address the open problems related to this study. When the original experiments reviewed in this paper were completed about 8 years ago, I thought of two future directions. One direction was to study dynamic adaptation schemes in which the robot has to learn about open environments in an incremental way rather than off-line. The studies have been conducted [15, 20] and are continuing focusing on how coherent and incoherent phases autonomously appear during the dynamic changes of the internal attractors. We proposed that such open-dynamics characteristics might explain the momentary self-consciousness discussed in the phenomenology literature. The other line of study is to consider articulation mechanisms of sensory-motor flows. The branching mechanism in the YAMABICO experiments was pre-programmed as described earlier. Then, we started to consider how "concepts" of branching or landmarks can be learned as articulated from the experiences of continuous sensory-motor flow in navigation tasks [19]. This question leads to further general questions of how behavior primitives could be self-organized[18] and how they can be combined to generate diverse behavior patterns[16]. However, I have to admit that these studies are still half-baked and there are many open problems left for future studies.

References

[1] Beer, R.: A dynamical systems perspective on agent-environment interaction. Artificial Intelligence **72** (1995) 173–215

[2] Crutchfield, J.: Inferring statistical complexity. Phys Rev Lett **63** (1989) 105–108

[3] Elman, J.: Finding structure in time. Cognitive Science **14** (1990) 179–211

[4] Gunji, Y., Konno, N.: Artificial Life with Autonomously Emerging Boundaries. App. Math. Computation **43** (1991) 271–298

[5] Harnad, S.: The symbol grounding problem. Physica D **42** (1990) 335–346

[6] Jordan, M., Rumelhart, D.: Forward models: supervised learning with a distal teacher. Cognitive Science **16** (1992) 307–354

[7] Kolen, J.: Exploring the computational capabilities of recurrent neural networks. PhD thesis, The Ohio State University (1994)

[8] Kuipers, B.: A qualitative approach to robot exploration and map learning. In: AAAI Workshop Spatial Reasoning and Multi-Sensor Fusion (Chicago). (1987) 774–779

[9] Mataric, M.: Integration of representation into goal-driven behavior-based robot. IEEE Trans. Robotics and Automation **8** (1992) 304–312

[10] Matsuno, K.: Physical Basis of Biology. CRC Press, Boca Raton, FL. (1989)

[11] Maturana, H., Varela, F.: Autopoiesis and cognition: the realization of the living. D. Riedel Publishing, Boston (1980)

[12] Pollack, J.: The induction of dynamical recognizers. Machine Learning **7** (1991) 227–252

[13] Rumelhart, D., Hinton, G., Williams, R.: Learning internal representations by error propagation. In Rumelhart, D., Mclelland, J., eds.: Parallel Distributed Processing. Cambridge, MA: MIT Press (1986)

[14] Tani, J.: Model-Based Learning for Mobile Robot Navigation from the Dynamical Systems Perspective. IEEE Trans. on SMC (B) **26** (1996) 421–436

[15] Tani, J.: An interpretation of the "self" from the dynamical systems perspective: a constructivist approach. Journal of Consciousness Studies **5** (1998) 516–42

[16] Tani, J.: Learning to generate articulated behavior through the bottom-Up and the top-down interaction processes. Neural Networks **16** (2003) 11–23

[17] Tani, J., Fukumura, N.: Embedding a Grammatical Description in Deterministic Chaos: an Experiment in Recurrent Neural Learning. Biological Cybernetics **72** (1995) 365–370

[18] Tani, J., Ito, M.: Self-organization of behavior primitives as multiple attractor dynamics by the forwarding forward model network. Trans On IEEE SMC-B (2003) in print.

[19] Tani, J., Nolfi, S.: Learning to perceive the world as articulated: an approach for hierarchical learning in sensory-motor systems. Neural Networks **12** (1999) 1131–1141

[20] Tani, J., Yamamoto, J.: On the dynamics of robot exploration learning. Cognitive Systems Research **3** (2002) 459–470

[21] Tsuda, I.: Toward an interpretation of dynamic neural activity in terms of chaotic dynamical systems. Behavioral and Brain Sciences **24:5** (2001) 793–848

[22] Wiggins, S.: Introduction to Applied Nonlinear Dynamical Systems and Chaos. New York: Springer-Verlag (1990)

Forward and Bidirectional Planning Based on Reinforcement Learning and Neural Networks in a Simulated Robot

Gianluca Baldassarre

Computer Science Department, University of Essex,
CO4 3SQ, Colchester, United Kingdom
Institute of Cognitive Sciences and Technologies,
National Research Council of Italy (ISTC-CNR)
Viale Marx 15, 00137, Rome, Italy
baldassarre@ip.rm.cnr.it

Abstract. Building intelligent systems that are capable of learning, acting reactively and planning actions before their execution is a major goal of artificial intelligence. This paper presents two reactive and planning systems that contain important novelties with respect to previous neural-network planners and reinforcement-learning based planners: (a) the introduction of a new component ("matcher") allows both planners to execute genuine taskable planning (while previous reinforcement-learning based models have used planning only for speeding up learning); (b) the planners show for the first time that trained neural-network models of the world can generate long prediction chains that have an interesting robustness with regards to noise; (c) two novel algorithms that generate chains of predictions in order to plan, and control the flows of information between the systems' different neural components, are presented; (d) one of the planners uses backward "predictions" to exploit the knowledge of the pursued goal; (e) the two systems presented nicely integrate reactive behavior and planning on the basis of a measure of "confidence" in action. The soundness and potentialities of the two reactive and planning systems are tested and compared with a simulated robot engaged in a stochastic path-finding task. The paper also presents an extensive literature review on the relevant issues.

1 Introduction

Since its birth, from the first experiments with the robot Shakey [8] to current research in planning [1] and behavior based robotics [6] [2], a major goal of artificial intelligence has been building intelligent systems that are both capable of acting in a reactive fashion and planning the course of action before its execution. In fact reactive behavior allows agents to quickly react to dynamic and unpredictable events. In order to be "tuned" with reality, the reactive behavior has to incorporate forms of "implicit anticipation" (cf. Butz et al. in this volume). On the other hand, the capacity to predict future states of the world on the basis of its regularities allows agents to be

M. Butz et al. (Eds.): Anticipatory Behavior ..., LNAI 2684, pp. 179-200, 2003.
© Springer-Verlag Berlin Heidelberg 2003

flexible and taskable, i.e. to re-use knowledge to pursue different goals (see below, and see the concept of "state anticipation" of Butz et al. in this volume).

Traditionally, artificial intelligence planning systems have been based on logical information representations built a-priori by the designer [1]. When these planning systems are used to control robots, sensors' readings are converted into logical representations, the control is implemented in terms of manipulations of these representations, and the outcome of this processing is again converted into effectors' commands. This approach has difficulties as the time-consumption of logical reasoning about the effects of low-level actions is too expensive to generate real-time behavior [21].

At least initially, behavior-based robotics proposed to eliminate logical deliberation and planning from control altogether [6]. Reactive behavior was implemented through numerical functions and/or rules that "directly" linked the sensors' numerical patterns to the effectors' numerical patterns. More recently, behavior-based robotics has attempted to integrate reactive behavior based on numerical representations and planning (usually) based on logical representations (see review in [2]). However, the resulting systems have an important limitation: they imply a *double recoding* of information, from a numerical format to a logical format and vice versa, that is difficult to implement, slow and prone to errors.

The motivation of this research was building reactive and planning systems that rely only on numerical representations. This novel approach should allow coping with noisy and unpredictable environments through reactive behaviors, having the flexibility of planning, and at the same time avoiding the problem of the interface between different information representation formats. Given the level of development that they have reached (cf. 9), neural networks have been chosen for this purpose. The attempt of this research is challenging and interesting at the same time. In fact its solution implies answering a number of questions of the following type: What kind of information representations can be used to plan with neural networks? What is the origin of this information (hand-coded, experience)? How implementing the loop "decision of action – prediction of consequences – decision of action - ..." required by planning with neural networks? How can the neural planning process be used to influence future action? What are the advantages and disadvantages of using neural networks versus logic-based algorithms to implement planning?

The systems presented here are based on dynamic programming and reinforcement learning methods. Roughly speaking, dynamic programming [20] is based on a model of the world that indicates which states and rewards, e.g. rewards associated to goal states and costs associated to other states, follow the execution of one of the available actions. Dynamic programming repeatedly and systematically explores all the states of the model of the world to associate an "evaluation" (e.g. a number between 0 and 1) with each of them on the basis of the rewards and the states' "contiguity" in time. This evaluations, that are gradually higher (or lower) for states closer to the goal states, form a gradient field. Actions are selected so as to ascended (or descend) this gradient field. The way this is done is called "action-selection policy". As dynamic programming, to which they are closely related, reinforcement learning methods build a gradient field of evaluations over the states of the world [25]. However, differently from dynamic programming, they usually do not use a model of the world ("model-free reinforcement learning methods"). Instead, they learn the evaluations by directly executing the actions in the world and by observing the consequences in terms of new states and rewards (costs) experienced. Dynamic programming and reinforcement learning methods are probably the most suitable framework currently available to implement planning with neural networks. In fact, even if developed for whole state representations, many interesting applications in these fields "break" the representation of states into feature patterns. Feature patterns, being distributed and parallel information representations, can be readily processed by neural networks. Moreover, both approaches are based on state evaluations that can be easily represented with neural networks (see [3] and the systems presented here).

Sutton [26] has integrated planning based on dynamic programming and reinforcement learning into a class of architectures called "Dyna" (from "dynamic programming"). The central idea of Dyna architectures is to use a model of the world to generate "simulated experience" for reinforcement-learning training (e.g. [14]). The algorithms proposed here are inspired by Dyna-PI architectures, a class of Dyna architectures based on actor-critic reinforcement-learning methods [25] ("PI" stands for "policy iteration", the process at the basis of actor-critic methods). Actor-critic methods, differently from other reinforcement-learning methods that select actions on the basis of the evaluations, use a data structure that stores the action-selection policy in the form of probabilities of selecting the actions in correspondence to different states.

To the best of the author's knowledge, so far Dyna architectures have been only used to speed up planning (e.g. [14], [19] and [26]) and not to implement genuine "taskable planning". Let us define a goal as a state to reach. A system is capable of executing taskable planning if it possesses a model of the world and can use the goal-independent information stored in it to autonomously reach a goal, never reached previously, more efficiently than a purely reactive system (see below). Dyna architectures are not taskable because they need a model of the environment that incorporates the rewards (or costs) associated with the goals [26]. As a consequence, when Dyna architectures are assigned a goal, they either need to learn the evaluations associated with that goal (but to do this they need to experience the goal several times, so they cannot be efficient since the first time they reach the goal) or the evaluations have to be hardwired by the designer into the model of the world (but in this case the systems are not autonomous). The systems proposed here are endowed with a computationally simple but theoretically important new component in order to overcome this problem: the "matcher". This is a simple neural network that compares the goal assigned to the system with the current state and *internally and autonomously* generates a reward if they are similar (cf. [4] for an hypothesis on the possible brain structures that might correspond to the matcher). Once capable of producing rewards internally, the systems can engage in planning by generating "simulated experience" to train their reactive components. Simulated experience is generated by producing several "chains of prediction". To generate a chain of predictions the systems select an action in the current state, predict the state that would follow that action's execution, suspend the execution of this action, select another action starting from the predicted state, and so on. After doing this, when the systems act again in the world they can reach the goal with high efficiency compared to a system made up by the reactive components only.

There is an interesting issue related to the use of neural networks for planning. The predictions generated by the neural networks of the model of the world, are affected by noise. As a consequence one would expect to observe an accumulation of noise if chains of predictions are generated. Surprisingly, we will see that there are some mechanisms for which this noise is filtered out so that long chains can be generated that maintain a correspondence with the images of the world.

While generating the chains of predictions through the model of the world, the systems learn by reinforcement learning *as if they were acting in the world*, and then they use the knowledge acquired in this way to act. Two novel algorithms are proposed to control these processes. These algorithms suitably interleave acting and planning according to the systems' "confidence" in action, defined on the basis of the action selection probabilities.

In a given moment, the systems know not only the current state, but also the goal state. The second planner proposed (bidirectional planner) alternately generates forward chains from the current state and backward chains from the goal. The bidirectional planner is more efficient than the forward planner because it is quicker in learning the evaluations and is more focused on the states relevant for the task while exploring the world's model.

Section 2 presents the reinforcement learning framework, the scenario and the simulated robot used to test the algorithms. Section 3 presents the forward planner and the results of the simulations run with it. Section 4 presents the bidirectional planner and compares its performance with the performance of the forward planner. Section 5 presents an extensive literature review and highlights the novelties of the algorithms presented here. Finally, section 6 presents the conclusions and the future work.

2   Reinforcement Learning Framework, Scenario of Simulations, and Simulated Robot

The reinforcement learning framework, based on "Markov Decision Problems", is sketched here (see [3] and [25] for details). A Markov decision problem implies that an agent interacts with the world at discrete time steps $t = 1, 2, 3, \ldots$ On each time step t the agent perceives a state of the world $s_t \in S$, and on the basis of this selects an action $a_t \in A$. In response to each action the world produces a reward r_{t+1} and a new state with probability: $p^a_{ss'} = \Pr[s_{t+1} = s' \mid s_t = s, a_t = a]$ for all $s, s' \in S$ and $a \in A$. The planners introduced below learn a simplified model of the world M: $(s_t, a_t) \rightarrow s_{t+1}$. This model is deterministic and does not take into considerations the rewards that are universally modeled by the matcher on the basis of the goal and the current state.

An (action selection) policy π is defined as a mapping from the states to the action selection probabilities, $\pi: S \times A \rightarrow [0, 1]$. For each state s a "state-evaluation function" $V^\pi[s]$ is defined that depends on the policy π and is calculated as the expected discounted future reward from s:

$$V^\pi[s] = E[r_{t+1} + \gamma\, r_{t+2} + \gamma^2\, r_{t+3} + \ldots] = \Sigma_{a \in A}[\pi[a, s]\, \Sigma_{s' \in S}[p^a_{ss'}\, (r_{t+1} + \gamma\, V^\pi[s'])]] \qquad (1)$$

where $\pi[s, a]$ is the probability that the policy selects a in s, $E[.]$ is the mean operator, and $\gamma \in [0, 1]$ is a discount coefficient. The agent aims at finding an "optimal policy" that maximizes $V^\pi[s]$ for all $s \in S$.

The algorithms presented here apply to a restricted class of Markov decision prob-·lems, the "stochastic shortest-path problems" (cf. [5]). In these problems the reward is equal to 1 for a *unique* "goal state" s^g and equal to 0 for all other states. The agent aims at finding a policy that leads from a start state s^s to the goal state s^g with the minimum number of steps. This restriction is introduced to allow the systems pre-sented here to generate an internal reward in correspondence to the goal through the matcher. This simplification does not reduce the generality of the systems: more complex situations might be dealt with by using more complex internal reward gen-erators (e.g. a component that generates costs when actions are selected).

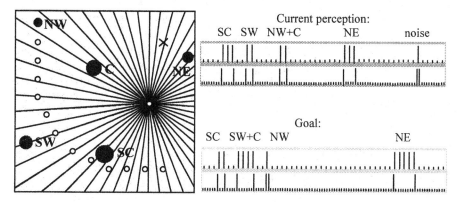

Fig. 1. *Left*: The scenario of the simulations containing five landmarks (*black circles*), the scope of the simulated robot's 50 visual sensors (*delimited by the rays*), the simulated robot (*circle at origin of rays*), the 12 start positions (*white circles*), and the goal (*marked with an X*) used in the tests. Letters indicate different landmarks. *Top right:* The sensory pattern generated by the simulated robot's sensors (at the position shown in the graph on the left) and the corre-sponding contrast pattern. Letters indicate the landmarks that affect the different portions of the image. *Bottom right:* The sensory pattern and contrast pattern corresponding to the goal

Now the scenario and the simulated robot used to test the algorithms are described. The scenario (Fig. 1) is a square arena with sides measuring 1 unit and with 5 circular landmarks inside. These landmarks are also obstacles for the simulated robot. The simulated robot can see the landmarks with a one-dimension horizontal retina cover-ing 360 degrees. Notice that the simulated robot cannot perceive distances, cannot see a landmark that is behind another landmark, and perceives just a "big" landmark if there are two or more landmarks that are contiguous in sight. This implies that differ-ent states might generate the same image. Problems of with these features, called "partially observable Markov decision problems", might raise difficulties for rein-forcement learning that are not considered here (see [10] and [13] on this).

Markov processes are based on whole state representations. However here, as in the most interesting applications of reinforcement learning, the simulated robot per-ceives the world through sensors that return a feature-like pattern (vector) x in corre-spondence to the state s (this also implies that s^s and s^g become x^g and x^s). More in details, the simulated robot has a retina made up by 50 units x. Each unit x_i activates with 1 if a landmark is in its scope, with 0 otherwise. This activation is affected by noise (0.01 probability of flipping for each sensor). The signals coming from the retina are always aligned with the magnetic north through a simulated compass before being fed to the systems. This facilitates the task by making the images rotation invariant. The reading of the compass is affected by Gaussian noise (0 mean, 1 degree variance). Before being sent to the controllers, the retina signals are re-mapped into a vector $\mathbf{y}$ of 100 binary units representing the image "contrasts" (contrasts between the landmarks, perceived as "black", and the "background", perceived as "white"). Two contiguous retinal units give unit activation to one contrast unit y_j if they are respectively on and off, to another contrast unit if they are respectively off and on, and to no contrast units if they are both on or both off. This simple re-mapping implements an expansion of the input space that allows the systems to work properly by using simple two-layer networks for the controller [3] (more complex tasks would require a more powerful preprocessing). At each cycle of the simulation the system has to select one of eight actions, each consisting of a 0.05 step in one of eight directions aligned with magnetic north (north, northeast, etc.). The outcome of these actions is affected by Gaussian noise with 0 mean and 0.01 variance. If the simulated robot moves against the arena's boundaries or the obstacles it "bounces back", i.e. it is set at the previous position.

3   Reinforcement Learning and Forward Planning

Fig. 2 shows the reinforcement-learning (reactive) and the planning components of the bidirectional planner. The forward planner can be obtained from the bidirectional planner by not considering the components necessary to generate the backward prediction chains (these components have a dotted border in the figure). The forward planner is made up by the reinforcement learning components (bold border in the figure) and by the components used to generate the forward prediction chains (thin border in the figure). Now the components of the forward planner are analyzed in detail.

3.1   Actor Critic Reinforcement Learning

The reactive reinforcement learning components of the system are the critic (composed of the TD-critic and the evaluator), the actor, and the stochastic selector. In the introduction it was mentioned that the matcher is used to generate an internal reward signal while the system is planning. However, in order to facilitate comparisons, it will also be used to generate this signal when the planner will be used as a share reactive system in order to show that it does implement genuine taskable planning. For this reason the matcher is analyzed here together with the reactive components. The matcher is a hand designed neural network with 200 input units and 1 output unit. It returns 1 as output when the first part of its input (100 units encoding the goal, i.e. the contrast image of the landmarks from the goal position) has at least 94% of units with the same activation of the corresponding units of the second part (100 units encoding the current input contrast pattern, or the current prediction). Otherwise it yields 0 (cf. [3] for the details).

The "actor" is a two-layer feed-forward neural network that is fed with the input pattern y_t and has eight sigmoid output units that locally encode the actions. To select an action, the activation m_l (the "action merit") of the output units is sent to a selector that implements a stochastic winner-take-all competition. The probability Pr[.] that a given action a_g becomes the winning action a_w is given by:

$$Pr[a_g = a_w] = m_g \, / \, \Sigma_l \, m_l \qquad (2)$$

In the scenario studied here the actors' sigmoid output units plus this selection method yield a behavior more robust than the popular soft-max function (cf. [3] and [25]).

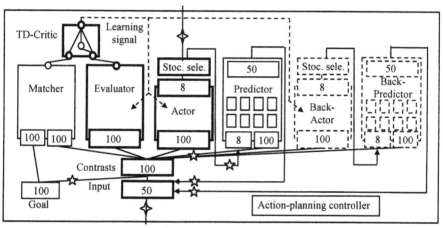

Fig. 2. The systems' architecture. Networks with a *bold, thin* and *dashed border* implement reinforcement learning, forward planning, and bidirectional planning respectively. *Arcs* and *arrows* respectively indicate forward and backward connections that "copy" a pattern from one layer to another. The *four and five spike stars* indicate the channels respectively set open and closed by the action-planning controller when acting (vice versa when planning). *Dashed arrays* indicate the learning signal used to update the weights of the evaluator, actor and back-actor.

The "evaluator" is a two-layer feed-forward neural network that is fed with the input pattern y_t and that returns the estimation $V'^\pi[y_t]$ of the evaluation $V^\pi[y_t]$ with its linear output unit. Similarly to what has been done for the states s, $V^\pi[y_t]$ is defined as the expected discounted sum of future reinforcements r, given the current actor's policy π (cf. equation 1; γ is set at 0.95 in the simulations).

The "TD-Critic" is a neural implementation (see [3] and [4] for details) of the computation of the Temporal-Difference error e defined as (see [25]):

$$e_t = (r_{t+1} + \gamma \, V'^\pi[y_{t+1}]) - V'^\pi[y_t] \qquad (3)$$

Notice that $V^\pi[y_t]$ is set at 0 if $y_t = y^g$ because a "trial" ends [3, 25]

The evaluator is trained with a Widrow-Hoff algorithm [31] that uses the error signal e_t. In particular, its weights w_j are updated as follows:

$$\Delta w_j = \eta\, e_t\, y_j = \eta\, ((r_{t+1} + \gamma\, V'^\pi[\mathbf{y}_{t+1}]) - V'^\pi[\mathbf{y}_t])\, y_j \qquad (4)$$

where η is a learning rate (set at 0.1 in the simulations). This rule implies that the evaluator's estimation $V'^\pi[\mathbf{y}_t]$ is made closer to the target value $(r_{t+1} + \gamma\, V'^\pi[\mathbf{y}_{t+1}])$. This target is a more precise evaluation of $\mathbf{y}_t$ because it is expressed at time t+1 on the basis of the observed r_{t+1} and the estimation $V'^\pi[\mathbf{y}_{t+1}]$, usually closer to the goal than $V'^\pi[\mathbf{y}_t]$.

The actor is also trained with a Widrow-Hoff algorithm according to the TD-Critic's error signal. The updating of the merit values is done by adjusting the weights of the neural unit corresponding to the winning action a_w (and only this) as follows:

$$\Delta w_{wj} = \zeta\, e_t\, (4\, m_j\, (1 - m_j))\, y_j = \zeta\, ((r_{t+1} + \gamma\, V'^\pi[\mathbf{y}_{t+1}]) - V'^\pi[\mathbf{y}_t])\, (4\, m_j\, (1 - m_j))\, y_j \qquad (5)$$

where ζ is a learning rate (0.1 in the simulations) and $(4\, m_j\, (1 - m_j))$ is the derivative of the sigmoid function multiplied by 4 to homogenize the actor's and critic's learning rates. When a new goal is assigned to the simulated robot, the weights of the evaluator and actor are randomized in [-0.001, +0.001], so the evaluations expressed by the evaluator's linear output unit are around 0, and the merits (probabilities) expressed by the actor (stochastic selector) are all around 0.5 (0.125). This implies that initially the simulated robot randomly explores the world (or the model of the world). Afterwards, the evaluator shapes the evaluations on the basis of the rewards, and at the same time the actor shapes the action probabilities on the basis of the evaluations. The parallel training of the evaluator and actor is called "policy iteration" [25] and gives the name to the Dyna-PI architectures (cf. [26] and see below).

3.2 Forward Planning

This section illustrates the components added to the reinforcement learning components to have forward planning. The "predictor", i.e. the system's model of the world, is a set of 8 feed-forward two-layer networks, called "experts". Each expert corresponds to one action. Each expert takes $\mathbf{y}_t$ as input, and is specialized to predict the following sensors' activation $\mathbf{x}_{t+1}$, if the action corresponding to it is executed, with its sigmoid output units. To have a binary output, the output of each sigmoid unit is squashed to 0 if below 0.5, and to 1 if above. A hand-designed selector chooses the expert corresponding to the selected action to yield the output of the predictor. The experts are trained while the robot navigates randomly in the environment for 200,000 cycles. This brings the "quadratic error" (computed as $(\sum_i[(x'_i - x_i)^2]/n))^{1/2}$ where x_i and x'_i are the actual and predicted activations of the n sensors) to about 0.24. This training is done before the simulations illustrated below. At each cycle, the contrast pattern $\mathbf{y}_t$ and the input pattern $\mathbf{x}_{t+1}$, observed after the execution of one action, are respectively used as input and as teaching output to train the expert that selected the action with a Widow-Hoff rule [31]. Notice that the predictor yields deterministic predictions that tend to be the average of the $\mathbf{x}_{t+1}$ observed. This is a simplification since the world is stochastic and the model should produce stochastic predictions.

The action-planning controller is a hand-designed algorithm (Fig. 3) that decides the planning or acting mode of the system and generates the forward (and backward) chains. Fig. 4 shows the details of one cycle of action and one cycle of (forward or backward) planning. The whole algorithm of Fig. 3 and Fig. 4 is executed sequentially *one time at each cycle of the simulation.*

```
//Initialisations
01 IF(NewGoalHasBeenAssigned)
02    MaxStepsPlan := 1
03    ConfThresh := MaxConfThresh
04    ForwardPlanning := TRUE
05    StepPlan := 0
06    InputFromWorld := TRUE
//Decision about planning or acting
07 IF(InputFromWorld)
08    System gets input x_t (y_t) from the robot's sensors
09    Actor gets y_t and gives m_t
10    Confidence is computed on the basis of m_t
11    IF(Confidence < ConfThresh)
12       Planning := TRUE
13    ELSE
14       Planning := FALSE
15       ConfThresh := MIN(MaxConfThresh, ConfThresh + Gain)
//Control of forward chains' length and interruption
16 IF(Planning)
17    StepPlan := StepPlan + 1
18    ConfThresh := ConfThresh - Decay
19    IF(ForwardPlanning)
20       IF(InputFromWorld = TRUE)
21          InputFromWorld := FALSE
22       ELSE
23          System uses predictor's output y_t as input
24       IF(GoalReached OR StepPlan = MaxStepsPlan)
25          IF(StepPlan = MaxStepsPlan)
26             MaxStepsPlan := MaxStepsPlan + 1
27          ELSE
28             MaxStepsPlan := MIN(MaxStepsPlan, StepPlan * 2)
29          InputFromWorld := TRUE
30          IF(BidirectionalPlanning)
31             ForwardPlanning := FALSE
32             ForwardSteps := StepPlan
33             GoalAsInput := TRUE
34             InputFromWorld := FALSE
35          StepPlan := 0
//Control of backward chains' length and interruption
36    ELSE
37       IF(GoalAsInput = TRUE)
38          System uses goal y^g as input
39          GoalAsInput := FALSE
40       ELSE
41          System uses back-predictor's output x_t (y_t)as input
42       IF(StepPlan = ForwardSteps)
43          ForwardPlanning := TRUE
44          InputFromWorld := TRUE
45          StepPlan := 0
```

Fig. 3. Pseudo-code of the action-planning controller. ":=" is the assignment operator, "MIN(., ., ...)" is a function that returns the minimum argument, "//" indicates a comment

If the variable "BidirectionalPlanning" is set at "FALSE", then the algorithm implements the forward planner. Let us see how the forward planner works. When a new goal is assigned to the system, some variable settings are done (line 1 to 6). The system's planning or acting mode (variable "Planning") is decided each time the system receives an input from the world (line 7) on the basis of the system's "confidence" (line 8 to 14).

```
//Networks' activation in one forward chain cycle
46 IF(Planning)
47   IF(ForwardPlanning)
48     Evaluator gets yₜ and gives V'π[yₜ]
49     Actor gets yₜ and gives mₜ
50     Stochastic selector gets mₜ and gives aₜ
51     Predictor gets yₜ, aₜ and gives xₜ₊₁ (Yₜ₊₁)
52     Matcher gets y_g, yₜ and gives rₜ
53     TD-Critic gets V'π[yₜ₋₁], V'π[yₜ], rₜ, gives eₜ₋₁
54     Evaluator gets yₜ₋₁, eₜ₋₁ and learns
55     Actor gets yₜ₋₁, mₜ₋₁, aₜ₋₁, eₜ₋₁ and learns
56     IF(BidirectionalPlanning)
57       Back-Actor gets yₜ and gives mₜ₋₁
58       Back-Actor gets yₜ, mₜ₋₁, aₜ₋₁(actor), eₜ₋₁ and learns
//Networks' activation in one backward chain cycle
59   ELSE
60     Back-actor gets yₜ and gives mₜ₋₁
61     Back-stochastic selector gets mₜ₋₁ and gives aₜ₋₁
62     Back-predictor gets yₜ, aₜ₋₁ and gives xₜ₋₁ (Yₜ₋₁)
63     Evaluator gets yₜ₋₁ and gives V'π[yₜ₋₁]
64     Matcher gets y_g, yₜ and gives rₜ
65     TD-Critic gets V'π[yₜ₋₁], V'π[yₜ], rₜ and gives eₜ₋₁
66     Evaluator gets yₜ₋₁, eₜ₋₁ and learns
67     Back-actor gets yₜ, aₜ₋₁, eₜ₋₁ and learns
68     Actor gets yₜ₋₁ and gives mₜ₋₁
69     Actor gets yₜ₋₁, mₜ₋₁, aₜ₋₁ (back-actor), eₜ₋₁ and learns
//Networks' activation in one action cycle
70   ELSE
71     Evaluator gets yₜ and gives V'π[yₜ]
72     Actor gets yₜ and gives mₜ (already done in line 9)
73     Stochastic selector gets mₜ and gives aₜ
74     Matcher gets y_g, yₜ and gives rₜ
75     TD-Critic gets V'π[yₜ₋₁], V'π[yₜ], rₜ and gives eₜ₋₁
76     Evaluator gets yₜ₋₁, eₜ₋₁ and learns
77     Actor gets yₜ₋₁, mₜ₋₁, aₜ₋₁, eₜ₋₁ and learns
78     System executes aₜ in the world
79     IF(BidirectionalPlanning)
80       Back-Actor gets yₜ and gives mₜ₋₁
81       Back-Actor gets yₜ, mₜ₋₁, aₜ₋₁ (actor), eₜ₋₁ and learns
```

Fig. 4. Pseudo-code of the activation of the neural networks components in one cycle of forward planning, backward planning and action

The system's confidence is defined as the highest of the actions' probabilities measured at the position currently occupied by the simulated robot. If the confidence is above a certain threshold the system acts in the world and the predictor is not used (cf. Fig. 4). If the confidence is below the threshold, the action-planning controller "disconnects" the robot from the world (line 11, 12, 16, 22 and 23; see 4 spike stars in

Fig. 2), in the sense that it starts to generate simulated experience by using the predictor and the matcher (cf. Fig. 4).

Each chain of predictions starts when the system switches from the acting to the planning mode (line 7, 8, 11, 12), and starts from the image that corresponds to the position currently occupied by the simulated robot (line 7 and 8). Then the chain continues with "rings" each made up by a prediction from the predictor (one for each cycle: line 19 to 23). Notice that chains of predictions tend to be different between them since the system selects actions stochastically (Fig. 4, line 49 and 50,). If one chain terminates without encountering the goal, the succeeding chain gets one "ring" longer (line 2, 24 to 26), otherwise it tends to get shorter (line 25, 27 and 28). While planning, the confidence threshold decreases (line 18). This prevents the robot from getting stuck in places in which the system is unable of becoming "confident" enough to start to move (for example, without this mechanism the simulated robot got stuck between the arena's border and the northwest landmark). While acting, the threshold increases again and reaches the maximum level without ever exceeding it (line 15). This guarantees that the simulated robot tends to move only when the confidence is above the maximum level of the threshold (MaxConfThresh). In the simulations the parameters that regulate these aspects are set as follows: Decay = 0.000001, Gain = 0.01, MaxConfThresh = 0.15. Each time a chain of predictions is terminated (either because the goal has been encountered or because it has reached a maximum length, line 24) the system "connects" again to the sensors (line 24, 29, 7 and 8), updates the mode (line 9 to 14), and starts to act or to generate another chain of predictions.

Fig. 4 shows the activation of the neural networks executed in one cycle of action when the variable "planning" is "FALSE" (line 46, 70 to 78). When planning is "TRUE" the predictor produces one of the predictions that make up the chain of predictions (line 51) and the matcher checks if the chain encounters the goal (line 52). Moreover, while the chains of predictions are generated the actor and the evaluator are trained with reinforcement learning as if the robot were acting in the world (compare line 48 to 55 with line 71 to 77). This allows the evaluator to improve its evaluating capacity and the actor to shape the action probabilities. As a consequence, when the system stops planning and acts in the world (line 78) it reaches the goal following a path that tends to be straight.

3.3 Reactive Behavior and Forward Planning: Results and Interpretations

The first simulation has tested the taskability of the reactive-planning system. This has been done by comparing its performance with the performance of the system made up by the reactive components only. The simulated robot is set at the southeast start, and its task is to reach the goal. Each time the simulated robot reaches the goal it is set at another *randomly-drawn* position of the arena. This is done for 50,000 cycles.

Fig. 5 reports the results of this simulation (averaged over 10 simulations run with different random seeds). For both the reactive and planning controllers the number of moves taken to reach the goal was measured and plotted against the cumulated cycles (this measure was sampled every 100 cycles, and then smoothed with a 10-step moving average). Each cycle reported in the graph implied the execution of one action and eventually, if the controller was planning, a variable number of planning cycles. In the case of the planner, the graph also reports the number of planning cycles per move.

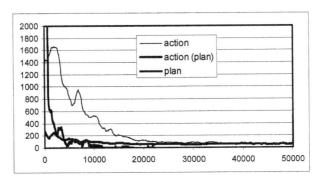

Fig. 5. *Y-axis*: moves per success taken by the reactive system (*action*), and moves per success (*action (plan)*) and planning cycles per success (*plan*) taken by the reactive-planning system, measured against the cumulated simulation cycles (*x-axis*). The measures have been sampled every 100 cycles and have been smoothed with a 10-step moving average, and are the average of the results of 10 simulations run with different random seeds

When assigned the (new) goal, the reactive system reaches it, by random walk, in about 1600 steps. When assigned the (new) goal, the reactive-planning system reaches it in about 200 steps *from the first time it pursues it*. This result is achieved through a considerable amount of planning processing: on average (10 random seeds) it takes 40,116 planning cycles to reach the goal the first time. During this planning activity the skills of the evaluator and actor improve so that when the system decides to act in the world it can achieve the goal by following an almost straight line. If the confidence threshold is set at 0.25 (a value higher than the previous 0.15), the performance of the planner improves to about 50 moves (see Fig. 6). These results demonstrate that the reactive-planning system implements genuine taskable planning. In fact, by using the goal-independent information stored in the predictor, the planner outperforms the corresponding reactive system from the first time it pursues the goal.

Fig. 5 shows another important property of the reactive-planning system. With repeated trials the system needs progressively less planning cycles, until it is capable of reaching the goal reactively without planning in about 40 moves from any position of the arena (the optimal path, not considering noise and obstacles, is about 15 moves). Interestingly, direct observation of the simulated robot's behavior also showed that once the robot has planned from a given position and then starts to act, it stops acting only if it reaches regions that are far away from the path between that position and the goal. This happens because planning process is focused on the states between the position where planning takes place and the goal. These results indicate that the algorithm nicely interleaves action and planning on the basis of "confidence", and that while planning it develops the skills of the evaluator and actor only for relevant states.

The quality of the chains of predictions generated by the predictor is a critical aspect of the system. Recall that when the predictor was trained, the square error per output unit did not go below 0.24, a quite high level if one considers that the error made by drawing a prediction randomly is about 0.5. On the basis of this, one could reasonably expect that the chains of predictions would have become completely random after few steps. In fact if a noisy prediction is fed back into the predictor, one would expect that the new prediction accumulates a double amount of noise, and so on for the further predictions generated along the chain. Interestingly, this is not the case.

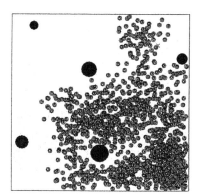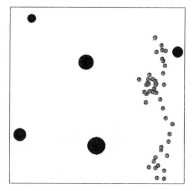

Fig. 6. The positions occupied by the simulated robot the first time that it reaches the goal by reinforcement learning (*left*) and by planning and acting (*right*)

A simulation was run to verify this. The simulated robot was set at the southeast start. The action "selected" was fixed to northwest by suitably changing the simulation program. A sequence of 29 successive predictions was then recorded (see Fig. 7). The results show a quite surprising capacity of the predictor to anticipate the consequences of actions. For example the predictor is capable of coping with noise, is capable of anticipating the appearance of landmarks from behind other landmarks, or the disappearance of them, and the overall chains of predictions are quite similar to the corresponding image sequences (compare the left and right parts of Fig. 7).

Though good, the predictor makes some mistakes (see Fig. 7 and cf. [3]; incidentally, these mistakes are quite interesting). For example it "loses" some landmarks, predicts the appearance of non-existing landmarks, tends to keep fixed images of landmarks at the middle of the scene (this is a strong regularity learned in the world: when the simulated robot moves toward a landmark, the image of it stays still while the images of all other landmarks move to the sides).

This robustness to noise of the predictor probably depends on the neural-networks' generalization property, and on the non-linearity of the output units of the predictor's experts (recall that these are sigmoid units whose output is compared with a 0.5 threshold and made binary accordingly, see above). When the predictor is trained, the images used as teaching output are those that correspond to the views of the world. As a consequence, even when fed with some images corrupted by noise the predictor's output will tend to be close to an image that corresponds to a view of the world since noise will tend to be filtered out by the non-linear output units.

This section is closed with some remarks on the noise robustness of the reactive system (the noise robustness of the planners is comparable). The system is robust with regards to the noise that affects actions and to the noise that causes the pixels of the retina to flip their values: doubling the variance of the former (from 0.01 to 0.02 of variance: recall that the step measures 0.05) or the probability of the latter (from 0.01 to 0.02) does not disturb the system. On the contrary, the system is very sensitive to the compass noise. Doubling it (from 1 degree to 2 degrees of variance) makes the system fall in local minima. Probably the reason of this sensitivity is that the rotation invariance of images is very important in the absence of a rotation-invariant pre-processing and with the simple one-layer neural networks used here.

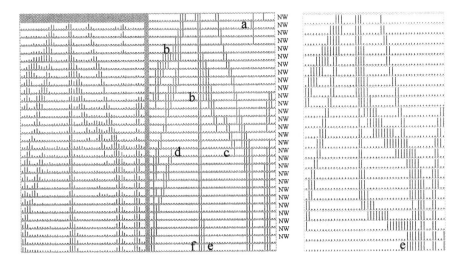

Fig. 7. The graph on the left reports an example of chain of predictions from the southeast start toward northwest (the action selected has been kept fixed to northwest). For each step the continuous output of the predictor (*left part of the graph on the left*), the binary prediction (*right part of the graph on the left*), and the selected action (*column of letters*) are shown. The succession of the predictions is reported from the top to the bottom of the graph. The first binary image corresponds to the image that the simulated robot perceives at the position currently occupied in the world, while the other images are generated by the predictor. *Graph on the right*: true images perceived by the simulated robot while moving from the southeast start toward northwest. It has been reported to allow the evaluation of the quality of the chain of predictions (reported in the left graph). Notice that this graph has been expanded to have a better correspondence with the chain of predictions: this means that the chains of predictions tend to be made up by more steps than the sequences of real images. Notice that the predictor is capable of coping with noise (*a* indicates a noisy activation that is suppressed after some time), is capable of predicting the appearance of a landmarks from behind other landmarks (*b*) and the disappearance of landmarks behind other landmarks (*c*); however, the predictor also produces distorted images, for example it generates non-existing landmarks (*d*) and it has biases, e.g. the prediction chain leads toward the north west landmark and not to the left of it as it should (*e*) probably because landmarks' images perceived at the center of the retina tend to persist (*f*)

4 Bidirectional Planning

We have seen that the forward planner needs a lot of planning cycles to reach the goal the first time. Is there a way to increase the planner's efficiency? One possible solution is to exploit the information given by the goal image. Next sections propose a bidirectional planner that exploits this information by generating chains of "predictions" both from the current position and from the goal image, and show that this planner has some important advantages in comparison to the forward planner.

4.1 The Architecture and Functioning of Bidirectional Planning

If the variable "BidirectionalPlanning" of the algorithm illustrated in Fig. 3 and Fig. 4 is set at "TRUE", the algorithm implements bidirectional planning. As the forward planner, the bidirectional planner decides if planning or acting on the basis of the measure of confidence at the position currently occupied by the simulated robot (line 7 to 14). The major difference between the two algorithms is that while planning the bidirectional planner generates prediction chains alternately forward from the current position image (line 47 to 55 implement one cycle of forward chain) and backward from the goal image (line 38; line 60 to 69 implement one cycle of backward chain). The length of each backward chain is the same as the last forward chain (line 32 and 42). Forward chains are executed as in forward planning. Backward chains are executed through the "back-predictor" and "back-actor" (Fig. 2).

The back-predictor is a network with the same architecture as the predictor. While the predictor is trained to produce the association y_t, $a_t \rightarrow x_{t+1}$, the back-predictor is trained to produce the association y_t, $a_{t-1} \rightarrow x_{t-1}$ (time indexes used backward) i.e. to *remember* (or guess) which *situation* x_{t-1} *led the system to the current situation* y_t after executing action a_{t-1} (line 62). Notice that each couple of experts of the predictor and of the back-predictor corresponding to a particular action could have been integrated in one bi-directional network associating $x_t \leftrightarrow x_{t+1}$ under action a_t. This has not been done since for simplicity only feed-forward networks have been used.

The back-actor has the same architecture as the actor, and is used to generate actions for the backward chains (the a_{t-1} of the association y_t, $a_{t-1} \rightarrow x_{t-1}$, see line 60 to 62). Before the tests shown below the back-actor weights are randomly drawn in the interval [-0.001, 0.001], so initially it selects actions randomly. During a back cycle that leads from y_t to y_{t-1} (from x_t to x_{t-1}), after the back-actor selects a_{t-1}, the merit of this action is updated according to the same formula used for the actor (see equation 3) and with the usual error $e_{t-1} = (r_t + \gamma V^\pi[y_t]) - V^\pi[y_{t-1}]$ (line 67). However, now the merit of the action is updated using y_t as input for the back-actor (and not y_{t-1} as for the forward actor). Notice that with this training the back-actor learns to generate actions that lead to states with the *lowest possible evaluation* $V^\pi[y_{t-1}]$, i.e. states *far* from the goal and visited few times. When backward chains are generated, the actor and evaluator are also updated using e_{t-1}. In particular the actor produces the actions' merit in correspondence to y_{t-1}, and then its weights are updated on the basis of those merits and the action a_{t-1} selected by the back-actor and back-stochastic selector (line 68 and 69). During forward planning and acting, the back-actor is also trained by using e_{t-1} (line 56 to 58 and 79 to 81). To this end, in each forward cycle the back-actor yields the actions' merit m_{t-1} in correspondence to y_t, and then its weights are updated on the basis of those merits and the action a_{t-1} selected by the actor and stochastic selector for y_{t-1}. The overall functioning of the bidirectional planning algorithm can be summarized as follows. The back-actor learns to yield backward chains that "escape" from the goal in "straight" lines, hence creating a big area of positive evaluations around the goal. This area is easily "found" by the forward planning chains that, as a consequence, expand the same area toward the position occupied by the simulated robot. At the same time the actor becomes competent in the area where positive evaluations diffuse, and soon gets ready to act efficiently in the world.

Some remarks about backward planning are due. Updating the evaluator when the back-actor is selecting the actions may cause some problems. In fact actor-critic methods require that state evaluations reflect the expected reward averaged over the actions *selected by the current policy* (actor) in that state (notice that one could avoid this problem using "off-policy reinforcement learning methods", cf. [25]). Notwithstanding this, the choice made here should not be impairing because the actor's policy and the back-actor's "back-policy" should be quite similar. In fact: (a) the actor and the back-actor have the same architecture and are trained an equal number of times with the same errors; (b) the probability that the back-actor selects the action a_{t-1} at y_t (x_t) is similar to the probability that the actor selects the same action at y_t (x_{t-1}) (yielded by the back-predictor on the basis of y_t and a_{t-1}) since x_t and x_{t-1} tend to be perceptually very similar; (c) the direction of the maximum slope of the evaluation gradient field built by the actor and by the back-actor tends to be the same (i.e. toward the goal). Incidentally, notice that these observations suggest that maybe it is possible to integrate the actor and back-actor in a unique network. It remains to be ascertained which are the domains different from navigation where these assumptions still hold.

Backward planning should present two important advantages vs. forward planning. The first advantage is exploration. Updating the evaluations backward from the goal brings immediately to change the evaluations of states close to the goal. On the contrary forward planning starts to update the evaluations only after the goal is encountered the first time. Since the first search of the goal is usually done by random walk (but cf. [30]), that event can take very long to occur (the expected time is exponential in the number of steps separating the start from the goal, cf. [30]). The second advantage is in terms of propagation of evaluations. This is particularly fast if done backward from the goal because newly updated evaluations of states are used to update the evaluations of other states [14] [30].

4.2 Forward and Bidirectional Planning: Results and Interpretations

The forward and bidirectional planners have been tested with the scenario illustrated in section 2. The task assigned to the simulated robot was to reach the goal several times each time departing from one of the 12 starting positions showed in Fig. 1 in a sequence, beginning with the northwest one (after the southeast starting position was used, the sequence repeated until the end of the test). The results show that both planners implement genuine taskable planning. In fact they respectively reach the goal for the first time in 245 and 186 moves in comparison to the reactive system that takes 1432 moves (averaged over 10 random seeds). Both planners suitably interleave planning and reactive behavior (Fig. 8 and Fig. 9): when they repeatedly reach the goal from the same start, the performance improves both in terms of planning cycles and moves; moreover, the experience gathered while planning to reach the goal from a given starting position is transferred to the starting positions close to it (Fig. 8).

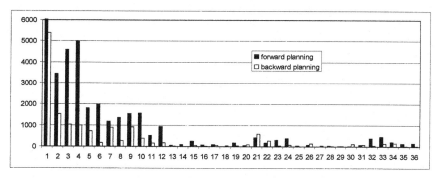

Fig. 8. Number of cycles (*y-axis*) spent planning for each success (*x-axis, 36 successes*) of the forward and bidirectional planners. Average over 10 random seeds. For graphical reasons the vertical axis is cut at 6000 (forward planning took 52,923 cycles to reach the first goal)

If the performance of the planners is compared, the following differences become apparent. The bidirectional planner is much more "goal oriented" than the forward planner. The forward planner spends nearly ten times the planning cycles used by the bidirectional planner (52,923 vs. 5,397 cycles) to reach the goal for the first time. After the first success in the world, bidirectional planning maintains its superiority for the following successes (Fig. 8). This difference is due to the fact that the forward planner takes several cycles to find the goal the first time (18,892 cycles on average). In comparison bidirectional planning is particularly efficient: in 6 tests out of 10 it reaches the goal in the world (for the first time) without having ever reached it in (forward) planning mode.

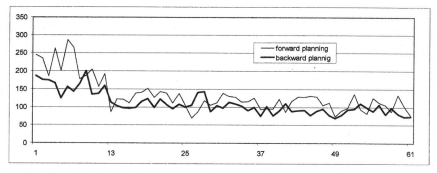

Fig. 9. Number of cycles (*y-axis*) spent acting for each success (*x-axis, 61 successes*) by the forward and the bidirectional planners. Average over 10 random seeds

This efficiency in exploration implies a fast "propagation" of the correct evaluations and policy updating through states. Fig. 10 shows the evaluations produced by the two planners in 20×20 positions of the arena after some cycles of action and planning. The bidirectional planner yields evaluations much closer to the optimal ones (equal to γ to the power of the number of steps to the goal) than the forward planning.

The bidirectional planner is also more effective than the forward planner in propagating the evaluations back from the goal. Direct observation of the dynamics of the graph of Fig. 10 drawn during the simulation, shows that, at the beginning of the simulation, in the case of the forward planner the evaluations tend to fall again to 0 between a (planned) reaching of the goal and the next. This happens because the goal is reached rarely and the evaluations get close to 0 due to the decay coefficient that, on the average, lowers the "targets" of the evaluations' updating (see formula 3). In the case of the bidirectional planner, positive evaluations are continuously "injected" in the graph from the goal state, and from there rapidly diffuse to contiguous states.

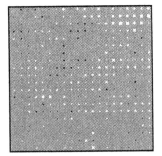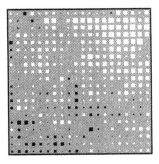

Fig. 10. Evaluations produced by the forward (*left*) and bidirectional (*right*) planners when put in 20×20 different positions of the arena. The area of the *white squares* (positive evaluations) and the *black squares* (negative evaluations) is proportional to the absolute evaluation produced at that position. *Gray* in correspondence to a position indicates an evaluation close to 0. For the forward planner, the evaluations have been recorded at the cycle after which the goal was reached the first time with a chain of predictions. For the bidirectional planner, evaluations were recorded after 13,340 cycles: this is the average number of cycles taken by the forward planner to reach the goal for the first time with a chain of predictions. Negative evaluations, that should not be present since the task does not involve negative rewards, are due to the insufficient number of degrees of freedom of the evaluator and the generalization properties of networks (negative evaluations are more pronounced in the case of the bidirectional planner simply because it is in a more advanced stage of learning in comparison to the forward planner)

5   Related Literature and Novelties of the Paper

The actor-critic part of the planners is similar to the actor critic models implemented with neural networks in [14] and [25]. The only difference is the way the actions are selected stochastically (see equation 2).

The idea of implementing planning as a form of (reinforcement) learning within a model of the world has been proposed with the Dyna models [26] (also compare [5], on "trial-based real-time asynchronous dynamic programming" applied to path finding problems, with the planners presented here). It is important to stress that previous works (e.g. [14] and [26]) used Dyna architectures exclusively to speed up learning, not to implement genuine taskable planning as done here. The reason is probably that no device like the matcher was used. The idea of generating simulated experiences on the basis of the current policy, called "trajectory sampling", was investigated in [5] and [25] but this is the first time that long prediction chains are generated with neural networks and their properties are investigated. The idea of increasing the depth of the path generated during planning resembles the "iterative deepening search" applied to deterministic problems by problem solving research [12]. However, the algorithms proposed here are new since they deal with stochastic problems.

The forward and bidirectional planning algorithms that control the flow of information between the components of the systems and generate the prediction chains are new. The idea of implementing the predictor (model of the environment) with a feed-forward neural network trained with experience has already been used in [14] and [18]. Notice that these works, as here, use deterministic neural networks to implement a model of a stochastic world. An alternative would have been to use stochastic networks, such as the feed-forward stochastic networks proposed in [17]. In this paper the predictor was trained while the simulated robot randomly explored the world. More sophisticated ways of exploring the world for model building have been proposed, e.g. in [23] and [30]. The idea of using expert networks for the predictor, each specialized to predict the consequences of one action, is proposed and used in [13].

The general idea of planning backward from the goal is widely used in the literature on classic planning [1]. However, its application to stochastic worlds, as done here, implies completely new problems, so the algorithms proposed here are new in this respect. The idea of updating the evaluations backward from the rewarding states has already been investigated within the reinforcement learning literature. In particular [13], [14], [19] and [30] have shown that this is a powerful strategy because state-evaluations are updated on the basis of recently updated evaluations. However, all these works used memory structures to store sequences of states that led to rewarding states in order to use them for iterated "backward" backups. If one wants to use neural networks, this strategy raises the problem of how implementing these memory structures and how using the information stored in them. With this respect, the back-actor, back-predictor and matcher used here are new since they allow generating rewards and states backward from the goal at will. Prioritized sweeping [7] [16], by updating states or state variables whose evaluations would change a lot if updated, often propagates evaluations backwards from states close to the goal.

There are two other important branches of research related to planning with neural networks. One is activation-diffusion planning (see several examples in [15]). These models are based on maps of Kohonen-like units [11]. Each of these units stores a "snapshot" of the world in its weights. While the agent randomly navigates in the world, lateral connections are formed/strengthened between units corresponding to places that are contiguous in space. Planning is implemented by "injecting" activation into the unit corresponding to the goal, allowing it to diffuse through lateral connections with a progressive attenuation, and selecting the actions so as to ascend the activation gradient field formed over the units. The model presented here differs from these models because: (a) places are represented by *patterns of active cells* instead of *patterns of weights*; (b) planning is based on focussed active exploration of the possible actions' consequences; (b) distributed representations are used allowing one unit to participate to the representation of several places (while activation diffusion planning needs to use one unit for each place represented).

The second branch of research is represented by neural planners based on gradient descent processes ([22], [28] and [29]). These planners formulate planning problems in terms of differentiable cost functions (and eventually differentiable actions) so they have a limited applicability. Plans are found by minimising these functions. Two other works relevant for the issues tackled here are [18] and [27]. They present two systems that use neural models of the world respectively to improve the performance of an agent controlled by a neural network evolved with genetic algorithms, and to allow a simulated robot to learn the world's regularities at multiple levels of abstraction.

6 Conclusion and Future Work

This paper has presented two new reactive and planning systems implemented with neural networks and inspired by Dyna-PI reinforcement learning methods. These planners present important novelties. Contrary to Dyna architectures, the planners are capable of executing genuine taskable planning using the goal independent information stored in the model of the world. This allows them to reach the goal with few moves from the first time they pursue a goal. This ability, that relies on the capacity to anticipate future states, is of great advantage if experience is costly or risky. Both planners are capable of nicely interleaving planning and action. They can decide to plan when their "confidence" in action, a novel concept introduced in the paper, is low, and to act reactively when they have enough experience about the goal and the current state. In the future, a better measure of confidence will be based on the concept of "entropy" applied to the actions' probabilities [24]. The model of the world is a compound neural network trained with experience to predict the states deriving from the execution of actions. The tests of the planners run with a simulated robot engaged in a stochastic path-finding task have indicated that the chain of predictions generated on the basis of this model of the world are more robust than expected, due to the non-linearity and generalization capabilities of neural networks. Since the model of the world is the core component of each planning system, this interesting result will be further investigated in the future. The bidirectional planner, that exploits the knowledge of the goal to generate "prediction" chains backward from it, has shown to be more efficient in exploring the state space and quicker in propagating the evaluations back from the goal than the forward planner. Unfortunately, these advantages have been obtained at the cost of a rather complex architecture. In the future a simpler planner will be studied that generates forward chains as the forward planner presented here, and updates the goal state's evaluation with a target of $1+\gamma$ when a forward chain terminates without encountering the goal ($1+\gamma$, and not simply 1, will be used since the states encountered one step before reaching the goal are updated on the basis of the undiscounted reward of 1). Given the generalization property of neural networks, this should change the evaluation of several states around the goal state, creating an extended area with positive evaluations. This would facilitate the forward chains in finding the goal state, allowing one to have some advantages similar to those obtained with the bidirectional planner, but with an architecture as simple as the forward planner's one (see [3] for details).

Acknowledgements

The Department of Computer Science of the University of Essex funded the author's research. Special thanks are expressed to my supervisor Prof. Jim Doran (University of Essex) who encouraged the exploration of the idea of bidirectional planning. The Institute of Cognitive Science and Technologies of the National Research Council of Italy (ISTC-CNR) funded the author's research during the writing up of the paper.

References

1. Allen J., Hendler J., Tate A. (eds.): Readings in Planning. Morgan Kaufmann, Palo Alto Ca. (1990)
2. Arkin R.C.: Behavior-Based Robotics. The MIT Press, Cambridge Ma. (1998)
3. Baldassarre, G.: Planning with neural networks and reinforcement learning. Computer Science Department, University of Essex (2002) Ph.D. Thesis
4. Baldassarre G.: A modular neural-network model of the basal ganglia's role in learning and selecting motor behaviors. Cognitive Systems Research. 3 (2002) 5-13
5. Barto, A.G., Bradtke, S.J., Singh, S.P.: Learning to act using real-time dynamic programming. Artificial Intelligence. 72 (1995) 81-138
6. Brooks R.A.: A robust layered control system for a mobile robot. IEEE Journal of Robotics and Automation. 2 (1986) 14-23
7. Dearden R.: Structured prioritized sweeping. In: Proceedings of the Eighteenth International Conference on Machine Learning. (2001) 82-89
8. Fikes R.E., Nilsson N.J.: STRIPS: a new approach to the application of theorem proving to problem solving. Artificial Intelligence. 2 (1971) 189-208
9. Haykin S.: Neural Networks: A Comprehensive Foundation. Prentice Hall, Upper Saddle River N.J. (1999)
10. Jaakkola T., Singh S.P., Jordan M.I.: Reinforcement learning algorithm for partially observable Markov decision problems. In: Tesauro G., Touretzky D.S., Leen T.K. (eds.): Advances in Neural Information Processing Systems 7. The MIT Press, Cambridge Mass. (1995) 345-352
11. Kohonen T.: Self-organized formation of topologically correct feature maps. Biological Cybernetics. 43 (1982) 59-69
12. Korf, R.E.: Optimal path finding algorithms. In: Kanal L.N., Kumar V. (eds.): Search in Artificial Intelligence. Springer-Verlag, Berlin (1988) 223-267
13. Lin L.-J., Mitchell T.M.: Memory approaches to reinforcement learning in non-Markovian domains. Carnegie Mellon University (1992) Technical Report CMU-CS-92-138
14. Lin, L.J.: Self-improving reactive agents based on reinforcement learning, planning and teaching. Machine Learning. 8 (1992) 293-391
15. Meyer J.-A., Berthoz A., Floreano D. (eds.): From Animals to Animats 6: Proceedings of the Sixth International Conference on Simulation of Adaptive Behavior. The MIT Press, Cambridge Ma. (2000)

16. Moore A.W., Atkeson C.G.: Prioritised sweeping: Reinforcement learning with less data and less real time. Machine Learning. 13 (1993) 103-130
17. Neal R.M.: Bayesian learning for neural networks. Springer-Verlag, Berlin (1996)
18. Nolfi S., Elman J.L., Parisi D.: Learning and evolution in neural networks. Adaptive Behavior. 3 (1994) 5-28
19. Reynolds S.I.: Experience stack reinforcement learning for off-policy control. University of Birmingham (2002) Technical report CSRP-02-1
20. Ross S.: Introduction to stochastic dynamic programming. Academic Press, New York (1983)
21. Russell S., Norvig P.: Artificial Intelligence: A Modern Approach. Prentice Hall, Englewood Cliffs N.J. (1995)
22. Schmidhuber J., Wahnsiedler R.: Planning simple trajectories using neural subgoal generators. In: Meyer J.-A., Wilson S.W. (eds.): From Animals to Animats 2: Proceedings of the Second International Conference on Simulation of Adaptive Behavior. The MIT Press, Cambridge Ma. (1992) 196-202
23. Schmidhuber J.: Artificial curiosity based on discovering novel algorithmic predictability through coevolution. In: Angeline P., Michalewicz X., Schoenauer M., Yao X., Zalzala Z. (eds.): Congress on Evolutionary Computation. IEEE Press, Piscataway N.J. (1999) 1612-1618
24. Shannon C.E.: A mathematical theory of communication. The Bell System Technical Journal. 27 (1948) 623-656
25. Sutton R.S., Barto A.G.: Reinforcement Learning: An Introduction. The MIT Press, Cambridge Mass. (1998)
26. Sutton R.S.: Integrated architectures for learning, planning, and reacting based on approximating dynamic programming. In: Proceeding of the Seventh International Conference on Machine Learning. Morgan Kaufmann, San Mateo Ca. (1990) 216-224
27. Tani J., Nolfi S.: Learning to perceive the world as articulated: An approach for hierarchical learning in sensory-motor systems. Neural Networks. 12 (1999) 1131-1141
28. Tani J.: Model-Based Learning for Mobile Robot Navigation from the Dynamical Systems Perspective. IEEE Transactions in System, Man and Cybernetics, Part B. 26 (1996) 421-436
29. Thrun S. B., Moller K., Linden A.: Planning with an adaptive world model. In: Tourtezky D. S., Lippmann R. (eds.): Advances in Neural Information Processing Systems 3. Morgan Kaufmann, San Mateo Ca. (1991) 450-456
30. Thrun S.: Efficient exploration in reinforcement learning. Carnegie-Mellon University (1992) Technical Report CMU-CS-92-102
31. Widrow B., Hoff M.E.: Adaptive switching circuits. IRE WESCON Convention Record, Part IV. (1960) 96-104

Sensory Anticipation for Autonomous Selection of Robot Landmarks

Jason Fleischer[1], Stephen Marsland[2], and Jonathan Shapiro[1]

[1] Department of Computer Science
University of Manchester
Manchester M13 9PL, UK
{jgf,jls}@cs.man.ac.uk
[2] Imaging Science and Biomedical Engineering
Stopford Building
University of Manchester
Manchester M13 9PT, UK
stephen.marsland@man.ac.uk

Abstract. There are many ways to define what constitutes a suitable landmark for mobile robot navigation, and automatically extracting landmarks from an environment as the robot travels is an open research problem. This paper describes an automatic landmark selection algorithm that chooses as landmarks any places where a trained sensory anticipation model makes poor predictions. The model is applied to a route navigation task, and the results are evaluated according to how well landmarks align between different runs on the same route. The quality of landmark matches is compared for several types of sensory anticipation models and also against a non-anticipatory landmark selector. We extend and correct the analysis presented in [6] and also present a more complete picture of the importance of sensory anticipation to the landmark selection process. Finally, we show that the system can navigate reliably in a goal-oriented route-following task, and we compare success rates using only metric distances with using a combination of odometric and landmark category information.

1 Introduction

Landmarks are commonly used for mobile robot navigation, but there is little consensus about the definition of the term and many theories about what constitutes a good landmark. In this paper we consider *perceptual* landmarks, for which we take the usual definition – a robot's sensor values in a particular location. A good perceptual landmark is one that enables a robot to maintain an accurate estimate of its location while traversing an area that it has previously mapped. Landmark-based localisation in this sense forms the basis of many successful mobile robot navigation systems, see for example [5, 20].

The question of how to select landmarks, then, is an important one, yet it is often solved by avoiding the issue, simply taking sensory scans at regular

M. Butz et al. (Eds.): Anticipatory Behavior ..., LNAI 2684, pp. 201–221, 2003.

intervals, and taking every scan as a landmark. In many environments a large number of perceptions are almost identical – especially to the limited sensors available to a robot – so the majority of these landmarks do not aid navigation. A slightly better approach is to *select* landmarks in some way, for example by having an experimenter choose landmarks while looking at images taken by the robot [11], or by using some form of statistical learning, e.g., [17, 3].

In this paper we address the problem of learning to select landmarks so that a robot can reliably navigate a route without getting lost. The robot should select landmarks that will be perceivable each time it follows the route (consistent), easily distinguishable from nearby locations (distinctive), frequent enough to be useful for localisation, but not so frequent that localisation is computationally expensive. These landmarks are not required to be completely unique, so that a robot might know exactly where it is from single perception – this may well be impossible in many environments. We assume that a robot is following a route and must localise itself as it travels, rather than trying to localise from scratch against all possible locations.

One solution to this problem, described in section 2, is to learn a model of sensory anticipations in a typical operating environment. This model of normal sensory sequences is used to predict the robot's next perception at every time step, and a landmark is defined to be any place where the robot's anticipations are not accurate; in other words, those places that are not adequately described in the model. This is a form of novelty detection. A Kalman filter is used both to remove the effects of noise in the predictions and to maintain statistics used for the selection of landmarks. Landmarks are therefore based on the novelty of the current perception in relation to recent context, ensuring that they are distinctive compared to nearby perceptions. The landmarks are consistent because the Kalman filter helps to remove both the effects of robot path variation and small environmental changes that may not be detected every time a route is followed.

We also investigate an alternative to anticipatory landmark detection, based on perceptual novelty in relation to the entire perceptual history. We use a growing neural network that adds nodes if a perception is not well described by previous perceptual categories. The growing network is trained in the typical operating environment, in the same way as the anticipatory method, and then operated in the new environment, with the growth mechanism frozen so that new nodes cannot be added. Any location where the network would have added a new node during training is considered to be a landmark. This approach to landmark detection, described in section 3, is completely different; it has no mechanisms in common with the anticipatory method, landmarks are selected based on their similarity to all perceptions seen previously, without any information about current perceptions. The perceptual novelty approach selects a landmark because it is distinctive with respect to the robot's entire perceptual experience.

The drawback of the anticipatory landmark detector is that an assumption must be made that the model's learning ability is sufficient to predict well, but not so great that it never produces enough error to detect a landmark. Previous experience [14] has shown that single-layer neural networks of various types meet this assumption in office corridor environments. A drawback of both approaches is that they require selecting parameters that determine the amount of novelty a perception must posses before it is selected as a landmark.

In previous publications, the use of sensory predictions to automatically select landmarks in static straight corridors where the robot took sensor readings at fixed intervals has been investigated [14]. The anticipatory method has also been applied to robot navigation in more complex routes involving turns, and some simple methods of localisation based on following route descriptions was investigated [6]. The perceptual novelty method has never before been applied as a landmark detector for localisation.

This paper extends and corrects the previous analysis in [6] that investigated the performance of various landmark alignment methods and parameter values for the anticipatory landmark detector. Those analyses were based on an assumption that the scores were normally distributed, which has since been found to be false in certain cases. In addition, we utilised many pairs of t-tests, opening up the possibility of an erroneous conclusion occurring due to the sheer number of comparisons being made. This paper verifies the previous results with distribution-free ANOVA analyses [7] in sections 4.3 and 4.4. We also look at the use of different complexities of sensory anticipation model and the resulting differences in landmark alignment in section 4.5. We compare the perceptual novelty method of landmark detection to the sensory anticipation method in terms of landmark alignments in section 4.6, and we present the results of using automatically selected landmarks in a route-following task in an office corridor environment in section 4.7. Finally, we close the paper with a discussion (section 5) and conclusions (section 6).

2   Landmark Selection by Sensory Anticipation

Anticipations are predictions of future values that have an effect on the current behaviour of the agent. There are several types of anticipatory mechanisms [4]; the direct prediction of sensory values given recent perceptions is one of them. Sensory predictions can be used either passively, as an attentional mechanism, or actively, in a sensorimotor context.

In neuromotor control models, sensory predictions are generated from current sensory values by feeding the motor commands to be executed through a forward model of the system's kinematics. The system can therefore detect the difference between sensory changes induced by self-motion and those that occur due to external factors, and cancel those effects if necessary [19], a process known as reafference. An example robotic application of sensorimotor prediction is the use of predictions to increase the accuracy of optic flow measurements in an active perception framework [16]. A reafferent model has also been used to investigate cognitive models of sensory anticipation, investigating the possibility that internal simulation can take the place of sensory inputs, allowing the robot to act blindly for short periods of time [8].

Our approach is to use sensory anticipations passively, we do not attempt to correct perceptions given self-movement, as in the reafferent case. The anticipatory mechanism described in this paper can be viewed as focusing the attention of the robot onto a particular perception in the temporal sequence because of its novelty with respect to the recent perceptions. This is related to the topics of change detection and novelty detection [12, 13].

The landmark detection system consists of a sensory anticipation network and a method of detecting when the difference between the prediction of the next sensor values and the actual perceptions (the error) is large enough to be considered a landmark. The sensory anticipation model used in this paper is a single-layer artificial neural network, as previous work [14] has shown this to be a suitable method. The network is trained in a latent learning fashion, by allowing the robot to travel through the environment using a wall-following behaviour, collecting sensor data and training its anticipatory model, as described in section 2.1. Once trained, the error between the model's predictions and the real perceptions is used to select landmarks. The prediction error is sent through a Kalman filter to both remove the effects of noise and to use the estimated statistics to select landmarks. Because the Kalman filter is recursively updated and includes a predictive noise model it provides the estimation of context perhaps nearly as much as the sensory anticipation model. Finally, each landmark is categorised using a Self-Organising Map, as described in section 2.2.

2.1   The Sensory Anticipation Model

Our sensory anticipation model learns to predict the next values of the five sonar sensors that face the wall being followed by the robot. The side-facing sonars are preferable because (a) they do not register humans walking along the corridor, and (b) for route navigation in a corridor potentially interesting landmarks are usually doors and hallway branches that will be best detected when directly to the side of the robot. The following types of single-layer artificial neural networks are investigated for learning the sensory anticipation model:

Standard Each node represents a sonar value, current (normalised) sensor readings being used as inputs, with the network attempting to predict the sensor readings at the next time-step. The output nodes have a sigmoidal activation function and are connected to one or more input nodes and a bias input. The output layer is not fully connected because sonars to the rear of the robot have no use in predicting future values of more forward sensors. Thus, each output node is only connected to its own input node and input nodes that represent sonars further forward on the robot; see [14] for more details.

Lagged This network uses the same structure as the standard neural network, but the inputs values consist of normalised sonar values from both the current time and the previous time-step.

Recurrent This is similar to the lagged version, except that the inputs consist of the current sonar inputs and the network outputs at the previous time-step.

Naïve This is not a neural network, but merely assumes that the next perception will be identical to the current one.

These models are compared with each other for the landmark selection task in section 4.5 and with a non-anticipatory on-line landmark detection method in section 4.6.

The neural network is trained to acquire a model of the relationship between successive sensory perceptions by adapting the weights during a training phase, where the robot explores an environment and records its sensory perceptions. At each time-step the current sensory perceptions are used as network inputs and the previous sensory perceptions are used as training examples. The weights are adapted by a least-mean-squares learning rule and trained for 17 epochs on the data with a learning rate of 0.2. The number of epochs was determined by using early stopping on several sample datasets and picking a reasonable value for general use.

We reiterate that this model makes no attempt at performing reafference; it learns to minimise prediction error including robot motion. Instead, we employ a Kalman filter [9] to remove components of the prediction error signal that are due to robot orientation and sensor repeatability, and to provide a systematic method to determine when sensory predictions are sufficiently incorrect to be labelled as a landmark. More detailed discussions of our implementation can be found in [6, 14].

The Kalman filter is a method of recursively estimating the state of a discrete-time controlled process that is governed by a linear stochastic difference equation. In our case the filter estimates a single scalar variable, the sum-squared error of the prediction network, E, which would be a constant value, hopefully near zero, if the robot maintained perfect orientation to a perfectly even wall. However, there is noise in E from both repeatability (measurement noise) and robot orientation and position (process noise). The Kalman filter attempts to optimally re-estimate E at each step so as to remove the effects of both measurement and process noise, leaving behind only the prediction error actually produced by variation of the environment. The filter equations use a constant measurement noise model and a time-varying process noise model to compute a gain, K, that is in turn used to recursively update estimates of the true prediction error, $\widehat{E}$, and its variance, V. The variance estimate can be used to select as a landmark any perception where $\widehat{E}(t) > \bar{E}(t) + n\sqrt{V(t)}$. The parameter n provides a way of adjusting the required level of landmark distinctiveness, and therefore to some extent landmark frequency. Typically, we find that values of n between three and five work well.

2.2   Landmark Categorisation

Some of the localisation methods presented in section 4.3 will not align the current perception with a location in the route description unless they have the same landmark category. Categorisation in this sense is merely clustering together similar robot sensor perceptions. We used Kohonen's Self-Organizing Map (SOM) [10], an unsupervised neural network based on vector quantisation, to perform categorisation of variance normalised sonar sensor values from the robot. The SOM is trained by comparing each sensor input vector to every node in the network and selecting the best-matching node, in terms of Euclidean distance between the input and the node weights, as the winner. The winning node and other nodes nearby in the output topology are adjusted so that their weights are closer to the current input. The category of a perception is the identity of the winning node in the map for that perception.

We used the SOM Toolbox [18] to generate the SOMs used in this paper. The SOMs all have 5×4 toroidal network topology, and the toolbox was allowed to automatically select the training parameters, based on the number of inputs, the number of data points, and the principal components of the training set. There are several possible choices of SOM inputs – we investigated using either all 16 sonar sensors or only those 5 facing the wall being followed – the same sonars used for landmark selection (see section 2.1). We also tried training the SOM on every perception in the training environment, or just on those that are classified as landmarks. The various types of SOM are compared in section 4.4.

3   Landmark Selection Using Perceptual Novelty

In machine learning a novelty filter is an algorithm that detects when inputs differ significantly in some way from those that it has seen before. This is different to the standard classification problem, where each new perception is classified as belonging to a known category after the algorithm has been trained on many examples of all the different classes. We used a novelty filter known as the Grow When Required (GWR) network [15] (see also the similar FOSART algorithm [2, 1]) to learn a model of a training environment. The GWR network is a self-organising topology-preserving network that can dynamically create and destroy network nodes, allowing it to model dynamic datasets on-line. When the robot explores a test environment, any perceptions that the novelty detector highlights as novel inputs were labelled as landmarks.

Learning in the GWR network is driven by an insertion threshold $0 \le a_T \le 1$, which can be thought of as tunable generalisation; the amount to which the network generalises between similar perceptions is controlled by the amount of discrepancy between perceptions that triggers a new node. Lower values of the insertion threshold produce fewer, more general categories. In this paper, the network is trained on one run in a subset of the hallways that are used as the testing environment. Thereafter, the network is fixed, and if the network would normally add a new node then the input is considered novel, and therefore a landmark. Experiments with the GWR network can be found in section 4.6.

4   Experiments

4.1   Physical Setup

The experiments in this paper were performed on a Nomad Scout, a differential drive robot with 16 Polaroid sonar sensors capable of giving range information on objects between 15 cm and 6 m away from the robot. The sensor values were updated as quickly as the processing speed of the PC controller would allow, giving less temporal structure to the data than there would be if the sensors were only updated at fixed distance or time intervals. Each experimental run consisted of between 30 and 200 metres of travel in normal office building corridors that were not modified for the robot's use in any way. These runs were made during daytime hours with normal use being made of the corridors; no attempt was made to remove data anomalies due to people walking through the sensor range.

The robot used a hard-wired wall-following program to follow the left-hand wall at a constant distance. Noise in the sonar sensor values comes principally from the imperfect wall-follower, but also from both systematic and repeatability errors in the sensors themselves. Another source of noise is the odometry, which is used to help localise the robot. A rule of thumb for our robot type is that maximum odometry drift is roughly five percent of distance travelled. Therefore we assume that drift can be modelled as $\nu(d) = 0.25 + 0.05d$, where ν is the maximum possible error and d is distance travelled in metres.

In the landmark alignment experiments (sections 4.3–4.5) three different, adjacent, hallway sections were used for training the landmark detection algorithms. The robot travelled through each hallway three times, thus producing a total of nine different datasets for training the algorithms. The training data were used to train both the landmark detector and the landmark categoriser. The test data were produced by propping open the doors between the three hallways and collecting a further three data runs using a 200 m long route through all three hallways. To compare the alignment between landmarks the three test sets were paired in all six possible permutations (since order of comparison matters) and the measures described in the next section were calculated. In all cases training and testing were performed off-line, using the data collected by the robot.

4.2 Analysis Methods

We use four metrics to quantitatively evaluate landmark alignments, and therefore localisation quality, between two runs of the robot along the same route:

Category score The number of times that the aligned landmarks share the same category in the SOM, divided by the number of alignments.

Distance score The mean of $\exp \frac{-\delta_k^2}{2\nu_k^2}$ for all aligned landmarks, where δ_k is the difference between the recorded odometry distance for landmark k and the distance of its aligned partner, and ν_k is the possible odometric drift at that landmark.

Sequence score The fraction of alignments where the aligned partner of landmark $k + 1$ is the same distance, or further on, from the aligned partner of landmark k. Landmarks without assigned alignments are not counted as a break in the sequence.

Percentage of landmark alignments The number of alignments made over the total number of landmarks in the current run.

Note that the first three metrics do not penalise alignments that produce fewer matches of better quality. This is based on the assumption that navigation with many landmarks of uncertain quality is more difficult than navigation with fewer, better quality landmarks, i.e., landmarks that are found consistently. It is also worth noting that these metrics give only an approximation to the accuracy of the landmark matching. There is no way to ensure that matching landmarks really are in the same location, since there is no global metric localisation system such as differential GPS or external cameras available. The best possible evaluation of the alignments in these circumstances is the success rate of using the alignments to find a goal location, as shown in section 4.7.

These scores are evaluated using distribution-free statistical tests of parameters, see for example [7]. In particular, we use a Kruskal-Wallis ANOVA, which tests if the null hypothesis – that all choices of treatments (i.e., experimental parameters) are equal – is false. We also use a Freidman ANOVA, which does the same thing, but also seeks to cancel out the effects of other systematic differences by blocking together trials that are homogeneous with respect to other factors that might influence the results. If a difference does exist, then a Tukey-Kramer multiple comparison of means is used to try grouping together those treatments that are not statistically distinguishable at some confidence level, therefore allowing us to determine the best performing treatments.

4.3 Aligning Route Descriptions

In this section, we propose some simple methods of robot localisation on a route, and compare them to select the best one to use for subsequent experiments on landmark detection. We find that the best results come from the CSL alignment method, which uses a combination of landmark category, distance travelled, and an odometric error model. CSL not only produces some of the best scores, but just as importantly it produces alignments between runs that are qualitatively pleasing.

Each trip that a robot makes along a route is recorded in a combined metric and topological route description, consisting of the sequence of landmarks perceived along the route, along with the distances between them as measured by the robot's internal odometry. The perceptual category of each landmark generated by the SOM is also recorded. When a robot attempts to follow a route it has previously taken it attempts to match it's current landmark perception with the stored route description. We investigated six different methods of producing such alignments, reflecting the possible combinations of information stored in the route description (sequence, distance, and category) and an odometric error model. The alignment methods can be described as:

Sequence (S) Each landmark in the current run is aligned with the next unassigned landmark in the route description. If there is a mismatch between the number of landmarks in each run then there is no alignment for the excess landmarks.

Distance (D) The $x - y$ position odometry is transformed into a single dimension of distance travelled from the beginning of the run. Landmarks in the current run are aligned to the landmarks in the route description that are closest to them in distance travelled.

Category and Distance (CD) Each landmark in the current run is aligned to the landmark in the route description that is closest to the same distance travelled and is also a member of the same landmark category (given by the SOM). If there is no category match then no alignment is made for that landmark.

Category and Sequence (CS) Each landmark in the current run is aligned with the next unaligned landmark in the route description whose category it matches. If there is no category match then no alignment is made for that landmark.

Category and Distance with Limited Range (CDL) As in alignment CD, but matches are only allowed that are within $\nu(d)$ of the landmark at distance d in the current run.

Category and Sequence with Limited Range (CSL) As in alignment CS, but matches are only allowed that are within $\nu(d)$ of the landmark at distance d in the current run.

To evaluate the six methods quantitatively we looked at the each of the four scores taken over all the possible sets of test and training sets. The first question is whether there is any statistical difference at all between the various alignment methods. A Friedman ANOVA ($p = 0.05$, 32 blocks formed by the possible combinations of anticipatory models and choices of landmark detector settings, 54 trials/block) does reject the null hypothesis that all alignment methods are equally as good, on all four scores.

The next question is which alignment is best in each score. Table 1 answers that using a Tukey-Kramer multiple comparison of mean ranks ($p = 0.05$). The best looking alignments in terms of score look to be the D and S alignments. Closer inspection, however, soon shows them to be inferior in several ways. Although they do well on all other metrics, the lack of high category score is an indication that the landmarks are not actually being aligned to the same locations. This is obvious after a quick look at the algorithms, D and S aren't picky, they merely rely on the landmark detector being very consistent from run to run, and are likely to accumulate more error the longer the route. CDL has the highest scores of the remaining algorithms; importantly it has a high category score, indicating that it probably is aligning positions in the environment correctly.

A visual inspection of the example alignments in figure 1 illustrates the point, showing that the distance alignment makes many dubious landmark assignments because some landmark features are not detected on both runs through the environment. Looking at the same figure also demonstrates why the CDL alignment does best – it finds fewer landmarks, but those that it finds are detected more consistently and at qualitatively better locations. This alignment method is therefore used to make all other comparisons in this paper.

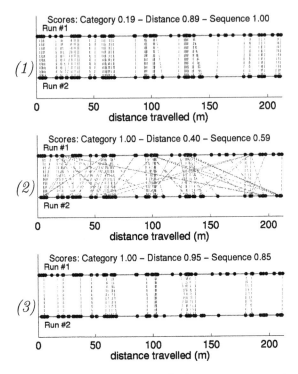

Fig. 1. Example alignments for a landmark detection system (standard single-layer network predictor, $n = 4$, using all sonars at landmark perceptions for SOM training) between two separate runs down the same route. The distance travelled along the route is the horizontal axis, landmarks are represented by black dots, and alignments by dashed lines between landmarks in the two runs. *(1)* Distance Alignment (D); *(2)* Category & Distance Alignment (CD); *(3)* Category & Distance Alignment with Limited Range (CDL)

It is easy to argue that the metrics proposed here are somewhat arbitrary and do not directly reflect the suitability of the algorithm to route navigation. Unfortunately it is not possible to externally check the correctness of the alignments since no accurate global position measurement is possible with these robots. However, these concerns are addressed by the experiments described in section 4.7, which show that CDL alignment is superiour to D alignment in real robot route navigation.

4.4   Effect of Parameters

This section investigates whether any particular choice of parameters for the anticipatory landmark detector are better than any others. Although the ANOVA finds that there is a difference somewhere in our choice of parameters, the comparison of means is unable to pinpoint it. Instead, we qualitatively observe a few

Table 1. Comparing alignment methods: mean scores, mean ranks, and statistically significant differences. See section 4.3 for a key of alignment abbreviations and an explanation of the scores. All statistically indistinguishable groupings, indicated by underlines, are due to a Tukey-Kramer comparison of rank means. ($p = 0.05$)

Category Score	CD	CS	CDL	CSL	D	S
rank mean	244	243	241	109	81	57
score mean	1.0	1.0	1.0	.26	.20	.15

Distance Score	CDL	D	CD	CSL	S	CS
rank mean	269	264	170	93	91	88
score mean	.87	.83	.51	.26	.26	.25

Sequence Score	S	D	CDL	CSL	CS	CD
rank mean	260	260	162	157	74	61
score mean	1.0	1.0	.82	.81	.70	.68

% LM Aligned	D	S	CD	CDL	CS	CSL
rank mean	274	214	205	120	120	42
score mean	1.00	0.92	0.92	0.77	0.77	0.47

trends in the scores, and prefer to use the following settings: $n = 4$, using left-side only sonars for categorisation, trained on all locations in the environment.

We investigate three parameters in the landmark detector: detection threshold n, the choice of whether to use all sonars, or just the left-side sonars used to categorise landmarks, and how the categoriser is trained. The first setting is obviously important, but the other two can conceivably have a large effect on alignment methods that use category information. A Friedman ANOVA was used to investigate the treatments, which is shown in table 2. The treatments were taken over blocks representing the various anticipatory models, for the CDL alignment only. The ANOVA was performed separately for the distance scores, sequence scores, and percentage of landmarks aligned (it was not necessary to run the analysis on category score since this would be 1.0 in all CDL alignments). The ANOVAs all rejected the null hypothesis at a significance level of 0.05.

In spite of the ANOVA results, the Tukey-Kramer comparison at $p = 0.05$ (see table 3) could not tell which treatments were different from the others. Qualitatively, though, we can see that the left-side sonar models seem to do particularly well for the distance and sequence scores. Although the significance of this effect cannot be shown, we postulate that better scores could result from less 'noisy' categorisation; the wall-following side sonars would not be as affected by people walking through the hallways or minor changes of robot orientation. In general, using all perceptions outperforms using only the landmark ones to train the categoriser, probably allowing a better categorisation of the world. We

Table 2. The treatments investigated in section 4.4 to investigate the effects of landmark detection parameters on performance metrics.

Treatment	Perceptions Used	Sonars Used	n
I	Landmarks	Left	3
II	All	Left	3
III	Landmarks	All	3
IV	All	All	3
V	Landmarks	Left	4
VI	All	Left	4
VII	Landmarks	All	4
VIII	All	All	4

also notice that the percentage of landmarks aligned does seem to decrease with increasing detection threshold. Therefore we generally prefer treatment VI for subsequent experiments.

Table 3. Comparing the effects of SOM inputs and landmark detection threshold: mean scores and mean ranks. No statistically significant differences could be found between settings using a Tukey-Kramer comparison of rank means at significance level 0.05. See section 4.4 for a key of the group labels and an explanation of the parameters that were varied

Distance Score	II	I	VI	V	IV	VIII	III	VII
rank mean	266	265	255	249	181	174	173	170
score mean	0.91	0.91	0.91	0.89	0.88	0.86	0.88	0.85

Sequence Score	VI	II	VIII	V	IV	I	VII	III
rank mean	256	243	227	227	221	208	186	163
score mean	0.79	0.81	0.77	0.76	0.78	0.77	0.74	0.74

% LM Aligned	II	VI	I	V	III	IV	VII	VIII
rank mean	302	290	277	255	165	162	145	135
score mean	0.65	0.63	0.61	0.59	0.46	0.46	0.43	0.41

4.5   Selecting an Anticipatory Model

In this section we investigate how varying the complexity of the anticipation model affects alignment. We find that although there are differences between the various networks, there is no network that is best in all scores. We also find that, in general, the naïve predictor not only scores poorly, but also produces landmarks that are not very useful when looked at qualitatively. In general, we prefer the standard anticipatory model due to it's combination of simplicity and good performance.

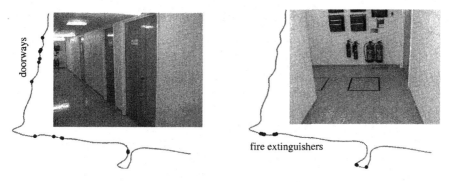

Fig. 2. An example of the differences in landmark selection between the *(left)* standard anticipatory model and the *(right)* GWR non-anticipatory model, $a_T =$ 0.85. All methods trained using the same parameters on the same data and tested on the same data (different from training set). Landmarks found are shown by black dots on the line of the route path. They correspond to physical features on the route: an alcove, fire extinguishers on the wall *(inset right)*, and a series of closed doorways *(inset left)*

Separate Kruskal-Wallis ANOVAs were run for each score, with the treatments being of the four anticipatory models described in section 2.1 plus the non-anticipatory GWR algorithm described in section 3. Discussion of the GWR results and the comparison of the two types of landmark detector is left to section 4.6. Blocking is no longer necessary since all tests are performed using the same alignment method (CDL, see section 4.3) and parameter settings (VI, see section 4.4). All ANOVAs rejected the null hypothesis at $p = 0.05$. A Tukey-Kramer multiple comparison of mean ranks at the same significance level can be seen in table 4.

Although all three networks were indistinguishable in terms of distance score, the naïve predictor had far worse distance performance. In sequence score, the naïve predictor comes out on top, with the networks behind. Finally, in terms of percentage of landmarks aligned, the naïve approach performs poorly compared to the other predictors. Because the naïve approach eliminates the neural network it lacks context both in terms of sensory prediction and also because the noise model of the Kalman filter changes, losing the prediction term. The combined effect leaves the naïve predictor finding landmarks only where very large single time-step changes in sensor values occur. Since such places are rare, it turns out that the naïve predictor finds about an order of magnitude fewer landmarks than the other anticipatory models. Additionally, the landmarks are very spread out; there are very few cases of such landmarks occurring in successive perceptions, and therefore little chance for the landmarks to be aligned out of sequence. Overall, the nature of the naïve landmark predictor is fairly unsuitable; the landmarks that it finds very large rapid changes of sensor value, such as cracks in the wall. Such landmarks are not sensed reliably in every run through the environment.

Table 4. Comparing the effects of different landmark selection models: mean scores, mean ranks, and statistically significant differences. Three versions of single-layer sensory anticipation network *(Standard, Recurrent, Lagged)* are compared along with a naïve anticipator that always guesses the current perception and a non-anticipatory landmark selector *(GWR)*. See section 4.5 for an explanation of the comparisons. All statistically indistinguishable groupings, indicated by underlines, are due to a Tukey-Kramer comparison of rank means $(p = 0.05)$.

Distance Score	Recurrent	Standard	Lagged	GWR	Naïve
rank mean	983	966	952	850	670
score mean	0.89	0.90	0.88	0.87	0.82

Sequence Score	Naïve	GWR	Lagged	Standard	Recurrent
rank mean	1378	1351	775	730	624
score mean	0.93	0.94	0.77	0.79	0.75

% LM Aligned	Recurrent	Standard	Lagged	Naïve	GWR
rank mean	1123	1096	951	479	236
score mean	0.56	0.55	0.49	0.29	0.18

4.6   An Alternative to Anticipation

In this section we compare a non-anticipatory, novelty filter approach (the GWR network described in section 3) to the previous landmark detectors. The novelty filter is shown to have a trade-off between generalisation and landmark detection. When the insertion threshold a_T is set high enough to ensure that at least a few landmarks are found, it tends to have too fine-grained a categorisation of the environment to reliably align itself.

The GWR network was investigated for several levels of node insertion threshold, between $a_T = 0.6$ and $a_T = 0.85$. The insertion threshold controls how similar two inputs have to be before the same node in the network will be used to represent them. A value close to one will require that each node matches the input very accurately, so that many nodes will be generated; a smaller value will generate fewer nodes to represent the same inputs. Informal experiments showed that low values of the insertion threshold do not discriminate between the sonar values seen during typical exploration by the robot, so that occasionally no landmarks at all are selected during a test run of over 200 m. We therefore used an insertion threshold of 0.85 to compare the performance of the GWR network with the various models described in the previous section.

The null hypothesis, that all anticipatory models and the GWR network perform the same at landmark detection, was rejected in section 4.5 and a comparison of score rank means can be seen in table 4. It can be seen that for distance and sequence scores the GWR performed comparably to the naïve predictor, but found very different landmarks. It is interesting to note that the naïve predictor finds very few landmarks, an average of 20, yet it is able to align them more often (% landmarks aligned) than the GWR network, which finds an average of 68 landmarks at $a_T = 0.85$. This may happen because the typical GWR size is 114 nodes, compared to just 20 in the case of the SOM. The GWR network finds a more fine-grained categorisation of the world, and therefore is less likely to find an exact category match for alignment. This hampers its performance in localisation using the CDL comparison from section 4.3. However, the best intuition as to what is going on can be had by examining figure 2. GWR only finds landmarks that have never been seen previously, like the fire-extinguishers on the wall. Naïve prediction (not shown in figure) finds only a single landmark at the entrance to the alcove, a place where there is a very large sensor value change between successive perceptions. But the full anticipatory model finds landmarks wherever perceptions change in relation to the recent history of perception, therefore landmarks are close enough together to be useful in correcting odometry drift.

4.7   Landmark Detection in Use: A Door-Finding Experiment

In this section we use the anticipatory landmark detector in a goal-oriented navigation task. We compare success rates at the task using D and CDL alignment methods for localisation, and show that CDL alignment performs better than D alignment, as predicted by section 4.3. We also note that the CDL alignment shows no obvious degradation of performance when the distance travelled is more than doubled.

The task is to start from either a close (about 22 m straight travel) or distant (about 68 m travel with two right-angle turns) starting point, navigating along a corridor with no possible branches along the route. At the open door, which is approximately 1.8 m wide, the robot must execute a 90 degree turn and attempt to find its way through the door using an obstacle avoidance behaviour. The robot performs a baseline run first, during which is was told to turn at the door by the experimenter. The robot then records the route description as the series of landmark categories perceived along the route with the relative distances between the landmarks and the action (such as 'wall-follow' or 'turn left') to perform at each landmark. The robot is then returned to the starting location and attempts to duplicate the route by using either the CDL or D alignment methods for localisation. In all, three baseline runs were performed in the short hallway, and three in the long. Each baseline run was followed by seven attempts at following the route directions to the door again. The number of successes in navigating through the door was recorded and the difference between the positions at which the robot stopped when going through the door were compared to the original training run.

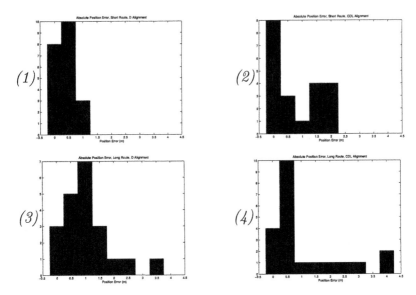

Fig. 3. Histograms of the absolute displacement error for each of the 21 following runs, for *(1)* D alignment over the short trip, *(2)* CDL alignment over the short trip, *(3)* D alignment over the long trip, and *(4)* CDL alignment over the long trip. A run with a displacement error of more than about 0.6 m would be very unlikely to make it into the doorway successfully

A Kruskal-Wallis ANOVA ($p = 0.05$) rejected the null hypothesis that the four possible treatments (CDL long, CDL short, D long, D short) were equivalent in success rate. A Tukey-Kramer comparison (see table 5) finds a significant difference between the long distance CDL and the long distance D alignments, which validates our earlier results that using category and odometry measurements aid robot localisation. In fact, this can be seen in the histogram of displacement errors in figure 3. The CDL alignment has a sharper peak near zero displacement error than the D alignment for both lengths of route. The amplitude of this peak is also unchanged with distance, unlike the D alignment. In both alignments though, the total spread of the errors is roughly proportional to the distance travelled, which is not surprising for the D alignment given our model of odometric error on the robot. In fact, one can fit a gamma distribution to the D alignment errors using maximum likelihood estimators and find that the 5% odometric drift model corresponds to roughly five standard deviations normalised against distance travelled. But the CDL alignment also displays this distance proportional error spread because if it misses making the category match it will continue until the odometry error model says the robot has gone past any reasonable hope of finding the doorway before stopping.

Table 5. A Tukey-Kramer comparison ($p = 0.05$) of rank means for the four cases of the door-finding experiment. Statistically insignificant difference between means are grouped together by the underlines. The CDL alignment method is found to function significantly better than the D alignment when the travel distance is longer.

Success rate	CDL long	CDL short	D short	D long
rank mean	50	46	42	32
score mean	0.66	0.57	0.48	0.24

5 Discussion

Given how much computational machinery is at work in the anticipatory land-mark selector, it is important to ask how much of it is actually necessary. There are three major components: a sensory anticipation model, a Kalman filter operating on the error of the sensory anticipations to detect landmarks, and a mechanism for producing categorisations of the resulting landmarks.

Firstly, the anticipation model needs to be of a useful complexity. Both here and in previous work [14, 6] it has been demonstrated that a single-layer network is sufficient for the sensor prediction. In fact, it is preferable to multi-layer networks, because the more complex model may learn to predict so well that there are never any landmarks [14]. In this paper, as in [14], it has been shown that variations in the network produce little difference in terms of score, but the network with lagged inputs does seem to learn to predict the environment better than the other two, producing fewer, better quality landmarks. In comparison, the naïve approach of always predicting the current perception is insufficient. It can only detect landmarks where perceptions change by a large amount very quickly; such landmarks are often not detected on every trip through an environment. Learning any kind of model of the environment allows for a better selection of landmarks.

The Kalman filter is necessary because peak detection on the model output produces an unwieldy number of poor quality landmarks [14]. The filter provides a way to remove error due to both sensor measurement variation and robot orientation. But could the Kalman filter operate directly on the sonar inputs, removing the need for the prediction network? The problem with this approach is that the sensor measurement variance must be modelled in the filter. In the anticipatory model the raw variance in the sensor measurements – an experimentally measurable quantity – is propagated through the non-linearity of the predictive network and the sum-of-squared errors to form a term of SSE variance due to sonar variation [6]. This value is always non-zero due to the summation of variance across all of the sonars. But a problem occurs if the filter operates directly on the sonars. The variance of the sonar perpendicular to the wall is almost zero, since sonars have very little measurement error for hard surfaces that have small angles of incidence to the beam. In the Kalman update equations, the gain will tend to unity and $\mathrm{Var}[\widehat{E}]$ will be close to zero, so that the filter would see every perception as a landmark unless some clever voting scheme were introduced that removed the effects of the sonar that is perpendicular to the wall.

Another good reason for using the filter is that both the recursive nature of the filter and the time dependency of the noise model (the sum-of-squared errors have terms at time t and time $t - 1$) provide additional context to the landmark detection system. The combination of context in the filter and context in the anticipatory sensor model dictate that landmarks selected with this system are distinct in comparison to recent landmarks, something that we believe to be especially important when using landmarks to localise. Assuming that the robot is in approximately the right area – something possible even with pure odometry over medium distances – accurate localisation is aided by individual landmarks being distinctive.

We used a SOM to categorise landmarks found by the anticipatory landmark detector because it is a well-known self-organising algorithm, but many other algorithms could be substituted to produce landmark categorisations. We could not find a definitive reason to prefer one choice of training perceptions or sensor inputs over another, in spite of the fact that using categories in the method of aligning and choosing landmarks radically improves their performance. It seems that the use of categories in alignment is more important than how those categories are created, although this has not been fully investigated. It does seem likely, however, that the number of categories has an effect on landmark alignment. A large number of categories would result in a very fine-grained categorisation of the environment, something equivalent to 'a T-junction with a scratch on the wall' rather than the more useful general category of 'T-junction'. Without a method to cluster similar categories the assumption that a landmark must exactly match in category will break down.

In general, the non-anticipatory GWR novelty filter performed on a similar level to the naïve anticipation model, but picked very different landmarks. One of the problems with this method of landmark selection is that when the insertion threshold is raised high enough to select a useful number of landmarks, the categorisation of the environment is very fine-grained, and therefore only a small percentage of the overall number of landmarks is actually aligned. As this method is not based on anticipation, any changes that the robot experiences that were not seen during training (for example, the robot travels closer to the wall that it did in training) will be detected as landmarks. An anticipatory system can deal better with these changes. Furthermore, this perceptual novelty system finds very different landmarks because it highlights locations that are different from the entire history of perceptions, rather than locations that differ given the recent context. This process potentially excludes a whole class of landmarks that could be useful: landmarks that are distinct compared to their surroundings even though they might be something that is familiar to the robot.

6   Conclusions

This paper has demonstrated that sensory anticipations, even very simple ones, can be valuable in selecting landmarks for a route-following system. It has presented both an anticipatory landmark detector that selects landmarks based on novelty compared to recent perceptions, and a non-anticipatory landmark detector that selects landmarks based on novelty compared to all previous perceptions. These two methods produce qualitatively and quantitatively different landmarks, resulting in different levels of landmark alignment when a robot tries to align it's current perceptions of a route with landmarks that it has seen when following the route previously.

In general, we would prefer to use the anticipatory landmark detector for a route-following task. It produces higher scores in our landmark alignments, perhaps because the perceptual novelty algorithm produces a finer-grained categorisation of the environment. The anticipatory method also has the theoretical advantage of being based on novelty in the local context, which can make the alignment of landmarks easier if the robot is looking for a match when it already knows that it is in the correct general area. We also find that the complexity of the predictive model must be greater than naïvely predicting the current perception, but we also know from previous experiments that a very complex model that can predict the environment very well will never have enough error to produce a landmark. The simple single-layer artificial neural networks presented here seem to offer a good compromise.

The paper also shows that combining category, distance, and odometric error information produces better alignments between landmarks in different runs – and therefore better route following – than merely following metric directions. This does not mean that more complex methods of localisation (e.g., multiple-hypothesis tracking or belief revision based on temporal evidence) are not necessary, but we do find that these simple methods can allow reasonable navigation success on routes on the order of 50 m.

Acknowledgements

Part of this work was supported by a Canadian & U.S. Citizens' Scholarship and an ATLAS grant from the Department of Computer Science at the University of Manchester. We would like to thank Tom Duckett, Ulrich Nehmzow, and David Gelder, who were involved in earlier publications on this work. We would also like to thank Susannah Lydon, Richard Craggs, and the anonymous reviewers for commenting on earlier versions of this paper.

References

[1] A. Baraldi and E. Alpaydin. Constructive feedforward ART clustering networks - part II. *IEEE Transactions on Neural Networks*, 13(3):662–677, 2002.

[2] A. Baraldi and P. Blonda. A survey of fuzzy clustering algorithms for pattern recognition: Part II. *IEEE Transactions on Systems, Man and Cybernetics - Part B: Cybernetics*, 29(6):786–801, December 1999.

[3] Eric Bourque and Gregory Dudek. On-line construction of iconic maps. In *Proceedings of the IEEE International Conference on Robotics and Automation (ICRA'00)*, pages 2310–2315, 2000.

[4] Martin V. Butz, Olivier Sigaud, and Pierre Gerard. Internal models and anticipations in adaptive learning systems. In *Lecture Notes in Artificial Intelligence*. Springer, this volume.

[5] Tom Duckett and Ulrich Nehmzow. Mobile robot self-localisation using occupancy histograms and a mixture of gaussian location hypotheses. *Robotics and Autonomous Systems*, 34(2–3):119–130, 2001.

[6] Jason Fleischer and Stephen Marsland. Learning to autnomously select landmarks for navigation and communication. In *Proceedings of the Seventh International Conference on Simulation of Adaptive Behavior*, pages 151–160. MIT Press, 2002.

[7] Myles Hollander and Douglas Wolfe. *Nonparametric Statistical Methods*. Wiley, New York, 1973.

[8] Dan-Anders Jirenhed, Germund Hesslow, and Tom Ziemke. Exploring internal simulation of perception in mobile robots. In *Proceedings of the Fourth European Workshop on Advanced Mobile Robots*, pages 107–113, 2001.

[9] R.E. Kalman. A new approach to linear filtering and prediction problems. *Journal of Basic Engineering*, 82:34 – 45, March 1960.

[10] Teuvo Kohonen. Self-organised formation of topologically correct feature maps. *Biological Cybernetics*, 43:59–69, 1982.

[11] David Kortenkamp and Terry Weymouth. Topological mapping for mobile robots using a combination of sonar and vision sensing. In *Proceedings of the Twelfth National Conference on Artificial Intelligence (AAAI'94)*, pages 979–984, Seattle, Washington, 1994.

[12] George Maistros, Yuval Marom, and Gillian Hayes. Perception-action coupling via imitation and attention. In *Proceedings of the AAAI Fall Symposium on Anchoring Symbols to Sensor Data in Single and Multiple Robot Systems*, pages 52–59, 2001.

[13] Stephen Marsland. Novelty detection in learning systems. *Neural Computing Surveys*, 3:157–195, 2003.

[14] Stephen Marsland, Ulrich Nehmzow, and Tom Duckett. Learning to select distinctive landmarks for mobile robot navigation. *Robotics and Autonomous Systems*, 37:241 – 260, 2001.

[15] Stephen Marsland, Ulrich Nehmzow, and Jonathan Shapiro. A self-organising neural network that grows when required. *Neural Networks*, 15(8-9):1041–1058, 2002.

[16] Volker Stephan and Horst-Michael Gross. Neural anticipative architecture for expectation driven perception. In *Proceedings of 2001 IEEE International Conference On Systems, Man & Cybernetics*, pages 2275–2280, 2001.

[17] Sebastian Thrun. Bayesian landmark learning for mobile robot localisation. *Machine Learning*, 33(1):41 – 76, 1998.

[18] Juha Vesanto, Johan Himberg, Esa Alhoniemi, and Juha Parhankangas. SOM Toolbox for Matlab 5. Report A57, Helsinki University of Technology, Neural Networks Research Centre, Espoo, Finland, April 2000.

[19] Daniel M. Wolpert and Zoubin Ghahramani. Computational priniciples of movement neuroscience. *Nature Neuroscience Supplement*, 3:1212–1217, 2000.

[20] Brian Yamauchi and Randall Beer. Spatial learning for navigation in dynamic environments. *IEEE Transactions on Systems, Man and Cybernetics Section B*, 26(3):496 – 505, 1996.

Representing Robot-Environment Interactions by Dynamical Features of Neuro-controllers

Martin Hülse, Keyan Zahedi, and Frank Pasemann

Fraunhofer Institute for Autonomous Intelligent Systems (AIS)
Schloss Birlinghoven, D-53754 Sankt Augustin, Germany
{martin.huelse, keyan.zahedi, frank.pasemann}@ais.fraunhofer.de
http://www.ais.fraunhofer.de/INDY

Abstract. This article presents a method, which enables an autonomous mobile robot to create an internal representation of the external world. The elements of this internal representation are the dynamical features of a neuro-controller and their time regime during the interaction of the robot with its environment. As an examples of this method the behavior of a Khepera robot is studied, which is controlled by a recurrent neural network. This controller has been evolved to solve an obstacle avoidance task. Analytical investigations show that this recurrent controller has four behavior relevant attractors, which can be directly related to the following environmental categories: free space, obstacle left/right, and deadlock situation. Temporal sequences of those attractors, which occur during a run of the robot are used to characterize the robot-environment interaction. To represent the temporal sequences a technique, called macro-action maps, is applied. Experiments indicate that macro-action maps allow to built up more complex environmental categories and enable an autonomous mobile robot to solve navigation tasks.

1 Introduction

There are now many attempts to increase the intelligence or cognitive abilities of autonomous systems like physical mobile robots. This seems to be desirable for many tasks one expects these systems to solve. Equipped with several types of sensors and with enough actuators they should be able to navigate and act in non-trivial changing environments. Sometimes they are assumed to develop also communication skills and some kind of social behavior which allows cooperative interactions - possibly with humans. In some sense they are often expected to mimic living systems.

This is of course a challenging perspective, and in general it is assumed that a better understanding of how such systems build up internal representations of their environment, how these representations can be modified during interaction with the environment, and how it can be adapted to a dynamical task management, are the prerequisites of this desired development.

On the other hand, using advanced dynamical neural networks for behavior control, it is quite unclear how an internal representation in a neuro-controller

M. Butz et al. (Eds.): Anticipatory Behavior ..., LNAI 2684, pp. 222–242, 2003.

will look like, i.e. on which level it will be implemented. It could be implemented as specific connectivity structure, as a weight matrix, as stationary states, or in terms of attractors of internal dynamical processes [25]. To approach these problems, it will be interesting to know, for instance, how motor commands can be mapped onto their sensory consequences or how a desired stream of sensor inputs can be accomplished by an appropriate sequence of motor commands [26], [15].

Following a modular neuro-dynamics approach, the basic assumption of this work is that cognitive performance is based on internal dynamical properties, which are provided by a recurrent connectivity structure of neural subsystems [2],[5],[11], [21]. An evolutionary robotics approach [18] is used to develop appropriate recurrent neuro-controller. But if we assume that higher cognitive abilities (e.g. planning tasks) need some kind of internal representation of the external world, one suggests to use those dynamical properties as basic elements for internal representations. The interaction of the robot is not only purely triggered by environmental conditions but is determined by specific internal dynamical features of its neural control system. This means in different situations different dynamical properties become active. Thus, it is suggested that those dynamical properties are the basic entities for a description of the robot's environment. Following this argumentation future prediction or expectations of robot behavior are state anticipations. If a state is interpreted as an specific "attractor of the robot-environment system", this implies that any goal-directed behavior can only be developed in a sensori-motor loop. Thus, the main focus of this work is to demonstrate how different dynamical features of neuro-controllers can be extracted during the interaction of the robot with its environment, how they can be used for classification of environmental properties, and how such categories can be utilized to encode and produce goal-directed behavior.

In the following investigations we concentrate, as a demonstration of method, on a simple example of robot behavior. An evolved recurrent neuro-controller is introduced which is able to endow miniature Khepera robots [16] with a robust obstacle avoidance behavior (section 2). The prominent feature of the used evolutionary algorithm ENS^3 [23] is its ability to evolve neural networks of general recurrent type without a specific connectivity structure determined in advance, which makes this algorithm similar to the GNARL algorithm [1]. It is thus mainly used for structure development, but it optimizes parameter values, like weights and bias terms at the same time. Only the number of input and output neurons of a controller are fixed according to a given sensor-motor configuration. Therefore resulting networks can have any kind of connectivity structure, including feedback-loops and self-connections.

In section 3 an implementation of a standard Braitenberg controller (BC) [7] and of an evolved neural network, called the minimal recurrent controller (MRC) [9], are used to analyze how the interaction of a robot with its different environments can be represented in terms of dynamical controller features. For this purpose methods like the *first return map* (FRM) of appropriate sensor and motor data are applied. In section 4 a method called *macro-action maps*

(MAM) is introduced to represent temporal sequences of specific dynamical features during the interaction of the robot. The specific dynamical features can be interpreted as "attractors of the robot-environment system", and the sequence of successively visited domains can be seen to represent relevant aspects of the environment and of the task, respectively. Some experiments are presented which illustrate how macro-action maps represent the robot-environment interaction and further on, how the elements of those maps can be used to build up more complex environmental categories. It is shown that those categories can enable an autonomous mobile robot to solve navigation tasks.

2   The Task

The task is to control a miniature robot, the Khepera, such that it can move collision free in a given environment with scattered objects; i.e., the classical obstacle avoiding task. To solve this task, eight infrared sensors can be used as proximity sensors, six at the front, two at the rear, and there are two wheels driven by two motors. The controllers only use two inputs, I_0 and I_1. At each time step t they serve as buffers for the average of the current values of the three left, respectively the three right front sensors. Data from the rear sensors of the robot are not used. Controllers will have two output units, O_0 and O_1, providing the signals driving the left, respectively right motor. The neurons of the controllers will be of the additive type; i.e. their dynamics is given by

$$a_i(t+1) = \theta_i + \sum_{j=1}^{n} w_{ij} \cdot f(a_j(t)) , \qquad (1)$$

where the activation a_i of neuron i at time $t+1$ is the sum of its bias θ_i and the weighted sum of the outputs $f(a_j)$ of the other neurons at time t, and w_{ij} denotes the strength of the connection from neuron j to neuron i. The transfer function f will be defined differently for both controllers. For the BC it is implemented as follows:

$$f(x) = \begin{cases} 0 & : & x < -1 \\ \frac{1}{2}(x+1) & : & -1 \leq x \leq 1 \\ 1 & : & x > 1 \end{cases}$$

For the MRC we use $f(x) = tanh(x)$ as transfer function. The connectivity of the controllers and a typical path of a simulated robot in one of the environments is shown in Figure 1.

2.1   The Braitenberg-Controller

In the case of the BC the sensor signals are pre-processed in such a way that controllers get input values $I_{0,1}$ between 0 and +1. They increase with decreasing distance between sensor and obstacle. In order to realize separate back- and forward movements for each wheel one needs positive and negative signals M_0

and M_1, driving the left and right motor. They are provided by a post-processing of controller outputs O_0, O_1 according to: $M_{0,1} = 2 \cdot O_{0,1} - 1$. Wiring and bias terms of the BC are inspired by [17]. Each controller output has excitatory connections to the sensors on its own side and inhibitory connections to the sensors on the opposite side. Thus, this implementation of the BC is a simple feed-forward network. The two bias terms realize a positive offset value for each motor, and therefore a forward motion is generated when the robot receives no sensor inputs. An input signal, I_0 say, steers the robot away from obstacles, as it inhibits the output unit O_1 and at the same excites the output unit O_0. 1.

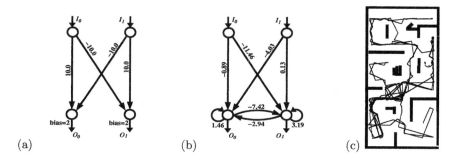

(a)   (b)   (c)

Fig. 1. Structure of the two controllers (a) BC and (b) MRC, which solve the obstacle avoidance task. Figure (c) indicates the behavior of a simulated Khepera robot controlled by the MRC.

2.2   The Evolved Minimal Recurrent Controller

The sensor signals for the MRC are pre-processed such that input values for I_0 and I_1 are mapped to the interval $[-1; +1]$. They increase with decreasing distance of an obstacle from the sensors. The interval $[-1; +1]$ for input values is chosen for consistency, because the MRC uses *tanh* as transfer function for the two output units. There are no bias terms, and no post-processing for the motor signals is necessary in this case.

Given these two input and output units, the evolutionary ENS^3-algorithm [23] is used to develop a controller achieving the desired properties without employing an internal unit. One of the results, shown in Figure 1(b), is used for the following discussions. This controller, the MRC, has positive self-connections (1.46 and 3.19) for its output units, which interact recurrently by inhibitory connections -7.42 and -2.94. It is well known from analytical investigations [20], [22] that single units with self-connections larger than $+1$ and 2-neuron loops with an even number of inhibitory connections can have co-existing fixed point attractors providing a hysteresis effect. The dynamical interplay of these three structures generates the advanced behavior of robots [9] as described in the following section.

2.3   Robot Behavior Controlled by BC and MRC

The BC and the MRC were introduced as networks which solve an obstacle avoidance task. But, the actual behavior is different. Figure 1(c) indicates a typical path of the robot generated by the MRC. It shows both, obstacle avoidance and exploration behavior. The behavior of the physical robot controlled by this network is comparable to that of the simulated one. Especially sharp corners and dead ends can be handled correctly, as indicated in the upper right corner of the environment in Figure 1c. In contrast to that the behavior of the BC is very different. Because it realizes an obstacle avoidance task in the sense that it turns to the right if an obstacle is detected on the left and the other way round. But in the case of deadlocks provoked by sharp corners or impasses it gets stuck and has no chance to escape autonomously from such situations. Hence one can say, the MRC shows a qualitatively improved behavior, since its interaction with the environment is context-sensitive in contrast to the pure reactive behavior of the BC. This better performance of the MRC is produced by the interplay of three different hysteresis effects generated by the already mentioned recurrent connectivity structure of the MRC. The hysteresis effect i.e. evoked by the positive self-connection on output unit O_0 (in the following denoted by LHE for *left-side hysteresis effect*) provides an appropriate turning angle for avoiding an obstacle on the right side. Whereas the self-connection on output unit O_1 generates a *right-side hysteresis effect* (RHE), which produces a right turn to avoid an obstacle on the left. The escape of the robot from dead ends is caused by the *extended hysteresis effect* (EHE), which lets the robot turn until it has a free space in front of it. The EHE results from an interplay of the LHE and the RHE which is mediated by the inhibitory ring of units O_0 and O_1 [9].

3   Sensori-Motor First Return Maps

In this section we will utilize the so called *first return maps* of sensor signals and motor signals, respectively, to represent the robot-environment interactions. These maps are defined as follows: The *sensor first return map* (S-FRM) plots the difference $\Delta_I := (I_0 - I_1)$ of the two inputs I_0, I_1 at time $t + 1$ over the difference at time t. Correspondingly, the *motor first return map* (M-FRM) plots the difference $\Delta_O := (O_0 - O_1)$ between the output signals O_0, O_1 to the left and right motors at time $t + 1$ over the difference at time t. We also make use of the sensori-motor map (SMM), which plots $\Delta_O(t)$ over $\Delta_I(t)$.

For these plots there are three characteristic points on the main diagonal: the lower left corner with coordinates $L = (-2, -2)$, the origin $(0, 0)$, and the upper right corner $R = (2, 2)$. For the S-FRM the points L and R represent a near obstacle at the right, and left side of the robot, respectively. The origin represents in general obstacle free space. For the M-FRM the points L and R represent left and right turns on the spot with maximal angular velocity, points on the main diagonal constant circular motion, and the origin represents straight movement along a line. Finally, a SMM represents the action following a sensor stimulus. The point L stands for a fast left turn of the robot if there is a near object at its right side, and R stand for a fast right turn if an obstacle appears at the left side. The origin codes the situation where there are identical left and right sensor inputs (they may, of course, be zero) and the robot reacts with a straight movement. Points in the upper left and lower right quadrants represent impossible situations like almost instantaneous jumps of objects from one side to the other (S-FRM), or instantaneous switching between left and right rotations (M-FRM), or undesirable reactions turns toward an object (SMM). Relevant quadrants for the discussion are therefore the upper right ($x > 0$, $y > 0$) and lower left quadrant ($x < 0$, $y < 0$), called R-quadrant and L-quadrant, respectively. Of course paths in the R-quadrant correspond to right turns, those in the L-quadrant to left turns.

For comparison, in Figure 2 these three types of first return maps are depicted for the Braitenberg controller and the minimal recurrent controller. These representations of sensori-motor signals clearly show the different control techniques applied by the two networks. Looking at the S-FRMs (figures (a)), one realizes that points accumulate around the main diagonal; i.e. successive differences in inputs change only gradually during time steps. According to the pre-processing of sensor data one observes that the BC receives smaller differences of input values than the MRC. Next consider the M-FRMs of the two controllers (figures (b)). The M-FRM of the BC appears as a more or less stretched version of its S-FRM in Figure 2(a), indicating roughly a pure reactive response of the controller to its sensor inputs. Whereas the M-FRM of the MRC shows a significantly different pattern. Both relevant quadrants show a sequence of points which can be described as leaf-shaped. These curves can be divided into upper and lower parts; i.e., parts above, respectively below the main diagonal. We will first concentrate on the L-quadrant.

Due to the obstacle avoidance task, for which the fitness function rewards straight movements of the robot, we expect that the evolved MRC tries to minimize the difference of the outputs as fast as possible. But a somewhat different behavior is observed and can be explained as a hysteresis effect. Whenever the output difference $\Delta_O(t)$ at time t grows (upper part), the output will grow steadily until it reaches its maximum. In fact, there are no shortcuts between the upper and lower path. Once it reaches the maximum difference, the MRC will steadily reduce the output difference until it reaches zero. Thus the origin and the point L are the only intersections of the upper and lower paths in the L-quadrant. The reason for this can be read from Figure 2(c) and will be discussed later in this section. If one looks at the R-quadrant, a very similar behavior is observed. The MRC steadily increases or decreases the difference of its output until it reaches one of the intersections points, R or the origin. But one can observe also an additional feature. The MRC allows a much stronger growth of differences in the R-quadrant. As will be shown later, it also keeps the maximum positive difference Δ_O for a longer time than the negative maximum difference. This is indication for the strategy to leave dead ends and sharp corners.

Next the SMMs of the controllers are compared (Figures 2(c)). The Braitenberg controller shows mainly three different activities corresponding to output

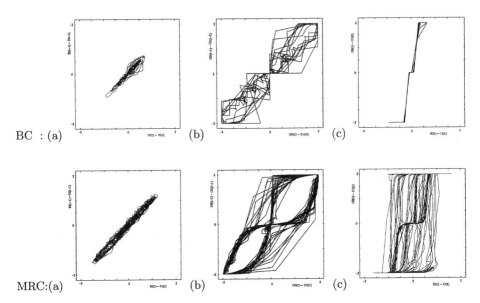

BC : (a)   (b)   (c)

MRC:(a)   (b)   (c)

Fig. 2. First return maps of the two controllers taken while the physical robot was moving in its environment: (a) first return map of the difference Δ_I of input values, and (b) of the difference Δ_O of output values; (c) the sensori-motor map plotting $\Delta_O(t)$ over $\Delta_I(t)$.

differences $\Delta_O = 1$, 0, and -1. We will start the analysis at the origin. If the input difference Δ_I increases to a certain threshold, the output difference Δ_O will jump to its maximum. If the difference of the input Δ_I is large and then slowly decreases until it falls below a positive threshold, then Δ_O will jump to 0. The same holds for negative input differences and the negative threshold value.

Analyzing the MRC, hysteresis effects are observed. The inner narrow loops in the R- and the L- quadrant correspond to hysteresis intervals for positive and negative input differences Δ_I. They appear as a "widening" of the jump lines of the BC. The figure shows, that for the MRC one has $\Delta_O \approx 0$ for a plateau of input differences Δ_I. If the positive input difference exceeds a certain positive threshold value, the Δ_O will jump to its maximum. If then Δ_I decreases, Δ_O will stay constant until Δ_I falls below a second, lower threshold value. For L-quadrant a similar behavior is observed. But there is a third hysteresis seen in Figures 2(c) for the MRC, which is remarkable. To see this, assume the system is in a state where obstacles are at equal distance from the robot and therefore it moves straight forward; i.e., we are in the origin of the SMM and the robot moves into a deadlock situation. If now the input difference Δ_I increases steadily the output difference Δ_O will then jump to a high value at some threshold, as discussed before. But if Δ_I is now decreasing, Δ_O can still remain stable at the high value, even if the input difference falls below the negative threshold value for the inner hysteresis. It is exactly this feature which is needed to handle dead ends and sharp corners. If the robot is in a sharp corner, it needs to turn in one direction, even if the difference in inputs decreases significantly. Else the robot would try to avoid the side on which its input is higher, thereby turning towards the other side with the lower sensor input. Then the higher input value would decrease, while the lower would increase, until the situations is reversed and the direction of rotation, correspondingly. The resulting behavior is shown by the Braitenberg controller. Hysteresis is the solution to this deadlock situations, and it is provided by the recurrent structure of the MRC as demonstrated in [9]. To emphasize this result a 3-dimensional plot of the SMM is plotted adding the time axis (Figure 3). The first two maximum differences (corresponding to L- and R-points) show how the MRC reacts to corners. The last maximum difference shows how the MRC reacts to a dead end or a sharp corner. The MRC provides maximum output difference Δ_O for a longer time (as can be seen from the 2-dimensional projection in the lower figure) until the robot can go straight ahead again. Observing the physical robot, one can see clear 180-degree turns in such dead end situations. Figure 4 demonstrates the robustness of the MRC. During the interaction of the robot most of the time the control signals lie in those domains, which produce precise actions like moving straight (40%), left turns (15%), and right turns (5%).

The comparison of the control strategies, as reflected by Figure 2, explains and supports our hypothesis, that the BC shows a purely reactive behavior, whereas the MRC shows a qualitatively improved behavior.

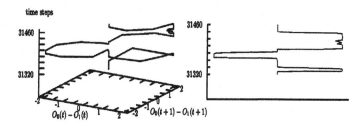

Fig. 3. The time development of the MRCs first return map of output differences $\Delta_O(t)$, and its projection onto the $(\Delta_O(t), t)$ plane (right).

4   Macro-Action Maps

As outlined in the last section, for the MRC there are four discernable robot-environment interactions which are clearly related to the internal dynamics of the controller. The two small hysteresis effects, corresponding to the co-existence of two fixed point attractors, become active during simple left or right turns. The third hysteresis effect with its extended input domain is activated in typical deadlock situations, whereas simple straight forward movements in obstacle free

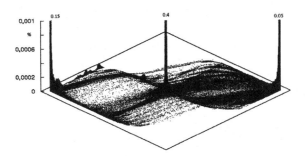

Fig. 4. The relative appearance of points in the MRCs first return map of output differences Δ_O, indicating three main actions represented by the points $\Delta_O = +2, 0, -2$.

space indicate a unique stable fixed point. Figures 2(b), 3, and 4 illustrate that these different interaction states are easy to distinguish if in addition to the difference of the motor outputs Δ_O also its time development is plotted. The following table shows the relations between features of the internal dynamics, difference in controller output, and the external situation:

internal dynamics	Δ_0	time steps	external world
unique stable fixed point	0	any	no obstacles
LHE	−2	< 20	obstacle on the right
RHE	+2	< 20	obstacle on the left
EHE	±2	> 20	sharp corners, impasses, etc.

Table 1. Relation between internal dynamics of the MRC and the external world.

4.1 Definition and Implementation

One possible way to implement a temporal segmentation of different robot-environment interaction states and its graphical representation is based on the following definitions.

A *macro-action* [{+/-}d, s], $d, s > 0$, describes a rotation of d time steps to the right (+) or left(-) followed by a positive straight movement of s time steps. We call {+/-} the *sign of the macro-action*. A sequence of turns with no straight forward movement between them are summarized to one turn. $\mathcal{A}$ is the set of all macro-actions. A *macro-action map* (MAM) is defined as a directed graph with nodes representing macro-actions and the directed edges indicate the temporal predecessor-successor relation between two macro-actions.

The development of such a MAM is inspired by the work on landmark-based navigation [12]. The robot segments its path through the environment according to landmarks. These landmarks are kept in a simple, self-organizing chain representation. Here, instead of landmark macro-actions are used. During interaction the emerging macro-actions are appended to the list of macro-actions. Usually this list would be temporally ordered. But before a new macro-action is added, it is tested if there exists a similar node in the list. If a similar node is found, values of the already existing node in the list will be updated with a value, which is the mean of the new and the already existing node. If there are more than one similar nodes in the list, only the first in the list will be changed. The predicate $S : \mathcal{A} \times \mathcal{A} \rightarrow \{0,1\}$ of similarity between two macro-action $[d_1, s_1]$ and $[d_2, s_2]$ is defined as follows:

$$S([d_1, s_1], [d_2, s_2]) = 1 \iff (\text{sign}(d_1) \cdot \text{sign}(d_2) = 1) \wedge (|d_1 - d_2| \le 5) \quad (2)$$
$$\wedge \left(|s_1 - s_2| \le 5 \vee |s_1 - s_2| \le \frac{1}{2}(s_1 + s_2) \right).$$

Finally a predecessor-successor relation will be established between the last node in the list and the already existing similar node. In such a way predecessor-successor relations between macro-actions are established. Iterating this procedure while the robot is moving in its environment, a MAM unfolds, which represents, on the level of attractor sequences, the interaction of the robot. This procedure is fulfilled as long the number of nodes does not increase anymore. Then the robot is stopped and nodes and relations in MAM are deleted if they are updated relatively seldom with respect to the total number of nodes in the MAM. Updates are counted for each node and relation. In the following experiments we delete nodes if their update rate was less than 5 %, and relations if their update rate is less than 2.5 %.

4.2   Experiments on Building Up MAMs

The following experiments will make clear how specific features of the robot environment are represented by the macro-action maps. Some examples of such MAMs and the related interactions are shown in Figure 5, 6, 7 and 9. The nodes of the graphs in these figures are labeled by the corresponding macro-action and a number, which is the absolute number of its occurrences during which the MAM was developed. Correspondingly, the numbers on the edges indicate the absolute number of occurrence of this sequence of macro-actions.

For the first experiment (Figure 5) the MAM consists of only one node and one edge. In this case the world is square shaped with no obstacles, and it can be argued, that the constant rotation angles lead to constant driving distances and vice versa. Therefore the interaction of this world with this MRC is characterized by a temporal constant sequence of two interaction states: turning and driving straight. They are characterized by the left hysteresis effect (LHE) and the unique fixed point attractor (compare Table 1).

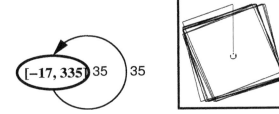

Fig. 5. Right: robot path controlled by the MRC in square shaped world. In this simple world a very simple macro-action maps consisting of only one node and one edge is developed (left side).

The same holds for the second experiment (Fig. 6) for which the world is chosen to be rectangular. In contrast to the first experiment there are now two different distances which the robot can move straight forward after a simple turn. Like in the first experiment the turning angle remains constant, but in this rectangular world the robot has to cover a short and a long distance until the next obstacle. Therefore the macro-action map representing the robot interaction in this experiment consists of two nodes and edges. This can be interpreted as the alternate appearance of the two macro-actions [-18, 57] and [-17, 423] during the interaction. The first indicates a turn to the left followed by a short forward move and the second also a turn to the left but followed by a longer forward movement.

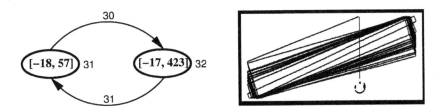

Fig. 6. Robot path controlled by the MRC in rectangular world (right side). The two different distances which can be covered by robot before an obstacle appears are indicated by the two nodes in the corresponding macro-action map on the left.

A more interesting experiment is shown in Figure 7. Here the robot is confronted with two sharp corners: the bottom left corner and the upper right corner. In the corresponding MAM these two corners are represented by one and the same node ([47,401]). This is because in the sharp corners only the EHE becomes active which produces the large turning angle with the same sign. Referring only to this single node the robot therefore can not discriminate by itself in which of these two sharp corners it is located.

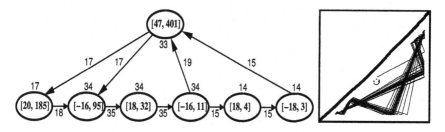

Fig. 7. Robot path controlled by the MRC in triangular world (right side) and the corresponding macro-action map on the left side. See text for explanations.

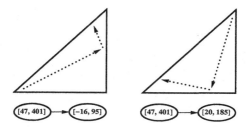

Fig. 8. Subpathes of the robots interaction in the triangular world an the corresponding macro-action sequences, which are also subsequences of the MAM in Figure 7. The first node / macro-action refers to both sharp corners, but the successor node indicates which corner exactly the robot comes from.

But if the temporal sequence of the macro-actions is taken into account then these two sharp corners became distinguishable. As it is pictured in Figure 8 the macro-action sequence [47, 401] [20, 185] indicates the robots way out of the bottom left sharp corner and the sequence [47, 401] [-16, 95] the way out of the upper right sharp corner. Because after the escape of the bottom left corner the next obstacle is always avoided by a turn to the left. Whereas after an escape of the upper right corner the next obstacle is always avoided by a turn to the right.

The last experiment in this section is an example for the fact that simple worlds can nevertheless produce complex MAMs. Figure 9 shows the environment of this experiment, which is a small corridor. One might suggest that the corresponding MAM contains only one node, as in the first experiment. But, the robot controlled by the MRC moves hardly parallel to the walls, and therefore it will oscillate between the two walls of the corridor. This means many turns and therefore many macro-actions are performed during the interaction. And finally it can be observed at the end of the corridor that the robot usually turns back in two steps. In the corresponding MAM this is represented by the two macro-action sequences [28,8] [18,21] and [-26,4] [-12,46].

Traveling through the corridor is characterized by oscillations as it is depicted in the macro-action map of Figure 9 by the loop between the two macro-actions

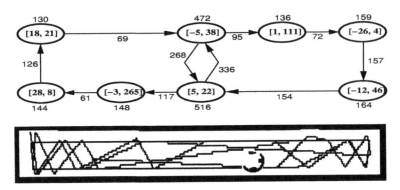

Fig. 9. Robot path controlled by the MRC in a small corridor (bottom) and the corresponding macro-action map on the top. See text for explanations.

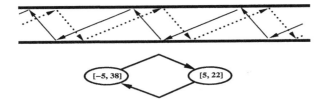

Fig. 10. Traveling through the small corridor is characterize by a zigzag move, because the robot moves hardly parallel to the walls. So, the robot-environment interaction in this experiment is mainly described by small turning angles followed by short straight forward moves. This is related to the substructure of the MAM in Figure 10 (bottom) with only two nodes and edges, because left and right turns are alternately performed.

[-5,38] and [5,22]. The relation between these two nodes and the movement through the corridor is demonstrated in Figure 10. The most frequent occurrence of this oscillation in this experiment is indicated by the large numbers of the corresponding nodes and edges in Figure 9.

Another point of the MAM in this experiment is, that the two nodes [-5, 38] and [5, 22] representing the zigzag move through the corridor are the only nodes, which have two successor nodes. It is easy to see in Figure 11 and 12, that for each node the second successor node reflects the "experiences" of the robot, that in narrow corridors smaller turning angles to avoid collisions lead to longer straight forward movements.

4.3 Experiments on Using MAMs for Exploration, Homing, and Navigation

In the previous section the reader saw how MAM can be used to characterize the robot–environment interaction. The next three experiments demonstrate how

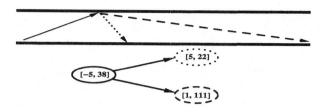

Fig. 11. In the MAM of the experiment with the small corridor (Fig. 9) the node [-5, 38] has the two successor nodes [5, 22] and [1, 111]. The paths which correspond to these macro-actions are shown, line style refers to the same node / macro-action in the subgraph at the bottom of this figure. The two successor nodes or macro-actions are mainly distinguished by their "turning angles" 5 and 1. It is easy to see in this figure, that a smaller turning angle leads in a corridor to a longer straight forward movement.

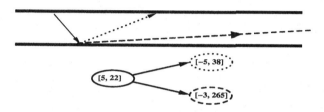

Fig. 12. Like Figure 11, but according to node [5, 22] the successor node [-3, 265] in comparison with successor node [-5, 38] reflects the fact that smaller "turning angles" lead to longer straight forward movements.

MAMs can be used to encode behavioral sequences, which overcome the limited interaction of the MRC. In the following it is shown that a MAM based categorization of environmental states can provide exploration, homing and navigation capabilities.

First of all a few definitions must be introduced in this section. As it was pointed out earlier, a MAM is a directed graph $G = (V, E)$, with $v \in V$ representing the macro-actions and $e \in E$ representing the predecessor-successor relationships between the macro-actions. A *path* of length k from a macro-action v_0 to a macro-action v_k in the directed graph $G = (V, E)$ of the MAM is a sequence $\langle v_0, v_1, \ldots, v_k \rangle$ of macro-actions such that $(v_i, v_{i+1}) \in E$ for $i = 0, \ldots, k-1$. The length of the path is the number of relations. We also have to define the similarity of macro-actions and paths of macro-actions. Two macro-actions are *similar* $(v_0 \approx v_1)$ if they satisfy the relation given in equation 2. A path p of macro-actions is a subset of a MAM $G = (V, E)$ if $\forall v_p \in p \; \exists v_i \in V : v_p \approx v_i$ and for each relation between to macro-actions in the path, there exists a corresponding relation between two similar macro-action in the MAM. A corresponding relationship does not have to have the same occurrence label; i.e.

$$p \subset G \iff \forall (v_p, v_{p+1}) \in p \; \exists e = (v_i, v_{i+1}) \in E : v_p \approx v_i, v_{p+1} \approx v_{i+1}, v_i \in V.$$

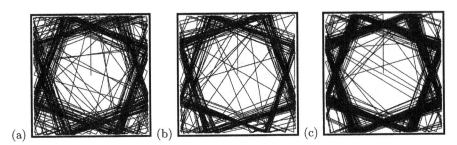

Fig. 13. Robot path controlled by the MRC, but including a random turn if the last two (a) / four (b) / ten (c) macro-actions are similar. (100,000 time steps are plotted in Fig. (a) and (b) and 250,000 in (c).)

The first experiment in this section demonstrates how the robot controlled by the MRC can develop an exploration behavior. We assume, that a good exploration behavior is shown, if the robot's path through the world covers up most of the free space. This is done using the square shaped environment as shown in Figure 5. As it was already mentioned the robots interaction in this world is represented by a MAM with only one node and one edge. That means, during the interaction only one macro-action emerges. During the interaction of the robot with the environment, a new path of macro-actions (in this case with $v_p = v_q, \forall v_p, v_q \in p$) is generated. To equip the robot with exploratory abilities the control of the interaction by the MRC is interrupted, if a path p of a certain length k of newly recorded macro-actions is a subset of the MAM. The interruption in this case means that the robot will turn a random angle and then be controlled by the MRC again. Figure 13 reflects the results of three experiments of this kind. The difference between these three experiments is the value, which determines how many of the last macro-actions have to be similar before a random turn is triggered.

The results of the three experiments show that after the robot is rotated by a random angle it does not take a long time to reach the old pathway, as shown in Figure 5. But in contrast to this figure the paths plotted in the Figures 13(a) – (c) indicate that there are two such stable pathways in this square shaped world. Depending from the start position and orientation the robot runs clockwise or counterclockwise through this world after a while. This determines in which of this two stable pathways the robot will end. Going back to the exploration abilities, one can see that in all three experiments the robot covers the whole area. The difference is according to the length k of the subset of macro-actions, that have to occur, before a random turn can be performed.

In the next experiment a homing behavior using MAM was implemented. In a triangular world (like Figure 7) the robot has to recognize only one of the two sharp corners as its home and wait there for a while. As it was already argued in the corresponding experiment of the pervious section, each sharp corner in this world is represented by the same node in the MAM. Therefore the robot has to distinguish the two sharp corners by a precedent sequence of macro-actions. In detail, additionally to the pure MRC control a *attention mode* is implemented, which is deactivated by default but becomes active, if a specific sequence of two macro-actions occurs. Finally, if the attention mode is active and a macro-action corresponding to an extended hysteresis effect (EHE) occurs, then the robot stops for a certain time. After the waiting time is over, the attention mode is deactivated and the robot is controlled again by the MRC. Thus, "home" is not determined by the single node representing both sharp corners but is distinguished by a macro-action sequence activating the attention mode. For example, if the upper right sharp corner is defined as "home" (compare Fig. 7), then a macro-action sequence similar to [47, 401][-16, 95] activates the attention mode. The sequence [47, 401][20, 185] defines the the bottom left corner as "home".

Fig. 14. Homing behavior experiment with the miniature robot Khepera in a triangular world, like Fig. 7. One of the two sharp corners in this world is defined as home. If the robot moves in this "home corner", it has to stay there for a certain time.

This experiment has been done with a simulated and a real Khepera robot testing each corner separately as home. A mpeg-video, which can be downloaded from [10], shows two experiments with the real Khepera robot. Figure 14 is a screen shot of the first experiment in this movie where the "upper right" corner (according to Figure 7) is defined as home. Both experiments show the following fact: the robot only stops in the correct corner, but it also passes the correct corner. This is because, to recognize the sharp corner, which is defined as home, a similar sequence of the corresponding macro-action has to occur. In detail this means, that the robot can not take a look at the currently appearing sensor values, because they are ambiguous. In fact, a sequence or a path of macro-actions must be recorded in order to find the correct corner. This means that a certain level of robot-environment interaction has to be performed, as the MRC has a limited sensory system. A robust discrimination between the two sharp corners can only be based on the anticipated macro-action sequence, which emerge from a specific robot-environment interaction.

In the last experiment of this section the MAM is applied to solve a navigation problem. Figure 15(a) shows the path of a robot through a small corridor, which is identical to the experiment in Fig. 9. As it was already mentioned in the pervious section, the movement through the corridor is mainly characterized by zigzag movements. But the MAM of this experiment reflects also "experiences" of the robot that small turning angles lead to longer distances before a new avoidance action has to be triggered. In this experiment it was tried to reduce the zigzag movements and bring the robot to straight movements through this corridor using the corresponding MAM in the following way.

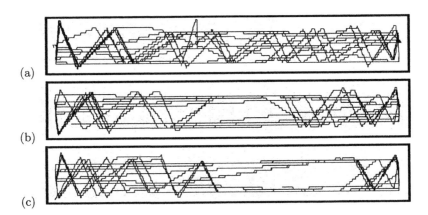

(a)

(b)

(c)

Fig. 15. Three robot paths through the small corridor. Figure (a) shows the incisive zigzag movements of the robot. The two other figures show more straight movements through the small corridor - mainly in the middle.

Like in the two experiments before the occurring macro-actions are monitored during the interaction. The control of the robot by the MRC is interrupted, if a macro-action occurs, which is similar to those macro-actions representing the oscillational behavior in the corridor (see Fig. 10). During this interruption the robot is turned 1 step to the right, if the last macro-action is similar to [-5, 38]. This is defined according to its second successor node [1, 111] in the MAM (Fig. 9 and 11). Because this second successor node refers to a longer distance before an avoidance behavior is triggered. Accordingly the robot is turned 3 steps to the left, if the last macro-action is similar to [5, 22]. Because in this case those 3 turn steps to the left refer to the second successor node [-3, 265] of node [5, 22] (Fig. 9 and 12), which again refers to a longer straight movement distance. After this rotation the control is given back to the MRC.

Figure 15(b) and (c) show the results of two runs of this experiment. The zigzag movements are still there, mainly at the ends of the corridor. This is not surprising, because first of all the specific macro-action must occur to "recognize" the small corridor and to trigger the smoother turns. In detail, the dynamics of the controller has to be in or near enough to the corresponding attractor. But in the middle of the corridor straight movements are the majority. Whereas in the experiment, where the robot is only controlled by the MRC (Fig. 15(a)), the zigzag movements are predominant.

5   Discussion

In this article a basic robot behavior - moving in a given environment by avoiding scattered objects - was chosen to demonstrate two methods for representing robot-environment interactions by dynamical features of neuro-controllers. The first method, based on first return maps of sensori-motor data, revealed a clear difference in control techniques between a standard feed-forward type of neuro-controller, the BC, and a more sophisticated one, the MRC, which makes extensive use of the recurrent connectivity of its motor neurons. The resulting differences in robot behavior becomes obvious especially in deadlock situations. In particular, the first return map (M-FRM) of the motor data and the combination (SMM) of sensor and motor data appeared to be most instructive. They already indicate the existence of four fundamental types of robot-environment interactions, which are constituted by the neural structure of the MRC.

Based on Table 1 the MAMs can be seen to indicate the temporal sequence of internal dynamical features of the MRC, which become effective during the interaction of the robot with its environment. Basically each macro-action for the MRC relates to a hysteresis interval of sensor inputs followed by a dynamics determined by a global fixed point attractor. The number of time steps of the actual rotation allows to distinguish between hysteresis effects over small and extended input domains. Finally, the sign of a macro-action indicates left and right turns; i.e., left and right hysteresis effects. Therefore one can use the MAMs as a representation of the objects or specific features of the world as they occur during the interaction of the robot with its environment. According to the four different dynamical properties of the MRC this representation can only refer to categories like *obstacle on the left*, *obstacle on the right*, *deadlocks*, and *free space*. This is an essential point, with respect to the landmark-based navigation approach. Landmark types are defined by human designers [12] whereas the MAM approach only takes into account the perception of the robot.

But the experiments in section 4 show that other categories of environmental conditions can be built, if substructures of MAMs are taken into account. The experiments in section 4.3 indicated that MAMs can provide homing and navigation tasks. The homing experiment demonstrated a method for distinguishing different objects in the environment based on macro-action sequences of a MAM. In addition the implementation of this experiment on a physical mobile robot shows that MAMs are able to handle noisy and discretized environments. Furthermore the navigation experiment introduced a MAM based classification of "small corridors". In addition, this classification was used to trigger specific actions in such a way that the number of turning actions (or zigzag movements) was decreased. Although in this case the improved action-selection process is "hard-wired", the trigger conditions and the triggered action were defined according to the underlying MAM. Therefore it is argued, that MAMs can be seen as basic entities for an anticipation process, on which planning tasks can be realized.

It was shown that the MAM approach implies the usage of regularities arising from the sensori-motor interaction of the robot-environment system. That is, because the basic entities of the MAMs, the macro-actions, are the behavior relevant dynamical features of the MRC, which become active in different environmental situation. But it was also shown that this implies, any object categorization (like a specific sharp corner or a small corridor in section 4.3) is a result of a performed interaction. An interaction is always specified by the environment, the robot as an embodied systems, and a controller. Therefore, the MAM approach is conform with approaches, which assume that perception and interaction can not be separated, and the generated behavior influences the perception process ([3], [4], [13], [15], [19]).

As was mentioned in the introduction, this contribution is intended to be a demonstration of method. In fact the presented MAM approach is here based only on the MRC. This controller was evolved for obstacle avoidance only and therefore the MAM is strongly dependent on this system and task. Especially the definition of the mapping (Table 1) between internal dynamical properties of the MRC and its output signals depend on this system. Using our modular neuro-dynamics approach to behavior control of autonomous agents this work is understood as a first step to equip autonomous agents with self-generated internal representations of behavior relevant properties of the external world. Other tasks and other robot platforms will in general require other controllers with different dynamical properties. Hence these systems will need other mappings to generalize temporal sequences of dynamical features during the interaction with the environment.

The experiments have shown that MAMs can become very complex, when representing the full sequences of the interactions. But the homing and the navigation experiments suggest also that a representation of the whole interaction sequence is not necessarily needed to solve a specific problem. In fact, homing and navigation tasks use only specific substructures of the underlying MAM. Therefore future work on MAMs has to include mechanisms which can focus on or separate relevant substructures of the MAM.

But the question which substructures of a MAM or which internal representation of the environment will be necessary and sufficient for the agent is then of course task dependent. Therefore the agent must be able to evaluate the MAM by itself with respect to a given problem or task. Following [6] we argue, that this evaluation of the agents internal representation has to include the agents anticipated and the actual interaction outcomes. The elements of the MAM are directly related to the external world and accessible by the agent and therefore a promising starting point to endow autonomous agents with self-generated internal representation to improve their behavioral repertoire.

References

1. Angeline, P.J., Saunders, G.B. and Pollack J.B. (1994), An Evolutionary Algorithm that Evolves Recurrent Neural Networks, *IEEE Transactions on Neural Networks*, **5**, 54-65.
2. Arbib, M.A., Erdi, P. and Szentagothai, J. (1998), *Neural Organization: Structure, Function, and Dynamics*, Cambridge, MA, MIT Press.

3. Bajcsy, R. (1988), Active Perception, *Proceedings of the IEEE*, 76: 996 – 1005, 1988.
4. Ballad, D.H. (1991), Animate vision, *Artificial Intelligence*, 48:57 – 86.
5. Beer, R., (1995) A dynamical systems perspective on agent-environment interaction, *Artificial Intelligence*, 72(1), pp. 173 – 215, 1995.
6. Bickhard, M. H., Treveen, L. (1995) *Foundational Issues in Artificial Intelligence and Cognitive Science*, Elsevier Scientific, Amsterdam, 1995.
7. Braitenberg, V. (1984), *Vehicles: Experiments in Synthetic Psychology*, MIT Press, Cambridge, MA.
8. Harnard, S. (1990), The symbold grouding problem, *Physica D*, 42, pp. 335–346.
9. Hülse, M., Pasemann, F. (2001), Dynamical Neural Schmitt Trigger for Robot Control, J. R. Dorronsoro(Ed.): ICANN 2002, LNCS 2415, Springer Verlag Berlin Heidelberg New York, pp. 783–788, 2002.
10. Hülse, M.: Implementation of homing behavior based on a recurrent neuro-controller and macro-action maps, MPEG video, http://www.ais.fraunhofer.de/INDY/aml/X/MRChoming.mpeg, 12.12.2002.
11. Krichmar, J. L., Edelman, G. M., (2002), Machine Psychology: Autonomous Behavior, Perceptual Categorization, and Conditioning in a Brain-Based Device, *Cerebral Cortex*, vol. 12, pp. 818 – 830, 2002.
12. Matarić, M.J. (1994), Navigating With a Rat Brain: A Neurobiologically-Inspired Model for Robot Spatial Representation, *Proceedings of the International Conference on Simulation of Adaptive Behavior: From Animals to Animats 3*, 282–290, 1994.
13. Metta, G., Fitzpatrick, P. (2002), Better Vision Through Manipulation, In: *Proceedings of the Second International Workshop on Epigenetic Robotics: Modeling Cognitive Developement in Robotic Systems*, Prince, C. G., Demiris, Y., Marom, Y., Kozima, H. and Balkenius, C. (Eds.), Lund University Cognitive Studies, 94, 97 – 104, 2002.
14. Michel, O., *Khepera Simulator* Package version 2.0: Freeware mobile robot simulator written at the University of Nice Sophia-Antipolis by Oliver Michel. Downloadable from the World Wide Web at http:// wwwi3s.unice.fr/ ∼om/ khep-sim.html
15. Möller, R. (1999), Perception Through Anticipation - A Behavior-Based Approach to Visual Perception. In: *Understanding Representation in the Cognitive Sciences* (A. Riegler; A. von Stein; M. Peschl, eds.), Plenum Press, New York, 1999.
16. Mondada, F., Franzi, E., Ienne, P. (1993), Mobile robots miniaturization: a tool for investigation in Control Algorithms. In *Proceedings of ISER' 93*, Kyoto, October 1993.
17. Mondada, F., Floreano, D. (1995), Evolution of neural control structures: Some experiments on mobile robots. *Robotics and Autonomous Systems*, 16:183 – 195.
18. Nolfi, S., Floreano, D. (2000), *Evolutionary Robotics: The Biology, Intelligence, and Technology of Self-Organizing Machines*, MIT Press, Cambridge.
19. Nolfi, S., Marocco, D., Active Perception: A Sensorimotor Account of Object Categorization, *Proceedings of the 7th International Conference on Simulation of Adaptive Behavior: From Animals to Animats 7*, 266–271, 2002.
20. Pasemann, F. (1993), Dynamics of a single model neuron, *International Journal of Bifurcation and Chaos*, 2, 271–278.
21. Pasemann, F. (1995), Neuromodules: A dynamical systems approach to brain modelling. In Herrmann, H., Pöppel, E. and Wolf, D. (eds.), Supercomputing in Brain Research - From Tomography to Neural Networks, Signapore: World Scientific, pp. 331-347.

22. Pasemann, F. (1995), Characteristics of periodic attractors in neural ring networks, *Neural Networks*, 8, 421-429.
23. Pasemann, F., Steinmetz, U., Hülse, M., and Lara, B. (2001), Robot control and the evolution of modular neurodynamics, *Theory in Biosciences*, 120, 311–326.
24. Tani, J. (1998), An Interpretation of the "Self" From the Dynamical Systems Perspective: A Constructivist Approach, *Journal of Consciousness Studies*, 5(5-6), 1998.
25. Tani, J., Sugita, Y. (1999), On the Dynamics of Robot Exploration Learning, *Proc. of 5th European Conf. of Artificial Life (ECAL99)*, 279-288, 1999.
26. Wolpert, D.M., Ghahramani, Z., Flanagan, J.R. (2001), Perspectives and problems in motor learning, *Trends in Cognitive Science*, 5(11): 487-494.

Anticipatory Guidance of Plot

Jarmo Laaksolahti and Magnus Boman

Swedish Institute of Computer Science
Box 1263, SE-164 29 Kista, Sweden
{jarmo,mab}@sics.se

Abstract. An anticipatory system for guiding plot development in interactive narratives is described. The executable model is a finite automaton that provides the implemented system with a look-ahead. The identification of undesirable future states in the model is used to guide the player, in a transparent manner. In this way, too radical twists of the plot can be avoided. Since the player participates in the development of the plot, such guidance can have many forms, depending on the environment of the player, on the behavior of the other players, and on the means of player interaction. We present a design method for interactive narratives which produces designs suitable for the implementation of anticipatory mechanisms. Use of the method is illustrated by application to our interactive computer game Kaktus.

1 Introduction

Interactive narrative is a form of entertainment that invites users to step into and interact with a fictive world. In contrast to traditional non-interactive narratives, participants in an interactive narrative are active in the creation of their own experiences. Through their actions, players interact with other agents (some of which might be artificial) and artifacts in the world—an experience that can be compared to role-playing or acting. Interactive narrative promises to empower players with a greater variety of offerings but also the capacity to deal with them [26], leading to deeper and more engaging experiences.

Stories may be told in many different ways, depending on the order in which events are disclosed to the player. Plot control concerns itself with deciding which event of a story to present next, not to create entirely new stories (cf. [29]). The player should have the feeling that anything may happen while being nudged through the story along various story arcs [11]. At the same time the player should feel that the choices she makes has a non-trivial impact on how the plot unfolds.

Plot guidance involves searching among a possibly huge amount of unfoldings for one that fulfills (some of) the author's intentions for the story (cf. [28]). The search of the state space quickly becomes intractable, however, as the number of scenes grow. For a scenario consisting of as little as 16 scenes (with a normal movie having 40-60 scenes) search would have to consider billions of states. Limiting the depth of the search can reduce the size of the state space but can also result in bad plot unfoldings (i.e., the player can get stuck).

M. Butz et al. (Eds.): Anticipatory Behavior ..., LNAI 2684, pp. 243–261, 2003.
© Springer-Verlag Berlin Heidelberg 2003

We hypothesize that anticipatory systems can provide an alternative means to efficient plot guidance. Robert Rosen suggests that the simplest way for anticipations to affect the properties of a dynamic system S through a model M is the following:

> Let us imagine the state space of S (and hence of M) to be partitioned into regions corresponding to "desirable" and "undesirable" states. As long as the trajectory in M remains in a "desirable" region, no action is taken by M through the effectors E. As soon as the M-trajectory moves into an "undesirable" region [...] the effector system is activated to change the dynamics of S in such a way as to keep the S-trajectory out of the "undesirable" region. [22] (p.247)

The system does not have to consider full trajectories but only those that start in the current state of S and stretch some time into the future. This gives the system a look-ahead similar to that provided by search. By contrast, Rosen's description suggests that the main concern for an anticipatory system is to detect and avoid bad trajectories instead of finding good ones. Hence, ideally the anticipatory system will not interfere with the object system's execution.

We will, in Section 2, present our own interactive narratives project, by means of an example. This example is then used as our object system S in an anticipatory system for guiding players through interactive plot development. The details of an automaton model M and effectors E of the anticipatory system are presented in Section 4. Since we present work in progress, we conclude by identifying topics for future research.

2   A Kaktus Scenario

Our ongoing Kaktus project aims to create truly interactive socio-emotionally rich stories in which the player actively participates and influences the plot in non-trivial ways. We strive to move away from the simplistic ways of interacting found in many games, and instead focus on social and emotional interaction.

The scenario centers around three teen-aged girls named Karin, Ebba, and Lovisa. The player of Kaktus acts as one of the girls while the system controls the others. We enter the story at a time when the girls are planning to organize a party for their friends. The story evolves over a period of several days prior to the party. During this period decisions have to be made concerning different matters related to the party, for instance, whom to invite, negotiating with parents about lending the house (or some other locale), what kind of food (if any) to serve, if alcohol should be allowed at the party, choice of music, etc. In addition to organizing the party, the game involves establishing and maintaining social relationships with the characters of the game. In order to be successful, the player must adopt the role of a teenage girl, be sensitive to the social and emotional cues in the environment, and act on the basis of those.

In the traditional arts, a story is described as a sequence of events that through conflict changes dramatic values of importance for the characters of the story (see e.g., [8,20]). Values are typically on a spectrum between counterpoints (e.g., love/hate, death/life, courage/cowardice, strength/weakness, good/evil). For example, the Kaktus scenario revolves around the values

- **love/hate** is illustrated by Lovisa's secret love for a boy in the local hockey team which may change due to events in the game.
- **friendship/enmity** is of great importance in the scenario. Plot unfolding—and ultimately success in arranging the party—depends on the players ability to interpret and manipulate the social configurations between characters.
- **boredom/exhilaration** is one of the main driving forces of the game. Boredom is the main reason for the girls to organize the party.

Story events are classified according to how much they change story values. Using the terminology of Robert McKee [20], the smallest value changing event is the *beat*. It consists of an action/reaction pair that causes minor but significant changes to at least one story value. A *scene* is built from beats and causes moderate changes to values. Scenes are the most fundamental building blocks of stories forming the arc of the story, or

> the selection of events from characters' life stories that is composed into a strategic sequence to arouse specific emotions and to express a specific view of life. ([20], p.33)

Next comes the scene *sequence* and *act*. Acts are strings of sequences causing major value reversals. Finally the *story* itself is a long irreversible change of values.

Using this definition of story events we can outline the story of our scenario through the following scenes:

q_1. Introduction of Karin, Ebba and Lovisa, their current state of mind, and conception of the party idea.
q_2. Karin and Lovisa find out that Ebba cannot afford to organize a party.
q_3. Lovisa's secret love for Niklas is revealed to Karin.
q_4. Plans are made to hold the party at Lovisa's house.
q_5. How to get hold of alcohol to the party is discussed.
q_6. Karin, Ebba, and Lovisa invite people to the party.
q_7. The girls decide not to have a party.

Each scene consists of beats detailing its content. Minimally a scene consists of beats detailing responses to input from the player. For instance, lines of dialog exchanged between characters and the player, or beats implementing (inter-) agent behaviors, such as behaviors for two agents to have a row. We will not go further into details regarding beats here.

Some events in a story are more important than others. *Kernels* [8] or *story-functional* events are crossroads where the plot of a story is forced to branch of in one direction or the other. By contrast *satellites* or *descriptive* events do not affect the plot in the same way. Instead, they serve as instruments for creating atmosphere, elaborating on kernels or fleshing out the story. In our scenario descriptive events are scenes where the user explores the narrative world she is immersed in, e.g., by chatting with the characters and getting to know them. Such events do not directly advance the plot but the story would nevertheless be impoverished without them.

3   System Overview

There is a fundamental conflict between interactivity and narrative. In narrative media experiences, such as suspense, comic effects or sympathy often depend on the ability to specify a sequence of events in the 'right order'. In interactive narratives, in which the player occasionally takes control of what will happen next, such effects are difficult to achieve. A player, in a few mouse clicks, may ruin an experience that has painstakingly been designed by an author. By limiting the amount of options, a player may be pushed along an intended path, but at the same time such a design will decrease her amount of influence over the story.

Stories are ultimately about characters, and hence agents have long been considered natural elements in many story-telling systems [2,1,11,13,21]. To date there has been a fair amount of research regarding various aspects of interactive narratives. However, most agent-based interactive narrative systems have adopted a locally reactive approach, where individual agents within the system ground their actions in their own perception of the current situation [18]. Comparatively little effort has gone into research on how to integrate local reactivity with a global deliberative plot control mechanism [19,28]. Selecting what events to recount and how to order them is what makes up the plot of a story. Kaktus adopts an agent-based approach to story-telling where characters in the guise of artificial agents or human players have a central role. Characters are the primary vehicles for conveying dramatic action and progress along different story arcs. As the scenario relies heavily on social and emotional interaction, artificial agents are required to simulate human behavior to some extent and be capable of 'reasoning' about emotions, interpersonal relationships, and consequences of actions.

3.1   Playing the Game

Game play in Kaktus is best described as a mix between role playing and simulation. It resembles role playing in that players will not act as themselves in the virtual world, but place themselves in the position of a fictive protagonist. The embodied agent will not be an instantiation of the user but rather a character. The player will act not so much as herself but as she thinks the character would act in any given situation.

Kaktus is also inspired by simulation games in that artificial characters have a (simulated) life of their own including emotions, goals, plans, and social relations which are all affected by players' choices and actions, as in The SIMS (http://www.thesims.com). In fact, we envision that Kaktus may be played in different ways. Instead of having party arrangement as the primary goal of a game session, some players may instead choose to, e.g., maximize the friendship relations between characters. In this mode of game play Kaktus can act as a social simulator where players can—within a limited domain—hone their social skills.

Players interact with the game mainly through dialog. Players type out text that is visible on screen while the artificial characters' dialog is synthesized speech. The discourse is real-time and players can use a number of objects available in the game (such as keys, diaries, candy, mobile phones, and buddy lists in those mobile phones) to accomplish tasks.

3.2   System Implementation

Kaktus uses a Belief Desire Intention (BDI) approach to model agents [5]. This approach gives agents a rudimentary personality in that they pursue subjective goals, based on equally subjective models of their environment. The core of our system is based on the JAM (for Java Agent Model) agent architecture [17] to which we supply an extension for doing anticipatory planning.

The anatomy of a JAM agent is divided into five parts: *a world model, a plan library, an interpreter, an intention structure,* and *an observer* (see Figure 1). The world model is a database that represents the beliefs of the agent. The plan library is a collection of plans that the agent can use to achieve its goals. The interpreter reasons about what the agent should do and when it should do it. The intention structure is an internal model of the goals and activities the agent currently has committed itself to and keeps track of progress the agent has made toward accomplishing those goals. The observer is a special plan that the agent executes between plan steps in order to perform functionality outside of the scope of its normal goal/plan-based reasoning.

A high level overview of the system is given in Figure 2. The system consists of four basic components:

- A graphical front-end/user interface
- An I/O manager
- A story manager
- An ensemble of JAM agents

The front-end is separated from the rest of the system according to a Model-View pattern. This facilitates having more than one type of interface to the system. Currently we are using an interface implemented in Macromedia Director but other interfaces, e.g., a web interface, are planned. Depending on the capabilities of the front-end some low-level functionality may be implemented in it, e.g., low-level movement primitives for agents.

The I/O Manager sits between the front-end and the story manager. The tasks performed by the I/O Manager vary depending on the type of front-end used. Minimally it consists of converting user input to internal formats and system output to front-end specific formats. For instance, text that the user has typed is converted to dialog moves and output is converted to formats such

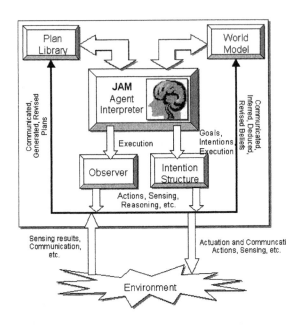

Fig. 1. The anatomy of a JAM agent. Picture taken from [16].

as Facial Animation Parameters (FAP), Body Animation Parameters (BAP) or ID's of animation files to play.

After conversion to an internal format user interface events are forwarded to the story manager. The story manager acts on the information it receives and continuously reflects the state of the system back to the front-end. Each character in the story is represented by a JAM agent–external to the story manager– with a distinct set of goals, plans and intentions. During game play agents act autonomously, but are from time to time given direction by the story manager.

The story manager embodies the anticipatory planning mechanisms of the system through the *anticipator* (cf. [10]). The anticipator monitors the progress of the story and decides on appropriate courses of action. The decisions are enforced by modifying the state of each agent—or tuning other system parameters—in order to accomplish the intended course of action. This includes, but is not limited to, adding or retracting information from an agent's world model, adding or deleting plans from an agent's plan repository and adding or deleting goals from an agent's list of goals.

We utilize the observer functionality of JAM agents to communicate with the anticipator giving it a chance to inspect and modify the state of an agent between each planning step. While this scheme does not provide the anticipator with uninterrupted insight into an agents mind it is sufficient for our needs, since in effect the anticipator can make changes to the agents state *before* the agent chooses and/or executes an action.

Other distributions of responsibility are also possible, e.g., the agent community could manage without a super agent deciding on appropriate courses of action through voting or the super agent can act more as an advisor, guiding agents when they are uncertain of what to do (cf. [4]).

The anticipator uses copies of each agent to simulate execution of the system. The simulation has no side effects, such as output to the front-end. In an interactive application such as ours, generating output and waiting for input are operations that typically make up the bulk of execution time. In a simulation these operations can, and should, be left out, thus freeing time for running the simulation itself. Hence we do not anticipate a need for creating simpler models of our agents, in order to achieve faster than real time performance for the anticipator.

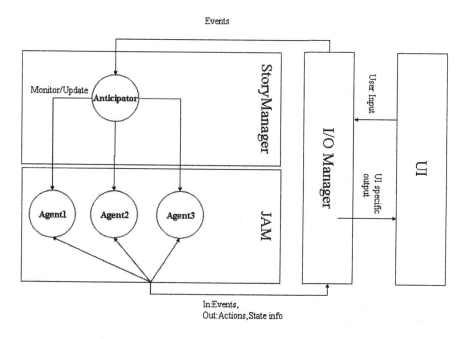

Fig. 2. Overview of the system.

In addition to the anticipator, the story manager contains a model of desired plot unfoldings. The plot model is encoded as a finite automaton which is described in detail in Section 4.1. The model includes bookkeeping aspects of the system state such as which scenes have been played, which scenes are currently playable, and which scene is currently active.

Each scene contains a set of beats encoded as JAM plans and goals. When a scene first becomes active agents in the scene are given goals and plans that enables them to perform the scene. Conversely, when a scene becomes inactive goals and plans specific to that scene are removed from agents' plan libraries and goal stacks. In this way the story manager portions out directions for the story in suitable chunks.

To illustrate how the anticipator works we will describe the sequence of events that takes place during execution of the system.

3.3   Example

After the system is started each agents *observer* plan will eventually be executed. The observer calls a synchronization method in the *anticipator* passing a reference to the agent as a parameter. Through the reference the anticipator has access to the agent's world model, plan library, and intentional structure, which together control the behavior of the agents.

The anticipator copies information from the agent, such as facts from its world model or the current goal of the agent, and stores it in a copy of the agent. The copy is later used for making predictions about the agents future behavior. If there are several agents in the system, the anticipator waits for all of them to call the synchronization method in order to get a snapshot of each agent's state, before relinquishing control.

Next, the anticipator starts a simulation of the system using the copied information. At regular intervals each agent observer plan calls the synchronization method again. If the anticipator, based on it's simulations, predicts that the system will end up in an undesired state, it searches for an appropriate effector to apply (see Section 4.2). In case there is more than one agent using the anticipator, it waits for all of them (or at least the ones that are affected by the chosen effector) to call the synchronization method in order to gain access to the entire system state, before it applies the effector.

Given that there is a user involved, providing input which may be hard for the system to predict, the synchronization can also act as a sensibility check of the anticipator's predictions. If the actual state does not evolve according to predictions the anticipator can discard the current predictions, gather new information, and start the cycle anew.

As an example consider the simple agent described in Figure 3. This agent has the single top-level goal of *ACHIEVE live* to pursue. The goal is achieved through displaying idle time behavior or, if the agent is on friendly terms with Karin, *gossip* about Lovisa's infatuation with Niklas.

Let us suppose that in the anticipator's simulation this agent has come to a point where it has selected *gossip* as its next plan to execute. However, suppose further that global story constraints dictate that for the time being this is not a good plan, since it leads to an undesirable state. The anticipator then starts a search for an effector (or effectors) that can prevent this situation from arising. In this case three effectors are found: lower the friendship value between the agent and Karin thus preventing the plan from becoming active, remove the plan from the agent's plan library (or replace it with a less harmful one) or give

```
GOALS:
  ACHIEVE live;
FACTS:
  FACT friends "Lovisa" "Karin" 1;
  FACT in_love "Lovisa" "Niklas";
PLAN: { NAME:
  "live"
GOAL:
  ACHIEVE live;
BODY:
  FACT friends "Lovisa" "Karin" $strength;
  OR
  {
    TEST( > $strength 1);
    ACHIEVE gossip;
  }
  {
    EXECUTE doIdle;
  };
} PLAN: { NAME:
  "gossip"
GOAL:
  ACHIEVE gossip;
BODY:
  RETRIEVE in_love "Lovisa" $who;
  PERFORM tell "Karin" "in_love" "Lovisa" $who;
EFFECTS:
  ASSERT knows "Karin" "in_love" "Lovisa" $who;
}
```

Fig. 3. An example of a JAM agent.

the agent a new goal with a higher priority. For simplicity let us suppose that the anticipator chooses to lower the friendship value between the agent and Karin. After this is done the agent resumes normal execution but with a new value on the friendship relation to Karin. At some later point in time when the agent would normally have selected the gossip plan for execution it now displays idle time behavior instead, and the undesired state is avoided. This cycle is repeated until the user stops playing or the plot has reached its conclusion.

The same basic procedure is also applicable when more than one agent is involved. However it is likely that different effectors will need to be applied to each agent, resulting in a larger number of applied effectors than in the single agent case. For instance, to initiate a fight between two agents regarding some matter, they will need opposing views of the matter at hand, different knowledge, different arguments, etc.

4   The Anticipatory System

After describing, in the classical Rosen sense (cf. [23]), the model and the set of effectors for the object system S (represented by the Kaktus scenario), we will concentrate on their significance to the design of interactive narratives.

Using the classification scheme introduced in [7] we can regard our system as a *state anticipatory* system. Through simulation, the anticipator forms explicit predictions about future states in order to avoid undesirable ones. Currently our model is limited in that we do not explicitly model players in any way. However, modeling users is notoriously difficult. Hence we will rely on empirical tests of the current system to indicate whether such an extension would be worth the added effort.

4.1   The Model M

Our model M is a finite automaton, in which each state corresponds to a scene. In Figure 4, the seven scenes of the Kaktus scenario are represented as states q_1 through q_7. Note that M only contains story-functional scenes as described in section 2. Descriptive events are accessible from most of the kernel events. However, since they do not influence the plot they are not included here. The start state is q_1, while the set of end states is $\{q_4, q_6, q_7\}$. The design method is top-down, in that M was completed only after the key scenes had been identified. Note that one may also proceed bottom-up, letting M depict sequential plot development from the start state to the end state of M. This entails that the design of M is more important than design of the plot. In this case, the states of M are compositional and scenes are identified as natural halts or crossroads in the evolution of the plot. Hence, if plot emergence is studied, a bottom-up design seems adequate. For conventional interactive game design, however, top-down is the default choice.

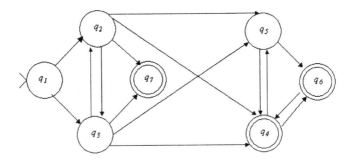

Fig. 4. A state diagram for the finite automaton M, with only desirable states.

Finite automata also lend themselves to non-sequential development of plot. Models have previously been restricted to directed acyclic graphs [28]. We argue that instead it should be regarded as an opportunity. Many forms of interaction build their narrative framework on plot cycles, e.g., the repeated mouse chases in Tom & Jerry cartoons.

We will show below that problems related to computational complexity and model design complexity can be avoided. In fact, we have harsh constraints for the execution of M, since the time complexity must be low enough to allow for faster than real-time execution.

The state transitions in M are labeled $a_0, a_1, \ldots$ indicating that each transition is different. (For clarity of exposition, we have not labeled the edges in Figure 4.) In practice, there might be simple scenarios or small parts of an interactive drama in which transitions may be reused. However, for scenarios of the size and complexity we consider, uniqueness can be stipulated, especially since this stipulation has no mathematical significance. Each transition a_m from q_i to q_j could be interpreted as (an ordinary Hoare triple with) pre- and postconditions. The precondition is the state described as q_i plus the added conditions described in a_m, and the postcondition is simply q_j.

Each transition a_m corresponds to a set of conditions that can be either *true* or *false*. Conditions are represented by one of the symbols $\{0,1,?\}$ denoting *false*, *true* and *either* respectively. A set of conditions is represented by a fixed length string where each position in the string corresponds to a particular condition. This means that transitions encode the desired value for all conditions in the system. However the wild card symbol '?' can be used in circumstances when the value of a condition is irrelevant to a transition. In our implementation conditions are represented by Java classes that are stored in a database. Each condition is evaluated at most once each cycle. The result of the evaluation is stored internally in the condition and can be reused later on. Currently we have implemented the following general purpose condition types:

- Range
- Boolean
- Greater
- Less
- Equal

These classes can be used to place conditions on any system parameters including story values, e.g., to prevent a transition from one climactic scene to another.

In addition we have implemented a set of conditions specific to agents that relate directly to facts, goals, plans, and emotions that an agent may have.

- Knows
- Feels
- HasGoal
- HasPlan

The language that M accepts is the set of words accepted. Each word is a sequence of names of transitions, e.g., $a_0 a_3 a_4 a_3 a_4 a_6$. Therefore, an ordinary finite automaton suffices. Should any manipulation be required in M, a finite state transducer (or if the language was very complicated, a Turing machine) would be required (see, e.g., [25]). The preconditions described in a_m are usually fairly complicated, and we will describe only one transition in the automaton pertaining to our Kaktus scenario.

The transition we will describe takes us from a state where Karin (the player) does not know about Lovisa's romantic interest in Niklas to one where Ebba tells her about it. This is a climactic scene where many story values temporarily are at their extremes. There are several preconditions that must be fulfilled for this transition to take place. Assuming that the transition is named a_m let us consider the set of preconditions $\{p(a_m)\}_j$.

- $\{p(a_m)\}_1$ states that Karin must be unknowing about Lovisa's interest in Niklas, or the transition will not make any sense.
- $\{p(a_m)\}_2$ tells us that Lovisa must be unwilling to have the party, since getting to know about and later inviting Niklas to the party is a way of persuading Lovisa.
- $\{p(a_m)\}_3$ requires that Ebba actually wants to have the party. If she does not, she may not have a strong enough incentive to break her silence and tell Karin about Lovisa and Niklas.
- $\{p(a_m)\}_4$ states that Ebba must be on speaking terms with Karin or she will be reluctant to tell Karin anything.

Preconditions also act as constraints on story values. For instance, Lovisa's love for Niklas can be expressed as an interval $(0\ldots9)$. Let us call an update of this interval a parameter update, and let us define a *radical parameter update* as a parameter update in which the value changes with at least five units. Given that a good dramatic arc should slowly oscillate between climactic scenes and quieter ones, we can consider, e.g., oscillations between states with at least one radical parameter update and states without such updates. The oscillatory sequence of scenes can now be achieved through preconditions constraining certain transitions.

4.2   The Set of Effectors E

In the following discussion we will, for simplicity, assume that all system parameters including agent goals, plans, intentions and relationships are stored in a global database.

An effector will in Kaktus be a function updating a parameter value in the database. We thus consider parameter updates to be atomic operations on S, or the environmental updates to S (cf. [22], p.248). In order to drag the plot from an undesired to a desired state, a single parameter update will hardly ever suffice. Depending on how much S needs to be altered any number of updates may be necessary. We amalgamate our terminology with that of McKee to achieve the following list of value-changing actions, in increasing order of importance to the plot:

- parameter update
- beat
- scene
- scene sequence
- act

A fairly small subset of E will typically constitute a beat. A larger subset of E is required for a scene change. Scene sequences and acts pertain to aspects of story design so domain specific that they will be left out of our discussion.

Below we give a short list of possible effectors in the Kaktus scenario:

- Simulate a player action, i.e. pretend that the player did something she actually did not.
- Filter out some player action.
- Introduce/remove a character.
- Alter the flow of time to bring about or delay an event.
- Start/stop a topic of discussion.
- Create a disruptive event such as throwing a temper tantrum or starting a row with another character.
- Give the player a hint or provide some information.

Effectors can have different functions. Weyhrauch divides his so-called Moe Moves, which roughly correspond to effectors, into *causers, deniers, delayers, substitutions,* and *hints.* Their respective function is to cause an event, stop an event from happening or deny access to an object, delay an event, substitute an event with combinations of other events, and finally give hints about unexplored paths.

Some effectors may not have any immediate impact on the plot e.g., turning on the light in an empty room or placing an item in that room. Such effectors can, however, create favorable conditions for alternative plot unfoldings in the long term. For instance, the player might find the item placed in the room and use it to overcome an obstacle at some later instant.

Other effectors can have a growing influence over time. Imagine for instance an effector instructing an agent to kill every other agent it encounters. At first such an act would have limited impact on the plot but as the agent roamed the world killing other agents, the effect would become increasingly noticeable.

It is important to remember that while the revision of story values during the drama describes the intended dramatic arc, these values are never directly manipulated. For instance, Lovisa's love for Niklas is never directly increased or decreased. Instead they are updated as a result of tuning other system parameters.

Finally, we wish to stress the importance of creating effectors that do not tweak parameters in a way that interferes with user experiences of the narrative. There should be no unexplained or unexplainable twists or turns to the story brought on by the application of any effectors. It is important that transitions from one state of affairs to another are made accessible and understandable to the player [24]. Hence the design of good effectors will likely require equal amounts of artistic work and engineering.

4.3   The Top-Down Design Process

Our top-down design process for the modeling of an interactive drama (after the initial cast of characters, rudimentary plot design, and means to player interaction have been determined) consists of the following steps.

1. Describe the entire scenario as a finite automaton
2. For some state/transition pairs, list the resulting state
3. Partition the class of states into desirable/undesirable states
4. Partition the class of desirable states into ordinary/end states
5. Review the graph of the automaton, and iterate from 1 if necessary

We will now review these steps in turn, in order to further explain M and E. The first step is ideally carried out by a team of authors. The team lists a number of scenes, and then for each scene, a number of beats. The scenes must then be linked into sequences by means of scene transitions. For each scene transition, a number of parameters and their required values are identified and arranged in a database. The first output of the team may be rudimentary, approximately at the level of our Kaktus scenario. The artistic part of the work carried out should affect the implementation, and vice versa, to some extent. For instance, detailing the parameters listed, such as when pondering whether life/death is a Boolean parameter, or an interval, has an artistic aspect as well as a direct influence on the implementation in the database of parameters and their values.

The really hard work begins in step 2. The reason step 2 does not read "For each state/transition pair, list the resulting state" is that this task is insurmountable and also unnecessary. It is insurmountable because the number of transitions is equal to the number of combinations of parameter values in the entire database, which for interesting portions of dramas will be huge. It is unnecessary because most of these combinations will not, and sometimes even cannot, occur. In fact, the objective of our anticipatory system is to steer clear of more than 99 per cent of these combinations. We will therefore happily leave many transitions non-specified, and seemingly also the execution of M undetermined. We will show why this is not problematic.

Since each state in M is a scene, listing only the interesting transitions between these scenes is likely to result in an automaton in which all states are desirable. Hence, much of the work in step 3 is already done. Steps 2 and 3 also reflect artistic vision and imagination, however, in that the authors should try to imagine interesting undesirable states too. Figure 5 shows a transition a_{16} leading to a situation in which Karin (alias the player) expresses an interest in Niklas. This is clearly an undesired state since we have no scene dealing with such a situation. Furthermore it prevents us from using scene q_3 where Lovisa's love for Niklas is revealed. However it is possible to recover from this situation, e.g., by revealing that Karin's declaration of love was only a joke. Here, transition a_{17} takes us back to desired territory. The necessity of explicitly listing undesirable states, and not simply stipulating that the listed states are desirable and the rest undesirable, is explained by the fact that the anticipatory system must be resilient. This resilience is a form of fault tolerance: if the changes to the interactive drama that M suggests requires too radical parameter updates to be performed within one scene, or if there is a latency in the real-time parameter updating, the drama may be in limbo for a limited time. This means that M will be in an undesirable state during that time. Note that analogously to the constraints placed on transitions from one climactic scene to another we can place constraints on the number and magnitude of parameter updates allowed by a single scene transition. If an undesirable state and its transitions leading back to a desirable state can be specified in advance, the drama can be led back on a desirable path. The player might in this case feel that the plot goes weird on her temporarily, but since things go back to normal, or at least come closer to what is expected by the player, after a short time, she might accept that she could not fully understand what was going on. The tuning of resilience procedures is a delicate matter and must be tested empirically. Since we do not explicitly model the player in the way Weyhrauch [28] and others do, we envisage such tests to be user studies. We recognize that there might be tacit iteration between steps 2 and 3, but their identification here as discrete steps is motivated by the fact that the number of such iterations will decrease with increased experience of the suggested design process.

Step 4 is relatively simple. Even if the preceding steps were only moderately successful, step 4 is meaningful already for automata with only desirable states; that M is in a desirable state does not entail that the drama could stop in any state. Instead, most desirable states will have transitions to one or more end states, in which the plot comes to a natural conclusion. If we for instance consider interactive computer games, M will normally have a single end state, in which the end credits are rolled out and all game processes are then killed. Note that an undesirable state cannot be an end state.

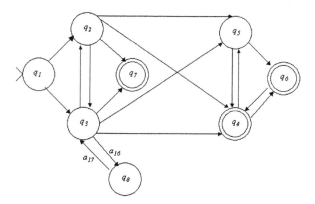

Fig. 5. A state diagram for the finite automaton M, with one undesirable state.

Step 5 is a return to the discussions that largely ran in step 1. The depiction of M as a state diagram is a pedagogical and instructive basis for self-criticism and revision. Since this step depends so largely on the form of interactive drama and on the working group, we will not pursue the reasons for iteration here.

4.4   Interaction between M and S

In order to execute M as a string acceptor as efficiently as possible, its design must be converted to canonical form. From the Myhill-Nerode theorem follows that each automaton can be minimized with respect to its number of states (see, e.g., [15]). There are several well-known efficient minimization algorithms, with the perhaps most widely used being Hopcroft's algorithm [14] which is included, e.g., in the PetC automata emulation tool (freely available at http://www.dsv.su.se/~henrikbe/petc/petc.html), developed by Henrik Bergström [3]. Hopcroft's algorithm requires as input a deterministic automaton. Hopcroft's algorithm runs in $\bigcirc(n \log n)$, but the computational complexity is not that important to us, since it is carried out in batch. In PetC, an algorithm due to Brzozowski [6] that can minimize non-deterministic automata is included. Brzozowski's algorithm has exponential worst-case complexity but is in practice often faster than Hopcroft's algorithm [27]. However, even small automata may experience its exponential nature [12].

Although there is no difference with respect to expressive power between the classes of non-deterministic automata and their subclass of deterministic automata, i.e. they accept the same class of languages, we must pinpoint which kind of non-determinism we allow for. There are three reasons for non-determinism in an automaton in general:

- Transitions are labeled with strings of length different from 1
- Undefined transitions
- Choices between transitions with identical labels

As explained earlier, our transitions can be labeled by fairly long strings, since each transition may have to fulfill a large number of preconditions in terms of plot parameter values. However, a string label $a_1 a_3 a_4$ from q_3 to q_4 for example, is easily converted to a sequence of transitions: from q_3 to a new state q_3' labeled a_1, from q_3' to another new state q_3'' labeled a_3, and from q_3'' to q_4 labeled a_4. Note that the designer would never see the two new states, as we are now describing a kind of compilation procedure, so the objection that $q3'$ and $q3''$ do not correspond to scenes is irrelevant. For completeness, we should also stress that we cannot allow for labels of length 0, i.e. the empty string. This is of no significance to our construction of M. Just like our string labels can be seen as shorthand, so can our incompleteness with respect to the state/transition pairs mentioned in step 2, which leads to undefined transitions. We simply stipulate that each transition not depicted in the state diagram of M leads to a dead state, i.e. a state with no edges directed to another state in M. This dead state is our garbage bin, and is naturally an undesired state. The only reason for non-determinism out of the three in the above list that we would like to recognize as significant is choice. We would like to allow for one out of several different scenes to follow a given scene stochastically. In an interactive drama with this possibility, the player may act in an identical fashion in two separate game rounds and still experience different unfoldings of the plot. This type of non-determinism is thus not only tolerated, but encouraged, in order for the drama to become less predictive.

So, at long last the drama begins. The appropriateness of simulated player versus real user tests depends on the maturity of the narrative. A special case is drama without human agents. In the early stages of development, this can provide ideas for plot, e.g., through plot emergence. At the other end of the spectrum is the special case of only human agents. If no artificial agents are involved in the drama, only the environment will be manipulated. As the number of human agents increases, so does in general the difficulty of maintaining a coherent plot, even though conversation between human agents might turn into an essential part of the drama, and so assume the role of a carrier of the plot. The instant the drama begins, so does the execution of M. The emulation of actions in the drama is coded in the parameter database and in the state of M. The automaton will monitor changes made to the parameter database via the operators implemented in the drama. In the other direction, the implemented agents will experience the effects of changes instigated both from their own actions and as a result of the look-ahead in M: for each detection of the possible transition to an undesirable state, M will make parameter updates so as to reduce the risk of, or eliminate, this possibility.

5   Conclusions and Future Research

We have described ongoing work on providing authors of interactive narratives with tools for subtle guidance of plot. The anticipatory system used for look-ahead is under implementation, and will rest in part on earlier computer science efforts to anticipatory learning [9,4].

Parts of our implemented system are being developed within MagiCster, an EC-funded project that explores the use of animated, conversational interface agents in various settings. Aspects of this work addressed within MagiCster include beat selection (or dialog management), a topic not covered in depth above. We have also omitted the demonstration of the efficiency of executing M. While it is well-known that the computational complexity of emulating execution of small automata is low, we still need to secure its faster-than-real-time properties. This will be done in connection with the implementation of our new version of the Kaktus game.

Acknowledgments

The authors would like to thank Paul Davidsson for comments. Laaksolahti was financed by the MagiCster and INTAGE projects, while the VINNOVA project TAP on accessible autonomous software provided Boman with the time required for this study.

References

1. Peter Bøgh Andersen and Jørgen Callesen. Agents as actors. In Lars Qvortrup, editor, *Virtual Interaction—Interaction in Virtual Inhabited 3D Worlds*, pages 132–165. Springer-Verlag, 2001.
2. Joseph Bates. Virtual reality, art, and entertainment. *PRESENCE: Teleoperators and Virtual Environments*, 1(1):133–138, 1992.
3. Henrik Bergström. Applications, minimisation, and visualisation of finite state machines. Master's thesis, Department of Computer and Systems Sciences, Stockholm University, 1998. Report no 98-02-DSV-SU.
4. Magnus Boman, Paul Davidsson, Johan Kummeneje, and Harko Verhagen. An anticipatory multi-agent architecture for socially acceptable action. In *The 6th International Conference on Intelligent Autonomous Systems*, pages 744–750. IOS Press, 2000.
5. Michael E. Bratman, David Israel, and Martha Pollack. Plans and resource-bounded practical reasoning. In Robert Cummins and John L. Pollock, editors, *Philosophy and AI: Essays at the Interface*, pages 1–22. The MIT Press, Cambridge, Massachusetts, 1991.
6. J.A. Brzozowski. Canonical regular expressions and minimal state graphs for definite events. In *Mathematical Theory of Automata*, volume 12 of *MRI Symposia Series*, pages 529–561. Polytechnic Institute of Brooklyn, N.Y., Polytechnic Press, 1962.
7. Martin V. Butz, Oliver Sigaud, and Pierre Gerard. Internal models and anticipations in adaptive learning systems. In *Proceedings of the Workshop on Adaptive Behavior in Anticipatory Learning Systems*, 2002.
8. Seymour Chatman. *Story and Discourse—Narrative Structure in Fiction and Film.* Cornell University Press, 1978.
9. Paul Davidsson. *Autonomous Agents and the Concept of Concepts.* PhD thesis, Department of Computer Science, Lund University, Sweden, 1996.
10. Paul Davidsson. A framework for preventive state anticipation. In *this volume*. 2003.
11. Tinsley Galyean. *Narrative Guidance of Interactivity.* PhD thesis, School of Architecture and Planning, MIT, 1995.
12. James Glenn and William I. Gasarch. Implementing WS1S via finite automata. In Darrell R. Raymond, Derick Wood, and Sheng Yu, editors, *Automata Implementation*, volume 1260 of *LNCS*, pages 50–63. Springer-Verlag, 1997.
13. Barabara Hayes-Roth, Robert van Gent, and Daniel Huber. Acting in character. In Paolo Petta and Robert Trappl, editors, *Creating Personalities for Synthetic Actors—Towards Autonomous Personality Agents*, volume 1195 of *LNAI*, pages 92–112. Springer-Verlag, 1997.
14. John Hopcroft. An $n \log n$ algorithm for minimizing states in a finite automaton. In Z. Kohavi and A. Paz, editors, *Theory of Machines and Computation*, pages 189–196. Academic Press, 1976.
15. John E. Hopcroft and Jeffrey D. Ullman. *Formal Languages and Their Relation to Automata.* Addison-Wesley, 1969.
16. Marcus Huber. *JAM Agents in a Nutshell.* Intelligent Reasoning Systems. Available at http://www.marcush.net.
17. Marcus J. Huber. Jam: a BDI-theoretic mobile agent architecture. In *Proceedings of the third annual conference on Autonomous Agents*, pages 236–243. ACM Press, 1999.

18. Brian Loyall. *Believable Agents*. PhD thesis, School of Computer Science, Carnegie-Mellon Univ, 1997. Tech report CMU-CS-97-123.

19. Michael Mateas and Andrew Stern. Towards integrating plot and character for interactive drama. In *AAAI Fall Symposium Technical Report FS-00-04*, pages 113–118. AAAI, 2000.

20. Robert McKee. *Story—Substance, Structure, Style and The Principles of Screenwriting*. Harper Collins, 1997.

21. Ana Paiva, Isabel Machado, and Rui Prada. Heroes, villians, magicians, ... : dramatis personae in a virtual story creation environment. In *Proceedings of the 6th International Conference on Intelligent User Interfaces*, pages 129–136. ACM Press, 2001.

22. Robert Rosen. Planning, management, policies and strategies—four fuzzy concepts. *International Journal of General Systems*, 1:245–252, 1974.

23. Robert Rosen. *Anticipatory Systems—Philosophical, Mathematical and Methodological Foundations*. Pergamon Press, 1985.

24. Phoebe Sengers. Do the thing right: an architecture for action-expression. In *Proceedings of the Second International Conference on Autonomous Agents*, pages 24–31. ACM Press, 1998.

25. M. Simon. *Automata Theory*. World Scientific, 1999.

26. Peter Voerderer. Interactive entertainment and beyond. In Dolf Zillman and Peter Voerderer, editors, *Media Entertainment: The Psychology of its Appeal*, pages 21–36. Lawrence Erlbaum Associates, 2000.

27. B.W. Watson. *Taxonomies and Toolkits of Regular Language Algorithms*. PhD thesis, Eindhoven University of Technology, 1995.

28. Peter Weyhrauch. *Guiding Interactive Drama*. PhD thesis, School of Computer Science, Carnegie Mellon University, 1997. Report CMU-CS-97-109.

29. Mads Wibroe, K.K. Nygaard, and Peter Bøgh Andersen. Games and stories. In Lars Qvortrup, editor, *Virtual Interaction: Interaction in Virtual Inhabited 3D Worlds*, pages 166–181. Springer-Verlag, 2001.

Exploring the Value of Prediction
in an Artificial Stock Market

Bruce Edmonds

Centre for Policy Modelling,
Manchester Metropolitan University
http://cfpm.org/~bruce

Abstract. An action selection architecture is described which uses two learning modules: one to predict future events (the PPM) and one to decide upon appropriate actions (the IALM). These are only connected by the fact that they have access to each other's past results and the IALM can use the predictions of the PPM as inputs to its action selection. This is instantiated in a model which uses GP-based learning for the two modules and is tested in an artificial stock market. This used to start exploring the conditions under which prediction might be helpful to successful action selection. Results indicate that prediction is not *always* an advantage. I speculate that prediction might have similar role to learning in the "Baldwin Effect" and that the "momentum" of the system might be a significant factor.

1   Introduction

There are two basic feedback mechanisms that are used in learning processes: *firstly*, from indicators such as pleasure, pain, or profit and, *secondly*, the (mis)match between what was anticipated to occur and what actually did occur, i.e. the error. These roughly correspond to the goals of *gaining utility* and *correctly predicting*. There are many ways of using these in different learning processes.

Planning systems attempt to *infer* what actions to take to achieve the former from models that accurately reflect their problem domain. Their use is dependent upon a raft of assumptions holding, including that:

1. A sufficiently correct model is known;
2. The domain is sufficiently static over the planning period;
3. And time, resources and the model structure make the inference feasible.

These conditions are onerous for many situations. Condition (1) means that either the model is knowable *a priori* or that some learning process using error-based feedback has had time to produce a sufficiently correct model. Condition (2) means that the domain must be pretty static and immune to unpredictable perturbation resulting from any actions taken. Condition (3) means that the model has to be pretty complete with respect to the domain, that the planning can be done without tight time constraints and that there are considerable computational resources available. Taken together these rule out almost all situations of interest to animals or animats. In particular they rule out any dynamic or tightly-interactive situations; they rule out

M. Butz et al. (Eds.): Anticipatory Behavior ..., LNAI 2684, pp. 262-281, 2003.

cases where any learnt model only captures restricted aspects of the domain; they rule out entities without the luxury of extensive off-line computation.

On the other hand learning systems that *only* use indicator-based feedback can be at a disadvantage. They are unable to prevent or prepare for any event before it happens, which can be critical in life-threatening situations (e.g. escaping from a predator, or being the first to acquire items of food). They may be 'locked-in' to local utility optima because they have no access to what *might* happen. Thus for some creatures it would seem that having some anticipatory ability would be highly advantageous (especially if it did not involve a high cost or 'time-out' to think).

However the exact conditions wherein an ability to anticipate future events might be advantageous are unclear. Many of the advantages of anticipation could be gained from learning over past sequences of events as to how to improve the appropriate indicator (e.g. avoidance of pain). Take an example: say event C tends to follow the event sequence A, B, and that it is in the entity's interest to take action X as (or just before) event C. Such an entity could approach this in two ways: *firstly*, it could learn to anticipate event C from the sequence A, B and thus learn to take action X, but *alternatively*, it could simply learn that taking action X after the sequence A, B was in its interest, i.e. the intermediate anticipation C could be eliminated from the process.

In this paper I use a simulation to compare the performance of entities with and without the ability to predict future events. This is thus a preliminary exploration of the conditions under which such anticipation is advantageous. I have chosen the environment of an artificial stock market as the test bed for the following reasons:

- it is inherently unpredictable and dynamic;
- actions taken by the agent change its environment;
- models of the environment (namely prices) have to be continually changed as part of a modelling "arms-race" with the other traders;
- accurate prediction (of future price of stocks) is cleanly distinguishable from utility (profit made by trading);
- it is possible, but unclear, that anticipation will be advantageous.

The agents employ learning algorithms based on Genetic Programming (GP) [7] – several learning process could have been used as long as they are sufficiently flexible and expressive. They do no planning and almost no inference – that is they do *no* formal inference and the action decision involves only fixed and limited steps that could be interpreted as inference (e.g. only attempting to buy when they have cash).

2   The Model Set-Up

The Environment

The environment is a stock-market. That is to say there are a fixed number of stocks which traders can buy or sell at the current price. There is only a limited quantity of each stock available in the market. There is one market-maker who sets the prices depending upon past demand. There are two such rules that I used: *firstly* what I call the "inflationary pricing mechanism" where if there has been a lot of recent buying the price goes up, and if there has been selling it goes down; and *secondly*, the "reverse pricing mechanism", where recent net buying means a drop in price and net selling a price increase. These pricing mechanisms are *not* realistic but serve to allow the market dynamics (for an account of how this actually happens in markets see [8]). The extent of the trading affects the extent to which the price changes. There is also some random noise added to the price changes. The market-maker is not anticipatory but has a fixed behaviour. All buying and selling transactions are subject to a transaction fee.

Each trading cycle each trader can seek to buy or sell each stock, as far as this is possible for it. Traders can only buy if: the market-maker has stock to sell; and the trader has sufficient cash. Traders can only sell if they have the stock. Also each trading cycle traders receive a dividend dependent upon the current dividend rate for the stock and their current holding of that stock.

The only 'fundamental' for this market is the dividend rate for each stock which wanders in a slow random walk. Almost all of the market dynamics result endogenously from the trading that occurs. These dynamics never 'quieten down' to anything like an equilibrium, because each agent is continually learning about the market, so if it settles down to a pattern this is quickly learnt; exploited and hence disrupted. The interplay of agents can be seen as a series of learning "arms-races" where by competing behaviours are being co-developed[1]. This can be characterised (albeit rather inadequately) as a version of the "minority game" [1, 3].

At the start each agent is endowed with 100 units of cash and a small amount of each stock. The initial price and dividend is random for each stock. The market maker has 100 units of each stock. For the first 5 trading cycles, the traders make small random trades.

The Agents

The agents perceive the current state of their environment and take a trading action each trading cycle. They do not have to trade. The actions they can take are: buying or selling an amount of each stock; or do nothing.

Their 'inputs' are: the current price of each stock; the current holding of each stock; the current holding of cash; the current stock index (an average of all prices); the last actions they took; the actions last trading cycle of other traders; any of the above a trading cycle ago (which can be iterated a number of times).

Each agent has two learning modules: the instrumental action learning module (IALM) and the price prediction module (PPM). Both are GP algorithms with very small populations (i.e. 10). The IALM attempts to learn the action strategy that will produce the most profit and the PPM attempts to accurately predict the prices of stocks. Each is given a rich vocabulary of nodes and terminals to build their tree-structures out of. The initial populations are generated at random out of these to a given depth.

---

[1] Since the learning is implemented by a GP algorithm the behaviours are literally co-*evolved*.

Each module does the following steps once each trading cycle:

1. they build a new generation of the population (generating new random trees as well as applying propagation and tree-crossover);
2. they evaluate their population over the past 5 trading cycles – the IALM according to the profit it would have resulted in if the strategy was used (presuming prices to be unchanged by any actions the agent took), and the PPM as to how well it would have predicted the prices;
3. they choose the best price prediction in PPM and interpret it to predict the prices;
4. they choose the best action strategy and interpret it to decide their attempted action.

It is important to note that the action strategies in IALM can include terminals that refer to the current predictions made by the PPM's best strategy. Thus the actions determined in stage (4) above can depend on the predictions in stage (3). The use of the predictions is *not* pre-programmed (e.g. by buying if it predicts the price will rise etc.) but is only provided as a possible input for the strategies learnt by the basic action learning module, the IALM. Thus it is quite possible that the IALM could learn to use the results of the PPM in strange ways, or even ignore it altogether. In fact, in the long run, one would *only* expect the IALM to preferentially evolve strategies that referred to the predictions *if* these provide it with some advantage in terms of profit.

Both the action strategies and the price prediction models being evolved can refer to *past* actions and predictions actually made as the result of the past selected best strategies and models. The current selected action strategy can utilise the present predictions about price. In order to test whether anticipation is advantageous to the agents we will set two versions of these entities in competition with each other. That is, the market will be composed of the market-maker plus two equally sized sets of traders – each set of traders made of the corresponding version of the entity. The only difference between the versions is that the first, *predictive*, version will predict the next price of each stock whilst the second, *non-predictive*, version will 'predict' only the present price of each stock. The alternative would be not allowing the action strategies to utilise the predictions as inputs in the non-predictive version, but this would not be a fair comparison because the trader could be using the price predictors as feature extractors and so increase the computational capacity of the entity. If this were done it would unfairly reduce the non-predictive traders some computational power. The predictive process in the PPM feedback mechanism is adapted so that in the predictive version the predictors are evaluated against past predictions of the *next* price (at those times) and in the non-predictive version the predictors are evaluated against past predictions of the *current* price (at those times). The two versions of the entity are illustrated in figure 1.

The non-predictive traders provide the control against which the predictive traders can be judged. The non-predictive traders' PPM modules are trying to 'predict' the current price, they can (and often do) learn that a 'predictor' of the current price is simply the current price! Thus, in effect, the non-predictive traders can be thought of as the predictive traders where the prediction of future prices used *is* the null one of the current price – i.e. the stable prediction.

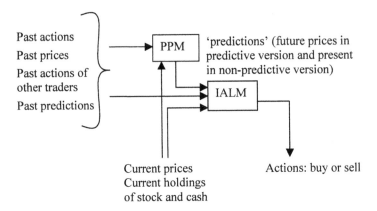

Fig. 1. An illustration of the learning structures of the traders

Thus the approach to anticipation used here is that the instrumental and predictive learning is done separately, but are loosely coupled by their ability to utilise (and hence adapt to exploit) the (best) results of each other. Each population co-evolves with the *results* of the other. This is in contrast to other approaches where the anticipation is directly associated with the action – in these approaches there is an anticipation of the effects of the possible action rather than, as here, an anticipation of environment. It is possible that in this model the PPM does evolve trees so that its predictions are dependent upon the previous actual actions if this was advantageous, but this is not necessarily so. Thus this algorithm is fundamentally different from the condition-action-anticipation of Tolman [11], Drescher [4] or Stolzmann [10] to a more modular structure where the anticipation and decision are done separately. In the final section I will briefly compare and discuss this approaches.

Obviously a major reason why I opted for the approach here is that it makes it easy to compare and test different adaptive mechanisms and analyse the results. When the action is bound tightly with the anticipation the feedback from the utility and the accuracy interact in ways that are sometimes difficult to detangle. I am not claiming here the superiority of my approach in any way, merely describing and investigating a series of approaches based on a loosely connected modular structure of learning processes.

3   Results

The results I describe here are preliminary. The time I had available and the slow speed of the simulations mean that I have not performed as many runs of the simulation exploring many different parameterisations as I would have liked.

Sets of runs were done for 8 simulation set-ups, each with an equal number of each kind of agent so that the relative success of each kind could be compared. The simulations came in two sizes: 3 agents of each kind and 7 agents of each kind. There were two numbers of stocks (4 stocks and 2 stocks); two pricing mechanisms (inflationary and reverse).

Inflationary Pricing Mechanism

Rising prices mean that agents learn to buy into stocks rather than keep their assets in cash. On the other hand, the fact that agents are always wishing to buy means that prices rise (under the inflationary pricing mechanism). Thus a pattern of rising prices is self-confirming. A typical pattern of exponential price rise from a typical run of the simulation is shown in figure 2 (note logarithmic scale).

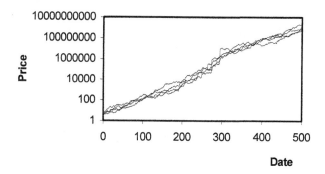

Fig. 2. Prices of stocks with inflationary pricing mechanism in simulation 1.

However, within this pattern of rising prices there is a lot of swapping between stocks, resulting in different stocks being the best buy at different times. The assets of the traders depend upon which stocks they choose to buy into at which stages.

Reverse Pricing Mechanism

In this version of the simulation the reverse pricing mechanism is used. This results in very different market dynamics – prices decline in the long run but with peaks caused by speculative bubbles in between. The prices of a typical run are shown in figure 3. Despite the appearance of less uniformity than in the simulations above with the inflationary pricing mechanism, the relative differences between stocks are more predictable being more stable. Also the price movements of the stocks are more obviously correlated.

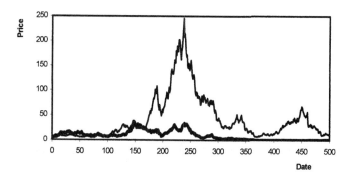

Fig. 3. Prices of stocks with reverse pricing mechanism (*non*-logarithmic price scale)

Table 1. Basic simulation set-ups

Number of agents of each kind	Number of stocks that can be traded	Pricing mechanism	Simulation label
3	2	Inflationary	3x2Agents 2Stocks, Inflationary
3	2	Reverse	3x2Agents 2Stocks, Reverse
3	4	Inflationary	3x2Agents 4Stocks, Inflationary
3	4	Reverse	3x2Agents 4Stocks, Reverse
7	2	Inflationary	7x2Agents 2Stocks, Inflationary
7	2	Reverse	7x2Agents 2Stocks, Reverse
7	4	Inflationary	7x2Agents 4Stocks, Inflationary
7	4	Reverse	7x2Agents 4Stocks, Reverse

The raw measure of the success of an agent is the total value of its assets, but since the market is roughly a zero-sum situation and the amount of money that it is possible to make depends upon the extent of the price turbulence. Thus it is more meaningful to compare the average success of each kind of agent against the average success of the other kind scaled by the standard deviation of the prices in that run. Figures 4 to 11 show the scaled difference of average assets for each of the 10 runs and the average of these. Lines above zero indicate that (on average) the predictive traders are doing better and below it the non-predictive agents.

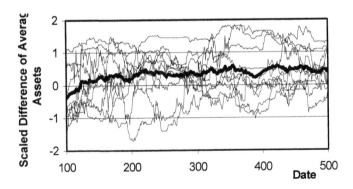

Fig. 4. Difference of average assets of predictive traders and average assets of non-predictive traders scaled by the standard deviation of each run in the simulation with label **3x2Agents 2Stocks, Inflationary** (the thin lines is the scaled difference for each of the runs, the average of these is the thick line).

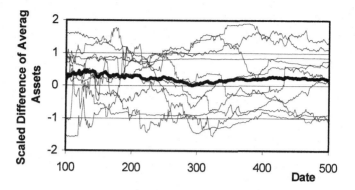

Fig. 5. Difference of average assets, **3x2Agents 2Stocks, Reverse**.

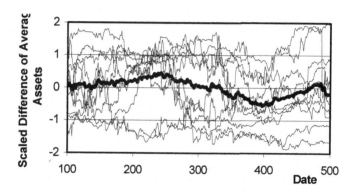

Fig. 6. Difference of average assets, **3x4Agents 4Stocks, Inflationary** .

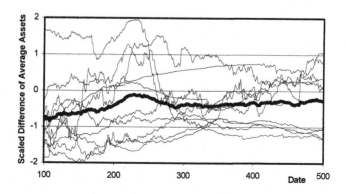

Fig. 7. Difference of average assets, **3x4Agents 4Stocks, Reverse**.

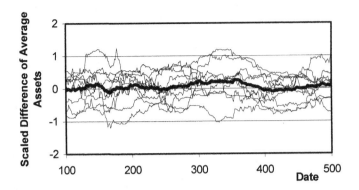

Fig. 8. Difference of average assets, **7x2Agents 2Stocks, Inflationary** .

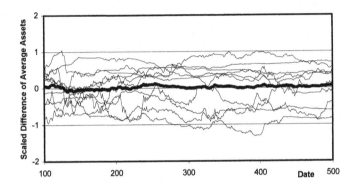

Fig. 9. Difference of average assets, **7x2Agents 2Stocks, Reverse**.

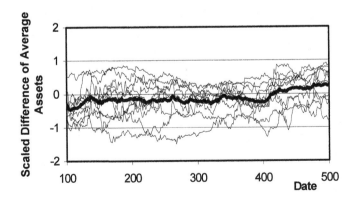

Fig. 10. Difference of average assets, **7x2Agents 4Stocks, Inflationary** .

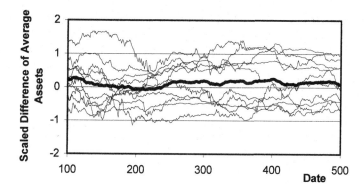

Fig. 11. Difference of average assets, **7x2Agents 4Stocks, Reverse**.

On the whole the above results do not indicate any solid evidence that prediction imparts any significant advantage over traders who do not predict. If any general trends are discernable the predictive traders did better when there where only 3 traders of each kind and only two stocks. It also did slightly better with the inflationary pricing mechanism than the reverse pricing mechanism. The table below shows the average of the average difference over the 500 time periods.

Table 2. A summary of the overall advantage (as the average over time of the average scaled difference of the average asset values of the two types of traders).

Number of agents of each kind	Number of stocks that can be traded	Pricing mechanism	overall advantage to predictors
3	2	Inflationary	0.25
3	2	Reverse	0.27
3	4	Inflationary	0.067
3	4	Reverse	-0.46
7	2	Inflationary	0.070
7	2	Reverse	-0.026
7	4	Inflationary	-0.098
7	4	Reverse	0.098

Momentum

I speculated that the reason that prediction did not seem to aid the agents was that there was no 'momentum' in the actions or their environment – in other words that everything can change direction instantly. It might be that prediction is more useful in circumstances where a significant level of momentum exists because then one has to anticipate the results – simple reaction will not be so effective because the effect of actions are delayed. In the next four sets of runs the set-up was the same except that the intentions of actions (that is the determined actions before they are moderated by what is possible for the agent) were delayed so that half the effect occurred immediately and half next cycle. That is each intended action is a 50/50 mix of those actions determined in this cycle and the last.

Four sets of runs were done with 3 agents of each type, with each pricing mechanism and with 2 or 4 stocks. The scaled difference of average asset values are shown in figures 12 to 15.

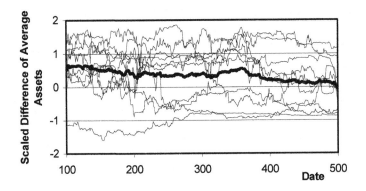

Fig. 12. Average assets of predictive traders minus average assets of non-predictive traders scaled in the simulation with action delay **3x2Agents 2Stocks, Inflationary**.

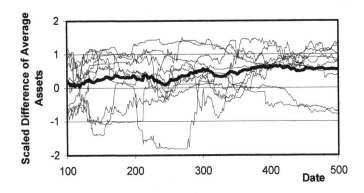

Fig. 13. Difference of average assets, **3x2Agents 2Stocks, Reverse**.

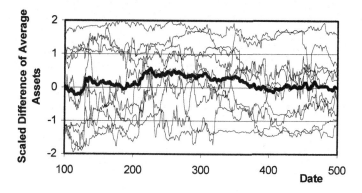

Fig. 14. Difference of average assets, **3x2Agents 4Stocks, Inflationary**.

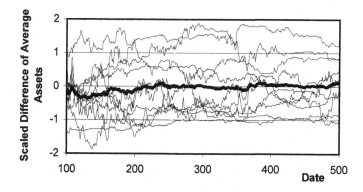

Fig. 15. Difference of average assets, **3x2Agents 4Stocks, Reverse**.

In each case the addition of momentum into the set-up did improve the overall performance of the predictive traders, but not by much. Table 3 below is the equivalent of table 2 for the set-ups with momentum.

Four Longer Runs

To see what happened after a longer period of time I ran the **3x2Agents 2Stocks, Reverse** set-up four times for 5000 time periods. I chose this set-up because this seemed to indicate that the predictive traders had the most advantage in this set-up. The results are shown in figure 16. Here one sees that the initial advantage slowly dissipates until there is none by time period 3000. One also observes that the agents gradually lock into their relative positions – this is beacause there is limited stock and hence, once this is all owned and nobody wants to trade the market becomes quietscent. These runs do not indicate any permanent advantage of the predictive traders over the non-predictive traders, but it may be that what we are observing here is a sort of Baldwin Effect (see discussion below).

Table 3. A summary of the overall advantage in the presence of momentum (as the average over time of the average scaled difference of the average asset values of the two types).

Number of agents of each kind	Number of stocks that can be traded	Pricing mechanism	overall advantage to predictors
3	2	Inflationary	0.36
3	2	Reverse	0.31
3	4	Inflationary	0.11
3	4	Reverse	-0.055

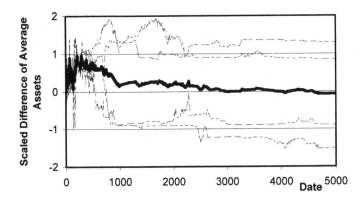

Fig. 16. Difference of average assets, **3x2Agents 2Stocks, Reverse**

4   Discussion

The simulations here do *not* indicate that prediction (in the form used here) always improves performance. Indeed this would be a surprising, even counter-intuitive result, if it had turned out to be the case. Rather the preliminary explorations highlight the importance of discovering the conditions under which prediction can aid adaptive action selection. Intuitively one would expect prediction to contribute to more successful decisions only if: (1) such prediction were feasible; and (2) that there was some advantage in knowing ahead of time what might occur compared to simply reacting to present events in the context of recent past events.

There are several reasons why prediction might not be significantly helpful here. It may be thought that prediction is simply not feasible in such markets – after all, if there is any pattern to the prices then a trader will learn to exploit this and the pattern will disappear. However examination of the predictive success of traders tends to discount this. Figure 17 below indicates the range of predictive success of the predictive agents over four stocks in a typical run. You notice that although they make a few spectacularly big errors, on the whole they do pretty well: always making some predictions more accurate than 10% and often being more accurate than 1%.

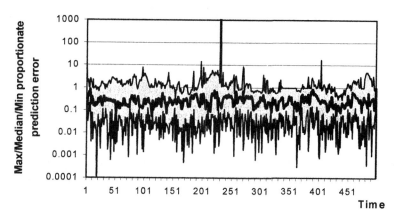

Fig. 17. The proportionate prediction error over three predictive traders and four stocks in a typical run of the simulation (reverse pricing, 3 traders of each kind, 4 stocks). Grey area indicates range of the errors, the bold line indicates the median error. A value of 1 indicates a 100% proportionate error, 0.1 a 10% error etc.

However in this environment success is very short-lived. What is successful in one time period may well be very misleading in the next. What I think is happening here is that although prediction is possible for short periods, any particular prediction strategy has a short 'shelf life' and will start to mislead after a while. Figure 18 below is a scatter graph of median prediction accuracy of the predictive traders against improvements (or otherwise) of the assets of predictive traders compared to those of intrumental traders over four runs of a market with 3 of each kind of agent, four stocks and the reverse pricing strategy. One notes that a low median prediction error is associated with some large increases in relative asset values but also with large decreases. It is notable that the correlation between the change in one time period and the next is –0.072, indicating that trading success does not last very long.

Thus the conclusion is that, although prediction is not possible in the long run, (since any prediction changes the market and the other traders soon learn to compensate for this) in the short term prediction is sometimes possible – one trader 'spots' a pattern in the prices of a stock that the other have not and exploits this. Both kinds of trader do worse than a trader who had a 'hard-wired' strategy of stable prediction (i.e. predict for the next price is the present price). Although they often 'chance' upon the fixed strategy they are 'mislead' into attempting to learn better predictive strategies due to the appearance of short-term patterns (which then quickly disappear). Clearly, in hindsight, the algorithm could be 'tuned' in order to improve its performance on this particular task. However this does not give one general information about the performance of learning with prediction. In particular it does not explain why the traders with prediction often did *worse* than those without prediction – if the prices were unpredictable because of some underlying randomness

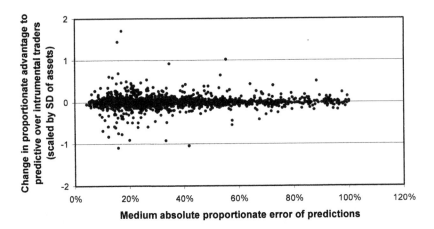

Fig. 18. A scatter graph of the change in proportionate advantage of predictive over non-predictive traders (scaled by standard deviation of assets) against the minimum predictive error made by predictive traders over the 4 stocks.

each kind would have done equally well. The fact is that anything that is learnt can be dangerous as well as advantageous when you are talking about prediction because you can never tell before hand which predictive strategy (if any) to rely on in dynamic environment. This contrasts strongly with the use of anticipation in static domains, and illustrates the danger of only using a restricted kind of test for one's technique.

It may be that we are observing something akin to the "Baldwin Effect" [2, 12], in which the advantage gained by learning is slowly incorporated into the genome over evolution. Here it might be that prediction is slowly learnt by the IALM module in a non-predictive fashion, cutting out the intermediate prediction and substituting a simple reaction to an observed pattern of events. This is not simply due to the traders trading less frequently with time as this does not account for the relative successes.

The results give some support for the hypothesis that one of the factors is the amount of momentum in the system, for the presence of such momentum would provide a clear need for prediction.

Fields tend to go through different stages in their development. Typically it runs like this: (stage 1) fields are started by isolated enthusiasts who know their new ideas have value and need to expend considerable energy in overcoming resistance from others; (stage 2) these then attract a close group of 'followers' who have come to appreciate these ideas and contribute to their study; (stage 3) some shortcomings and difficulties are raised causing initial resistance from the now established group; (stage 4) theories about the working of the technique and its conditions of application are mapped out. This paper seeks to play a part in moving the field to stage 3 and beyond – in essence, to go beyond the engineering approach of trying to make things work towards a scientific approach of understanding the root causes.

Whilst one can understand that one's intuitions may lead one to believe that correct anticipation is always useful, and thus that anticipation can always be usefully added when prediction is possible, such intuitions can lead one astray. When one is talking about a true anticipation (i.e. an anticipation *before* the event occurs) one can never *know* that the basis of one's anticipations are still valid. If one *knew* that the basis of one's anticipations remains valid then, of course, they can be usefully applied to gain leverage on the problem, but then this rules out the situated, dynamic and uncertain environment animals and animats inhabit. If one is ruling out such environments then one might as well attempt to learn the correct model of it and then use planning for one's actions.

That there are situations worse than random, where prediction is not only not advantageous but disadvantageous is illustrated by the results here. This should not be surprising – there are now many theoretical results (e.g. [13]) to show that any algorithm will only be useful with certain problems and that for every problem where a certain algorithm gives some advantage there will be another where it will be at a disadvantage. True progress can only come when one starts to distinguish each case.

Acknowledgements

Thanks to David Hales for comments, Scott Moss for discussions, Steve Wallis for writing SDML and to Mark Mills for discussions on how real markets work.

References

1. Arthur, B., Inductive Reasoning and Bounded Rationality. American Economic Association Papers, 1994. 84: p. 406-411.
2. Baldwin, J.M., A new factor in evolution. American Naturalist, 1996. 30: p. 441-451.
3. Challet, D. and Y.-C. Zhang, Emergence of Cooperation and Organization in an Evolutionary Game. Physica A, 1997. 246: p. 407.
4. Drescher, G.L., Made-up Minds - A Constructivist Approach to Artificial Intelligence. 1991, Cambridge, MA: MIT Press.
5. Edmonds, B., Modelling Bounded Rationality In Agent-Based Simulations using the Evolution of Mental Models, in Computational Techniques for Modelling Learning in Economics, T. Brenner, Editor. 1999, Kluwer. p. 305-332.
6. Edmonds, B., Developing Agents Who Can Relate To Us - putting agents in our loop via situated self-creation, in Socially Intelligent Agents, K. Dautenhahn, et al., Editors. 2002, Kluwer. 7.   Koza, J.R., Genetic Programming: On the Programming of Computers by Means of Natural Selection. 1992, Cambridge, MA: MIT Press.
8. Mills, M., Price fluctuations from the order book perspective: empirical facts and simple model. 1998. http://www.htcomp.net/markmills/order book and price fluctuation.pdf.
9. Palmer, R.G.e.a., Artificial Economic Life - A simple model of a stockmarket. Physica D, 1994. 75: p. 264-274.
10. Stolzmann, W. Anticipatory Classifier Systems. in Genetic Programming. 1998. University of Wisconsin, Madison, Wisconsin: Morgan Kaufmann.
11. Tolman, E.C., Purposive behavior in animals and men. 1932, New York: Appleton.
12. Turney, P., D. Whitley, and R.W. Anderson, Evolution, learning, and instinct: 100 years of the Baldwin effect. Evolutionary Computation, 1996. 4(3): p. iv-viii.
13. Wolpert, D. H. and Macready, W. H., No Free Lunch Theorem for Optimization IEEE Transactions on Evolutionary Computation, 1(1):67-82, 1997.

Appendix – Specification of the Simulation

Original Sources
The original agent-based artificial stock market was [9]. The rationale for the style of learning used in the learning modules may be found in [5]. The dual instrumental-predictive module structure was suggested in [6] but for a different purpose.

Static Structure
There is one market, where a fixed number of stocks are traded. In this market there is one market-maker who sets global prices and with whom all trades are made. There are a number of traders who buy and sell stocks with a view to increasing (the total book value of) their assets. Each of the traders is one of two types: predictive and non-predictive . There are equal numbers of each type of agent.

Each agent holds an amount of each stock and some cash. They have two learning modules: the IALM and the PPM. Each of these modules is a Genetic Programming algorithm with its own small population of tree structures. The population is of fixed size. Each member of the PPM population is a composite model with one tree for each stock. When the corresponding tree is interpreted it outputs the predicted price for that stock. Each member of the IALM population is a similar composite tree with a couple of trees as 'feature extractors' and then one tree for each stock which outputs how much the trader would like to buy and sell of that stock.

Temporal Structure
The model progresses via a sequence of discrete trading cycles. All the traders trade (or not) each cycle in parallel. In each trader there are a fixed number of stages that the cognition works through in order to make a trading decision.

Important Parameters
- Number of trading cycles
- Number of anticipatory traders
- Number of non-anticipatory traders
- Size of the GP populations
- Initial depth of the GP trees
- Proportion of new GP population created by propagation
- Proportion of new GP population created by crossover
- Proportion of new GP population newly generated
- Nodes and terminals of the populations (see below)
- Number of trading cycles over which the GP trees are evaluated
- Initial cash of the traders
- Initial amount of each stock held by the traders
- Initial amount of each stock held by the market-maker
- Level of random noise in the price setting process
- Number of initial trading cycles where traders trade randomly

The nodes and terminals of the trees are the following:

- *PPM Nodes*: AND, averageDoneByLast, averageIndexOverLast, averageOccuredToStockLast, boundedByMaxPrice, divide, dividendOf, doneByLast, F, greaterThan, IBoughtLastTime, IDidLastTime, indexLag, indexLastTime, indexNow, indexTrendOverLast, ISoldLastTime, lagBoolean, lagNumeric, lastBoolean, lastNumeric, lessThan, maxHistoricalPrice, minus, myMoney, NOT, onAvIBoughtLastTime, onAvISoldLastTime, OR, plus, presentStockOf, priceDownOf, priceLastWeek, priceNow, priceUpOf, randomBoolean, randomIntegerUpTo, randomPrediction, T, times, totalStockValue, volumeLastTime;

- PPM Terminals: F, indexLastTime, indexNow, maxHistoricalPrice, myMoney, onAvIBoughtLastTime, onAvISoldLastTime, randomBoolean, randomPrediction, T, totalStockValue, volumeLastTime *plus*: the names of the traders, stocks and a random selection of constants;

- IALM Nodes: AND, averageDoneByLast, averageIndexOverLast, averageOccuredToStockLast, avTraderPredictedIndex, boundedByMaxPrice, divide, dividendOf, doneByLast, F, greaterThan, IBoughtLastTime, IDidLastTime, indexLag, indexLastTime, indexNow, indexTrendOverLast, ISoldLastTime, lagBoolean, lagNumeric, lastBoolean, lastNumeric, lessThan, maxHistoricalPrice, minus, myMoney, NOT, onAvIBoughtLastTime, onAvISoldLastTime, OR, plus, predictionFor, presentStockOf, priceDownOf, priceLastWeek, priceNow, priceUpOf, randomBoolean, randomIntegerUpTo, randomPrediction, statResult, T, talkResult, times, totalStockValue, volumeLastTime;

- IALM Terminals: avTraderPredictedIndex, F, indexLastTime, indexNow, maxHistoricalPrice, myMoney, onAvIBoughtLastTime, onAvISoldLastTime, randomBoolean, randomPrediction, T, totalStockValue, volumeLastTime *plus*: the names of the traders, stocks and a random selection of constants;

There is not room to explain every single primitive, so I will give some indicative examples and summarise others.

- AND, OR, NOT, T, F are the obvious Boolean operators;
- times, divide, plus and minus are the normal arithmetic operators (except that divide by 0 results in 0);
- greaterThan and lessThan, a Boolean depending on the comparison;
- those of form random* produce a randomised output;
- those of form last* shift the evaluation one time period back;
- those of form lag* shift the evaluation back an indexed number of time periods;
- priceNow outputs the present price of a stock;
- indexNow outputs the present price index (the average of all prices);

- `donByLast` outputs the action of another trader for a given stock;
- `myMoney` outputs the agents amount of cash;
- `averageDoneByLast` outputs the average action over stocks;
- `boundedByMaxPrice` takes the min. of the number and the max. price;
- `dividendOf` outputs the current dividend rate of a stock;
- `IboughtLastTime` outputs a Boolean to represent whether the agent bought the given stock last time;
- `IdidLastTime` outputs the action for the given stock last time period;
- `predictionFor` outputs the prediction made for the current stock (PPM);
- `indexTrendOverLast` projects the next value given a linear trend over the last number of time periods;
- `presentStockOf` outputs the present level of a given stock ownedt;
- `totalStockValue` outputs the total value of the agent's stock.

The idea here is to provide agent with a rich set of primitives that include most of the operations they will need. If some turn out to be useless they will quickly be discarded by the learning modules. Below are a couple of examples.

- `[minus    [priceNow    'stock-1']    [maxHistoricalPrice 'stock-1']]` – sell if price is greater than the max. historical price otherwise buy;
- `[lagNumeric [2] [divide [doneByLast 'trader-2' 'stock-3'] [indexNow]]]` – do action of the action done by trader-2 for stock-3 divided by the price index 3 time periods ago.

Initialisation
The GP trees were initially randomly generated to a depth of 4 from the nodes and terminals available. For the first 5 cycles each trader makes small random trades, this has the effect of giving the GP process something to go on when it starts at cycle 6 and has the effect of damping down initial transient dynamics. The initial price of each stock and the initial dividend rate is randomly determined within a limited range.

Dynamics
Each trading cycle the following occurs in the market:
- Prices are set by the market maker, dividend rates are updated
- Each trader does the following (in parallel to other traders):
  1. Constructs new populations for its learning modules using propagation, crossover and new generation of trees (based on the fitness of trees last cycle).
  2. IALM evaluates its trees by how much its assets would be worth if the trading strategy of the tree was followed over a fixed number of past trading cycles (assuming historical prices).
  3. PPM evaluates its trees by their accuracy in predicting the next stock prices over a fixed number of past trading cycles (assuming historical prices). In the non-anticipatory version this is all done shifted one cycle back so that it is evaluated against predicting current stock prices.
  4. The best PPM tree is picked and evaluated to make predictions of next prices.
  5. The best IALM tree is picked and evaluated to decide intended decisions.

6. Trading decisions are moderated depending on what is possible for the trader (e.g. only buying what it can afford).

- The market maker deals with the bids to buy and sell it stock. And updates the accounts (how much stock and cash traders have).

Results Claimed as Significant and Intended Interpretation

That an ability to predict future prices is not always an advantage in this market. The simulation is intended to be interpreted in terms of how an animat might learn in a dynamic environment in which it was embedded.

Implementation Details Necessary to Get the Simulation to Run

The price setting mechanism of the market maker is basic and somewhat arbitrary. At the present it simply changes prices depending upon the net buying of the stock concerned during the previous trading period. A more sophisticated mechanism would be if the market maker tried to predict the net demand for each stock and guess the price necessary in order to fulfil the demand.

Implementation Language, Source Code, and Example Simulation Output

The simulation was implemented in the social simulation language: SDML, version 4.1. This is freely downloadable for academic use – see http://sdml.cfpm.. The source code is available as an SDML module at http://cfpm.org/~bruce/etvop/code. It requires SDML version 4.1 or later to run.

Generalized State Values in an Anticipatory Learning Classifier System

Martin V. Butz[12] and David E. Goldberg[1]

[1] Illinois Genetic Algorithms Laboratory
University of Illinois at Urbana-Champaign, IL, USA
{butz,deg}@illigal.ge.uiuc.edu
[2] Department of Cognitive Psychology
University of Würzburg, Germany
butz@psychologie.uni-wuerzburg.de

Abstract. This paper introduces generalized state values to the anticipatory learning classifier system ACS2. Previous studies showed that the evolving generalized state value in ACS2 might be overgeneral for a proper policy representation. Thus, the policy representation is separated from the model representation. A function approximation module is added that approximates state values. Actual action choice then depends on the learned generalized state values predicted by the means of the predictive model yielding anticipatory behavior. It is shown that the function approximation module accurately generalizes the state value function in the investigated MDP. Improvement of the approach by the means of further anticipatory interaction between predictive model learner and state value learner is suggested. We also propose the implementation of task dependent anticipatory attentional mechanisms exploiting the representation of the generalized state-value function. Finally, the anticipatory framework may be extended to support multiple motivations integrated in a motivational module which could be influenced by emotional biases.

1 Introduction

Anticipatory learning classifier systems (ALCSs) are learning systems that learn a generalized predictive model of an environment online. The environment is usually modeled by a Markov decision process (MDP). Predictive knowledge is usually stored in rules, called *classifiers*. This paper introduces an approach for the simulation of anticipatory behavior in an ALCS by the means of online generalized state values.

Previous studies have learned state values with other ALCSs [13, 11] exhibiting optimal behavior. In these approaches all encountered states are stored explicitly in a state list. Value iteration is applied over the stored states, learning an optimal policy represented in the state values. None of the approaches tried to generalize the state list, however.

ACS2 is an ALCS that recently showed to reliably evolve a complete, accurate, and compact predictive model in many typical MDP environments [6, 3].

M. Butz et al. (Eds.): Anticipatory Behavior ..., LNAI 2684, pp. 282–301, 2003.

The behavioral policy in ACS2 is directly encoded in the model by the means of reward prediction values (similar to Q-value in reinforcement learning [15, 22]). The direct encoding, however, was shown to possibly cause a *model aliasing* problem. In model aliasing, the learned predictive model is accurate in terms of the prediction of resulting states but inaccurate in terms of predicting reinforcement. This can result in incorrect reward propagation as well as misleading reinforcement values.

This paper introduces a generalizing state value learner to ACS2. The combination of ACS2 with this generalizing state value learner is termed XACS. XACS shows optimal behavior in a blocks world problem in which ACS2 was not able to learn an optimal behavior due to model aliasing. We also show that the generalization mechanism in XACS, currently implemented by a modified version of the classifier system XCS [24], works efficiently. We confirm that the generalization properly specifies task relevant perceptual attributes. Moreover, generalization enables learning of problems with larger state spaces. The implementation of other anticipatory mechanisms in XACS, such as task dependent attentional mechanisms, further interactions of the learning components, and multiple behavioral modules for the representation of multiple motivations (or needs) as well as the integration of an emotional component are suggested.

The paper starts with an introduction to the anticipatory learning classifier system ACS2. Next, the problem of model-aliasing is explained and the blocks world problem is introduced along those lines. Section 4 introduces the function approximation module, implemented by the accuracy based classifier system XCS, that approximates the state values for an optimal anticipatory behavioral policy. The derived system XACS is tested in a blocks world problem confirming optimal performance and proper generalization. Section 6 outlines how further anticipatory mechanisms could be included in XACS. Finally, summary and conclusions are provided.

2 ACS2—A Predictive Model Learner

ALCSs learn a predictive model of an encountered environment specifying the consequences of each possible action in each possible situation. Predictions are usually represented explicitly in each rule or classifier. That is, each rule consists of a condition-action-effect triple that specifies which perceptual effects are expected after executing the specified action provided the specified conditions are satisfied.

A first anticipatory learning classifier system, termed ACS, was published in [19, 20]. The learning mechanism in ACS is based on Hoffmann's learning theory of anticipatory behavioral control [14]. Essentially, ACS specifies encountered action-effect relations first and then considers conditional dependencies where necessary. The ALCS herein, termed ACS2, is an enhancement of Stolzmann's ACS including a genetic generalization mechanism. A complete description of the system including an algorithmic description as well as an experimental performance analysis can be found in [3]. This section gives a basic introduction to the system. The intention is to give the reader the idea of what the system is actually learning and how it basically works. For more details, the interested reader is referred to the cited literature.

2.1 Agent Architecture

ACS2 interacts with an environment or problem perceiving situations and reinforcement and executing actions. In reinforcement learning terms, ACS2 learns the state transition function of an MDP.

More formally, ACS2 perceives situations $\sigma \in \mathcal{I} = \{\iota_1, \iota_2, ..., \iota_m\}^L$ where m denotes the number of possible values of each environmental attribute and L the number of attributes. ACS2 executes actions $\alpha \in \mathcal{A} = \{\alpha_1, \alpha_2, ..., \alpha_n\}$ where n specifies the number of different possible actions in the environment and $\alpha_1, ..., \alpha_n$ denote the different possible actions. The environment is characterized by its state transition function f that transfers current states and actions into consequent states: $f : \mathcal{I} \times \mathcal{A} \to \mathcal{I}$. ACS2 learns an approximation of the state transition function. We assume the Markov property of a problem, that is, the consequences of any action in any situation only depend on the current perceived situation and not on the history. Moreover, we restrict ourselves to deterministic environments. Thus, any situation-action combination $(\sigma - \alpha)$ deterministically leads to a new situation σ'.

Additionally, the MDP provides reinforcement. That is, the environment provides scalar feedback ρ for specific situation, action, next situation combinations. The feedback ρ is determined by a reinforcement function $\nabla : \mathcal{I} \times \mathcal{A} \times \mathcal{I} \to \Re$. Thus, the environment represents an MDP with states $\mathcal{I}$, actions $\mathcal{A}$, a reward function ∇, and a state transition function f.

2.2 Knowledge Representation

Knowledge in ACS2 is represented by a *population* $[P]$ of condition-action-effect rules, i.e. the *classifiers*. A classifier in ACS2 predicts a complete resulting state. Each classifier further carries the following additional attributes. The *quality* q measures the average accuracy of the effect prediction of the classifier. The *reward prediction* r estimates the average discounted reward encountered. The *immediate reward prediction* ir estimates direct reward. Moreover, a classifier carries a *mark* M, a *GA time stamp* t_{ga}, an anticipatory learning process (ALP) *time stamp* t_{alp}, an *application average estimate* aav, an *experience* counter exp, and a *numerosity* counter num.

Condition and effect part consist of the values perceived from the environment and '#'-symbols (i.e. $C, E \in \{\iota_1, ..., \iota_m, \#\}^L$). A #-symbol in the condition, called *don't-care symbol*, denotes that the classifier matches any value in this attribute. A '#'-symbol in the effect part, called *pass-through symbol*, specifies that the classifier anticipates that the value of this attribute will not change after the execution of the specified action. Non pass-through symbols in E anticipate the change of the particular attribute to the specified value in contrast to ACS in which a non pass-through symbol did not require a change in value. Action parts specify any action possible in the environment ($A \in \mathcal{A}$). The measures q, r, and ir are scalar values where $q \in [0, 1]$ and $r, ir \in \Re$. The mark $M \in \{m_1, ..., m_m\}^L$ where $m_i \in \{\iota_1, ..., \iota_m, \#\}$ characterizes the set of situations in which the predictions of the classifier were inaccurate to enable directed specialization. The GA time stamp and the ALP time stamp record the time of the last genetic generalization and ALP application, respectively. The application average estimate estimates the average time between ALP applications. The experience counter specifies the actual number of ALP applications. Finally, the numerosity allows the representation of multiple identical classifiers in one (macro-)classifier.

2.3   Learning in ACS2

While interacting with an environment, the population of ACS2 increasingly approximates the state transition function of the MDP. Usually, the agent starts without any prior knowledge except for the knowledge implicitly included in the coding structure and the provided actions. Initially, classifiers are mainly generated by a covering mechanism. Later, an anticipatory learning process (ALP) generates specialized classifiers while a genetic generalization process produces generalized offspring. RL techniques are applied for the evolution of an optimal behavioral policy represented directly in the predictive model.

Figure 1 illustrates the interaction of ACS2 with its environment and its learning application in further detail. After the perception of the current situation $\sigma(t)$, ACS2 forms a match set $[M]$ comprising all classifiers in the population $[P]$ whose conditions are satisfied in $\sigma(t)$. Next, ACS2 chooses an action $\alpha(t)$ according to some action selection strategy and an action set $[A]$ is generated that consists of all classifiers in $[M]$ that advocate $\alpha(t)$. After the execution of $\alpha(t)$, classifier parameters are updated by ALP and reinforcement learning techniques and new classifiers might be generated as well as old classifiers might be deleted by ALP and genetic generalization.

The ALP is derived from the cognitive learning theory of anticipatory behavior control [14, 8]. The theory emphasizes that learning in "higher" animals and humans appears to establish primarily action-effect relations. Situational dependencies are only learned in a secondary process. Similarly, ACS2 first learns action-effect relations which are further conditioned on situational dependencies where appropriate. This is implemented in the ALP which first generates classifiers that represent action-effect relations. Situational dependencies are then established using the information derivable from the difference in the mark M and a successful situation.

In further detail, the ALP first updates classifier parameters in the current action set $[A]$ with respect to the subsequent perception $\sigma(t + 1)$. If a classifier predicted the encountered change correctly, its quality is increased by $q \leftarrow q + \beta(1 - q)$ whereas its quality is decreased by $q \leftarrow q - \beta q$ otherwise. If there was no representative present for the chosen action, a *covering classifier* is generated that specifies the encountered action-effect relation in its condition and effect part. Otherwise, specialized offspring is generated where necessary (indicated by

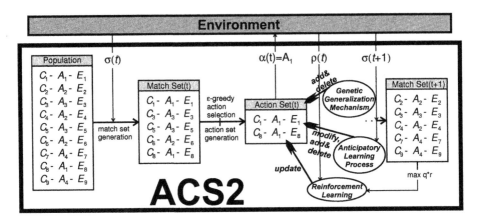

Fig. 1. During one agent/environment interaction, ACS2 forms a match set representing the knowledge about the perceived situation. Next, it generates an action set representing knowledge about the possible consequences of the chosen action. After executing the chosen action, payoff and the resulting situation is perceived. Classifier parameters are updated by RL considering payoff and ALP considering the resulting situation. The model evolves by the means of ALP and genetic generalization.

the mark M) and inaccurate classifiers are deleted. Parameter updates are done using learning rate β.

Despite the informed specialization process by the ALP, over-specializations can occur sometimes as studied in [3]. Since over-specialization can be the result of several distinct circumstances, a genetic generalization (GG) mechanism is applied. GG selects classifiers in action sets [A] proportionally to their fitness. Selected classifiers are reproduced and the conditions are generalized (by a generalizing mutation operator) and crossed over. Finally, classifiers are deleted in [A] if the action set size exceeds a certain threshold θ_{as}.

The interaction of ALP and GG results in the generation of a complete, accurate, and maximally general environmental model. While ALP continuously specializes over-general classifiers, GG continuously generates generalized offspring. Thus, there is a continuous generalization pressure that alleviates possible over-specializations. Moreover, possible over-generalizations are alleviated by the specialization pressure where necessary. Further balancing is achieved by a subsumption technique in which over-specialized offspring is subsumed by existing accurate, more general classifiers.

The applied RL technique is an approximation of Q-learning [23]. The reward prediction r of a classifier is updated according to the following equation:

$$ r \leftarrow r + \beta(\rho(t) + \gamma \max_{cl \in [M](t+1) \wedge cl.E \neq \{\#\}^L} (cl.q \cdot cl.r) - r) \tag{1} $$

Parameter $\beta \in [0, 1]$ denotes the learning rate biasing the parameters more or less toward recently encountered reward values. $\gamma \in [0, 1)$ denotes the discount factor biasing the reward values more or less toward future reinforcement values. The constrained that the classifier whose reward prediction is propagated does not have a completely general effect part $(cl.E \neq \{\#\}^L)$ is imposed to avoid self-propagation of reward or propagation of reward that actually results in no change in the environment. The reward prediction value r consequently estimates the average resulting discounted reward after executing action A in all possible situations σ in which the classifier is applicable following an optimal policy thereafter.

Action selection can now be biased more or less strong on the reward prediction values r. Usually, ACS2 applies a simple ϵ-greedy action selection strategy [22] in which an action is chosen at random with a probability ϵ and otherwise the best action is chosen. The action of the classifier with the highest qr value in a match set $[M]$ is usually considered as the best action in ACS2. Recently, an additional action selection bias was applied speeding up model learning [4]. Instead of choosing an action at random, with a probability p_b an action is chosen that promises the highest knowledge increase (represented either by the highest average application delay or the lowest average quality).

3   Model Aliasing

While the evolving behavioral policy represented by the reward prediction values of classifiers previously showed to solve different maze problems, it was also shown that a problem, termed *model aliasing*, can occur. The problem is that the classifiers in the evolving predictive model might be over-general to specify accurate reinforcement values. That is, although a classifier might accurately specify the perceptual effects in all situations its conditions are satisfied, its reward prediction value might be inaccurate. Similar problems have been reported in the reinforcement learning literature when generalizing the state value function [2, 15].

3.1   A Simple Example

An example with respect to model aliasing might clarify the problem. Imagine a game such as four-wins in which ACS2 learns the moves of the game. Actions would be to place a coin in either of the (usually) seven slots. After some time, ACS2 would have learned a proper representation of the effects of all possible actions (representing the moves of the other player in some way). To make accurate action-effect predictions the condition of a classifier must specify the current number of coins in the slot the specified drop-coin action in the classifier aims for. The conditions, however, would be over-general to specify an accurate reinforcement value since the classifiers are applicable nearly independent of the game situation. Thus, the reward prediction values would average over all possible actual game situations and would consequently be meaningless. The basic intuition behind this is that a representation of the rules of a game is usually insufficient for the representation of a good strategy for playing the game. An example in the investigated blocks world below further clarifies the problem.

3.2   Blocks World Problem

We investigate the behavior of ACS2 in a blocks world scenario introduced in [3] in which a model aliasing problem applies similar to the four-wins game. In the blocks world, b blocks are distributed over a certain number of stacks s. ACS2 is able to manipulate the stacks by the means of a gripper that can either grip or release a block on a certain stack. ACS2 perceives the current block distribution coding each stack with b attributes. One additional attribute indicates if the gripper is currently holding a block. Thus, the perceivable situations are a subset of $\mathcal{I} \subset \{*, b\}^{bs+1}$. Since there are $\binom{b+s-1}{s-1}$ possibilities to distribute b blocks over s stacks and in a current situation ACS2 might currently hold a block in its gripper or not, there are $\binom{b+s-1}{s-1} + \binom{b+s-2}{s-1}$ possible environmental situations.

Given the task to have exactly y blocks on the first stack, the predictive model is usually too general to represent an optimal behavioral policy. Figure 2 visualizes the problem in a simple $b = 4, s = 3, y = 4$ blocks world. To predict the effect of releasing a block on the middle stack accurately, the condition of a classifier needs to specify only that there is no block present on the middle stack and a block is currently gripped. This classifier matches in all four situations shown at the left of Figure 2. Thus, its reward prediction value will approximate an average value over all those possible situations estimating that $(8 + 6 + 4 + 2)/4 = 5$ steps are necessary to reach the goal. Releasing a block on the left stack in the left-most situation requires exactly 6 further steps to reach the goal. Thus, dropping a block on the middle stack would be preferred exhibiting the model-aliasing problem. Additionally, the model-aliasing problem is propagated by the means of the adapted $Q - learning$ mechanism so that the values might be even more misleading.

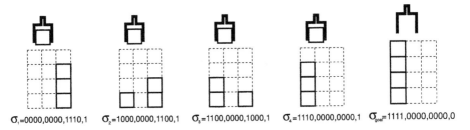

σ_1=0000,0000,1110,1 σ_2=1000,0000,1100,1 σ_3=1100,0000,1000,1 σ_4=1110,0000,0000,1 σ_{goal}=1111,0000,0000,0

Fig. 2. Model Aliasing in ACS2: To accurately predict the effects of a releasing action on a particular stack, classifiers in ACS2 evolve whose conditions only specify the state of the addressed stack. Thus, given the goal shown in the right-most scenario, the classifier for releasing a block on the first stack will have a lower reward prediction value than the classifier for releasing a block on the second stack in the left-most scenario since the latter classifier is applicable in all four scenarios on the left.

The next section introduces a function approximation module to ACS2 that approximates the optimal state value function making the policy representation independent of the predictive model representation.

4   Generalized State Values in ACS2

As the last section revealed, the RL approach in ACS2 is somewhat limited. The approach is only capable of solving Markov decision processes in which the learned predictive model is specific enough to represent accurate reinforcement values. This section is dedicated to remedy this drawback introducing a generalizing state value learner implemented by a modified version of the learning classifier system XCS [24].

A previous approach to the same problem in another ALCS using a list of all encountered states to learn a proper state value approximation was published in [13]. Our approach takes this idea one step further by using a generalized representation of states. We first formalize the actual state values that ACS2 learns to behave optimally. Next, we introduce XCS that learns a generalized representation of these state values.

4.1   State Values

In the MDP framework formulated above, reinforcement is provided in certain situation, action, resulting situation combinations. Since ACS2 learns a predictive model, i.e. the state transition function f of the MDP, dynamic programming techniques can be applied using the learned predictive model to approximate the following Bellman equation representing the state value function of an optimal behavioral policy.

$$V^*(\sigma) = R(\sigma) + \gamma \max_{\alpha} V^*(f(\sigma, \alpha)) \tag{2}$$

where

$$R(\sigma) = \frac{\sum_{\{(\sigma^{-1}, \alpha^{-1}) | f(\sigma^{-1}, \alpha^{-1}) = \sigma\}} \nabla(\sigma^{-1}, \alpha^{-1}, \sigma)}{|\{(\sigma^{-1}, \alpha^{-1}) | f(\sigma^{-1}, \alpha^{-1}) = \sigma\}|}$$

Hereby, $R(\sigma)$ denotes the average immediate reinforcement encountered when entering state σ. This notation differs from the usual state value definition. It enables us to disregard a separate representation of immediate reward.

The optimal policy can be defined based on the state transition function f and the optimal state values V^*.

$$\pi^*(\sigma) = \arg \max_{\alpha} V^*(f(\sigma, \alpha)) \tag{3}$$

The optimal strategy π^* evolves by iteratively learning the optimal state value function in the following way.

$$V(\sigma) = V(\sigma) + \beta(\rho(t) + \gamma \max_{\alpha} V(f(\sigma, \alpha)) - V(\sigma)) \tag{4}$$

Parameter β denotes the learning rate. In order to converge, it must be assured that ρ is sampled uniformly over all possible $(\sigma^{-1}, \alpha^{-1}, \sigma)$ combinations. However, in all environments (such as the one investigated below) in which payoff depends only on a state and not on the state-action combination that led to that state (that is $\nabla(\sigma^{-1}, \alpha^{-1}, \sigma)$ is equal for all combinations of σ^{-1} and α^{-1} that can lead to σ) this requirement is not necessary [5].

As explained above, ACS2 approximates the state transition function f of the environment. Once ACS2 represents the complete state transition function f accurately, the updates can be determined using ACS2's predictive model instead of the actual state transition function f. Since model and policy should be learned simultaneously, the state transition function is approximated by ACS2's best guess in a given situation-action combination. As the best guess, denoted by bg, we use the highest quality classifier of ACS2 of all classifiers in ACS2 that match in the given situation and specify the corresponding action.

$$\text{bg}(\sigma, \alpha) = \arg \max_{\{cl | cl \in [P] \wedge cl.C \text{ matches } \sigma \wedge cl.A = \alpha\}} cl.q \qquad (5)$$

Value updates in XCS are now applied in all encountered situations using the prediction of classifier cl, $p(\sigma, \alpha, cl)$, for the next-state prediction. Thus, updates are applied as follows:

$$V(\sigma) = V(\sigma) + \beta(\rho(t) + \gamma V(p(\sigma, \alpha, \text{bg}(\sigma, \alpha))) - V(\sigma)) \qquad (6)$$

These updates are applied to the state value learner XCS. The next section explains how V^* is approximated by XCS in interaction with ACS2. The implementation that stores all encountered situations explicitly and evaluates the situations according to equation 6 is referred to as ACS state value learner (ACS-SVL) in the remainder of this paper.

4.2   XCS Learns Generalized State Values

The learning classifier system XCS was introduced in [24]. XCS's learning is based on the accuracy of the reward prediction values of classifiers evolving a complete and accurate mapping of all possible situation action combinations in the environment. XCS was also used as a function approximation tool [25]. In our framework, XCS approximates the optimal state values V^* defined above. This section first gives an overview of the basic XCS structure and mechanisms. For further details the interested reader is referred to the cited literature as well as the recent algorithmic description of XCS [9]. In the following, it is explained how XCS interacts with ACS2 to evolve a proper function approximation.

XCS in a Nutshell Similarly to ACS2, XCS evolves a population of classifiers. In our framework, each XCS classifier has a condition part C_x with the same structure as conditions in ACS2's classifiers. Since XCS approximates state values in our framework and not values for state action combinations, the XCS classifiers have no action in this framework. Moreover, each XCS classifier has a reward prediction value p_x, a measure for the average error of the reward prediction value ϵ_x, and a fitness value F_x. The fitness in XCS measures the average accuracy of the classifier relative to the accuracies of all classifiers in the sets cl

participates. Moreover, each classifier in XCS comprises a set size estimate ss_x that estimates the average match set size in which cl participates, the GA time stamp ts_x, an experience counter exp_x that counts the number of parameter updates the classifier underwent, and a numerosity value num_x that counts the number of actual micro-classifiers this "macroclassifier" represents.

In interaction with a problem, XCS encounters situations σ and forms match sets $[M]$ that consist of all classifiers in the population $[P]$ that match σ. Moreover, XCS encounters reinforcement values ρ_x and updates its parameters as follows:

$$p_x = p_x + \beta_x(\rho_x - p_x)$$
$$\epsilon_x = \epsilon_x + \beta_x(|p_x - \rho_x| - \epsilon_x)$$
$$ss_x = ss_x + \beta_x(|[M]| - ss_x) \tag{7}$$

The fitness is updated by first determining its absolute accuracy κ_x of the classifier by a scaled function and next, its relative accuracy κ'_x with respect to the current match set. Finally, its fitness value is updated according to this relative accuracy.

$$\kappa_x = \begin{cases} 1 & \text{if } \epsilon_x < \epsilon_{0_x} \\ \alpha_x\left(\frac{\epsilon_{0_x}}{\epsilon_x}\right)^{\nu_x} & \text{otherwise} \end{cases}$$
$$\kappa'_x = \kappa_x num_x/(\sum_{cl\in[M]} cl.\kappa_x cl.num_x)$$
$$F_x = F_x + \beta(\kappa'_x - F_x) \tag{8}$$

Parameter ϵ_{0_x} denotes an accuracy threshold. Once the average error ϵ_x is lower than this value the classifier is considered accurate. Parameter ν_x further scales the inaccurate classifiers and α_x adds an offset to further distinguish accurate from inaccurate classifiers.

After value updates and offspring generation by the ALP in the current match set $[M]$, a GA might be applied if the average last GA application in $[M]$ exceeds the threshold θ_{ga}. If a GA is applied, two classifiers are selected in $[M]$ usually applying proportionate selection based on fitness. In our implementation, we use tournament selection with a tournament size proportionate to the actual match set size $[M]$. This selection method has shown to improve performance of XCS and make it more noise robust [7]. The selected classifiers are mutated with a niche mutation technique [9] that mutates an attribute in C_x with a probability μ_x. Moreover, the conditions of the two classifiers are crossed with a probability χ_x applying uniform crossover. The resulting offspring is added to the population. However, if there exists a more general, experienced ($\theta_{sub} < exp_x$), and accurate classifier ($\epsilon_0 > \epsilon_x$) in $[P]$, then an offspring classifier is not added but the numerosity of the subsuming classifier is increased by one. Moreover, classifiers are deleted in the population if the population size exceeds the threshold N_x. For deletion, classifiers are selected using proportionate selection with respect to the set size estimate of the classifiers, their experience, and their fitness.

Interaction with ACS2 For the interaction with ACS2, it is necessary to define when and how parameters are updated in XCS. Each time ACS2 perceives a situation σ and an immediate reward ρ, XCS is called.

The reinforcement value for reward prediction updates in XCS are determined by a combination of the current predictive model knowledge of ACS2 and the current state values of XCS.

$$\rho_x = \rho + \gamma \max_{\alpha} \frac{\sum_{\{cl|cl \text{ matches } P(\sigma,\alpha,\mathrm{bg}(\sigma,\alpha))\}} cl.p_x \cdot cl.\kappa_x \cdot cl.num_x}{\sum_{\{cl|cl \text{ matches } P(\sigma,\alpha,\mathrm{bg}(\sigma,\alpha))\}} cl.\kappa_x \cdot cl.num_x} \qquad (9)$$

This is similar to the prediction array calculation in XCS but uses the predictive model of ACS to determine next states. Moreover, the reward estimates are weighted according to the current accuracy instead of the current fitness of a classifier. The equation essentially adds the current immediate reward to the best state value prediction in all possible future state.

Once ρ_x is determined, all classifiers in the current match set $[M]$ of XCS can be updated according to the update rules defined in equations 7 and 8. Moreover, the GA can be invoked.

Restricted and Additional State Value Evaluation First experiments showed that XCS is not able to evolve a complete state value representation reliably. The problem is that XCS evolves classifiers with very general conditions, if the position in which immediate reward is experienced is not visited early enough nor frequently enough, recovery from this early over-generalization effect appears to be hard. Thus, two mechanisms are added to the XCS evaluation procedure. (1) XCS parameter updates and GA application are restricted. (2) An additional list of states is maintained on which internal evaluations are performed. Both mechanisms are explained in further detail below.

To prevent early over-generalization and disruption due to strongly biased sampling, learning is only triggered in XCS, if there exist useful prediction values. Moreover, learning is never triggered more than once in a row. Prediction values are considered useful if ACS2 is actually able to predict at least one consequent state and XCS has a state value representation of the predicted state. To prevent XCS from updating classifiers in one situation multiple times in a row, the last situation in which an update occurred is remembered. If the same situation is encountered in the next step, neither parameter updates nor genetic algorithm are applied.

Additionally, XCS keeps a list of states of size sl_x in which parameter updates with high average prediction error occurred. Each entry in the list stores a perceived situation together with its average error encountered in the state as well as the encountered prediction value ρ_x. After each actual parameter update and GA application, the state list is scanned. If the current state exists in the state list, the prediction value and average error value are updated. The entry will be deleted once the average error decreases below $10 \cdot \epsilon_0$. If the current state does not exist and the lowest average error in the list is smaller than the one encountered, the entry is replaced by the actual situation with corresponding values. Next, an internal update is triggered in which an entry is chosen from the maintained list proportional to the average error values. The update is applied using the stored ρ_x value. Next, the average error is replaced by the average error of the current update. Moreover, a GA might be triggered.

4.3 Resulting Behavioral Policy

With the definitions above, we have constructed a learning agent that learns a predictive model of its environment (i.e. the state transition function in an MDP) as well as the optimal state value function of the problem. The state value function is learned by using the model predictions of ACS2 and the value estimates in the predicted situations to update the reward predictions in XCS. In the remainder of this work, we will term the combination of ACS2 and XCS the extended anticipatory learning classifier system XACS.

What remains is to show how anticipatory behavior is applied in XACS. Given a current situation σ we defined the optimal policy π^* above given the state transition function f and the optimal state value function V^*. Since ACS2 learns the state transition function and XCS the optimal state value function, an optimal policy can now be defined by

$$\pi(\sigma) = \arg\max_{\alpha} \frac{\sum_{\{cl|cl \text{ matches } P(\sigma,\alpha,\text{bg}(\sigma,\alpha))\}} cl.p_x \cdot cl.\kappa_x \cdot cl.num_x}{\sum_{\{cl|cl \text{ matches } P(\sigma,\alpha,\text{bg}(\sigma,\alpha))\}} cl.\kappa_x \cdot cl.num_x} \tag{10}$$

since $f(\sigma,\alpha)$ is set equal to the prediction done by the best guess of ACS2 and $V^*(\sigma)$ is set equal to the accuracy weighted average prediction of classifier in XCS that match σ.

Essentially, this policy is an anticipatory policy since it predicts the result of each action before action execution and then chooses the best action with respect to the prediction. The major difference to traditional dynamic programming approaches is that the predictive model as well as the state values are learned online and generalized online.

4.4 Performance in the Blocks World

In order to evaluate the performance of XACS, we monitor its performance in the blocks world introduced above. XACS interacts with the blocks world by the provided s gripping and releasing actions. A reinforcement of 1000 is provided when the first stack reaches its specified height y. Moreover, the environment is reset to a random situation in this case. The experiments are divided into exploration and exploitation trials. During exploration, actions are chosen at random applying the biased exploration procedure published in [4] with a probability of p_b. All learning components are applied. Exploit trials are used for performance evaluation purposes only and consequently no parameter updates nor offspring generation is triggered. Actions are chosen according to policy $\pi(\sigma)$ defined in equation 10. A trial is limited to 50 steps in all the experiments. If the goal was not reached after 50 steps, the environment is reset to a random situation.

Performance is displayed by averaging the steps to the goal over the last 50 exploitation trials. Curves are averaged over fifty runs.[1]

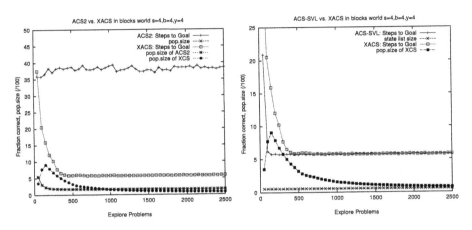

Fig. 3. As observable on the left hand side, ACS2 suffers from model aliasing while XACS is able to learn an optimal policy in the blocks world problem. XACS achieves similar performance to the state value learner without generalization (right hand side).

The left hand side of figure 3 shows that XACS is able to adapt its behavior in the $s = 4$, $b = 4$, $y = 4$ blocks world reaching optimal behavior after approximately 150 learning trials. The figure also displays performance of ACS2 clearly showing that ACS2 is not able to evolve optimal behavior in the problem. The right hand side of figure 3 compares performance of XACS with performance of ACS-SVL that does not generalize the state values but keeps a list of all encountered states propagating reinforcement similar to XACS over this state list. The comparison shows that XACS eventually reaches the goal as fast as ACS-SVL but learning takes slightly longer. The population size of the XCS module in XACS approaches the actual number of states in the problem. This shows that the generalization mechanisms does not cause disruption but it does not result in any gain, either, in this setting.

To see if the generalization mechanism was functioning at all, we took a look at the evolving classifier list of XCS. Table 1 lists the highest numerosity classifiers in the blocks world problem $s = 4$, $b = 4$, $y = 4$. The condition

---

[1] If not stated differently, parameters were set to: $\beta = 0.05$, $u_{max} = \infty$, $\gamma = 0.9$, $\theta_{ga} = 100$, $\mu = 0.3$, $\chi = 0.8$, $\theta_{as} = 20$, $\theta_{exp} = 20$, $\epsilon = 1$, and $p_b = 0.5$. XCS parameters were set to $N_x = 2000$, $\beta_x = 0.05$, $\alpha_x = 1$, $\epsilon_0 = 1$, $\nu_x = 5$, $\theta_{GA_x} = 25$, $\chi_x = 0.8$, $\mu_x = 0.01$, $\theta_{del_x} = 20$, $\delta_x = 0.1$, $\theta_{sub_x} = 20$, $p_{I_x} = 0$, $\epsilon_{I_x} = 10$, $f_{I_x} = 0$. Note that some of the parameters were not introduced above but are provided for completeness. The meaning of the parameters can be inferred from the cited literature on ACS2 and XCS.

part successively codes the four positions in each of the four stacks coding from left to right starting from the lowest position in each stack. Also the gripper state is coded. It can be seen that all classifiers are accurate and predict the correct state value. Moreover, nearly all classifiers are maximally general except for classifiers number 9,10, and 12. Classifier 9's specified problem subspace is more specific than the one of classifier 11 and should be eventually absorbed by it. Similarly, classifier 10 is more specific than classifier 7 and classifier 12 is more specific than classifier 8. Note that classifiers 9 and 10 are more specific in the blocks world problem than 11 and 7 although they specify less attributes. The listed classifiers in table 1 confirm that XCS successfully evolves a generalized state-value representation.

Table 1. A typical classifier list ordered by their numerosity values show that XCS properly evolves generalized state values in the blocks world problem.

	C_x	p_x	ϵ_x	F_x	num_x	exp_x	ss_x
1	.### #### #### #### A	478	0.001	0.992	199	27880	210
2	.### #### #### #### .	430	0.001	0.972	189	20625	204
3	A.## #### #### #### .	531	0.002	0.998	167	15695	181
4	###A #### #### #### #	1000	0.000	0.851	138	1565	187
5	A.## #### #### #### A	590	0.001	0.820	136	12964	180
6	##A# #### #### #### A	900	0.000	1.000	123	1582	157
7	#A.# #### #### #### .	656	0.004	0.803	95	8328	145
8	#A.# #### #### #### A	729	0.000	0.643	90	5750	169
9	##A# #### #### A### #	810	0.000	0.538	88	243	177
10	#A## #### #### #A## #	656	0.004	1.000	57	351	174
11	##A. #### #### #### .	810	0.000	0.638	57	4100	131
12	#A## A### #### #### A	729	0.001	1.000	53	1182	175

. . .

The left hand side of figure 4 shows the behavior of XACS in the larger blocks world problem where $s = 7$, $b = 7$, $y = 4$. In this problem setting, XACS actually evolves a state value representation with a smaller number of classifiers than states in the problem confirming that the generalization mechanism works properly. Moreover, XACS reaches the optimal policy faster than ACS-SVL. Applying XACS to the blocks world problem of size $s = 8$, $b = 8$, and $y = 4$, displayed on the right hand side of figure 4, shows that ACS-SVL fails to evolve an optimal policy during the 2500 trials due to the large state space. XACS does evolve an optimal policy with a state-value representation of less than 720 classifiers—much smaller than the actual 9867 states possible in the problem which are all specified by the explicit state value learner.

The results confirm that the combination of an online generalizing predictive model learner, represented by ACS2, with a state value function approximation method, implemented with the XCS mechanism, was successful. While it was shown in [3] that ACS2 is able to build a compact model of the underlying state transition function in the blocks world problem, the results herein show that it is also possible to learn a generalized representation of a state value function yielding improved performance, a compact representation, as well as anticipatory behavior. The next section puts the results in a broader perspective and compares the resulting system with other existing approaches.

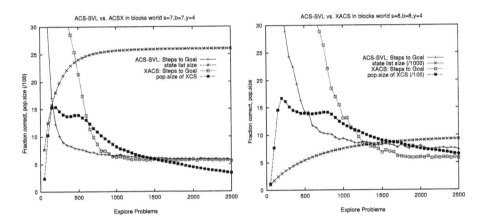

Fig. 4. Looking at larger blocks worlds, XACS outperforms ACS2 with a non-generalizing state value learner memory wise as well as policy wise.

5   Related Systems

The previous section confirmed that a combination of a generalizing predictive model learner with a generalizing state value learner can yield optimal performance with a more compact representation. This section focuses on the relation of this approach to other anticipatory learning systems.

As explained above, XACS essentially learns the state transition function of an MDP and evolves a state value representation of a Bellman equation that gives rise to an optimal behavioral policy. The basic underlying concept of evolving an optimal state value representation is that of dynamic programming (DP) in which a state value representation is usually generated iterating over all possible states of a problem. DP usually assumes that state transition function as well as payoff function are known beforehand. Thus, DP only learns the state value function. XACS, on the other hand, learns the state transition function, the payoff function, and the state value function.

The dynamical reinforcement architecture Dyna [21] is a general framework in which the state transition function as well as the payoff function is learned online. Dyna implementations represent optimal policies by reinforcement values for state-action combinations or state values. Most Dyna implementations do not generalize online state transition function nor state-value function.

More closely related are two other ALCSs that take a similar approach to XACS: (1) Yet Another Classifier System (YACS) [13, 12] and (2) the modular anticipatory classifier system (MACS) [11]. Both systems learn the state transition function of a problem similar to ACS2. YACS basically has the same predictive model representation as ACS2 but uses more informed mechanisms to learn the model. MACS evolves separate predictive rules for each perceptual attribute being able to exploit additional regularities in some environments. Both approaches keep a list of all encountered states using DP techniques similar to the one in ACS-SVL to evolve an optimal policy represented in state values for each state in the list. YACS represents immediate reward directly in the predictive rules which could result in a model aliasing problem for immediate reward predictions. MACS evolves identical state values to the ones in ACS-SVL and XACS. Neither YACS nor MACS attempt to approximate the state values with a function approximation mechanism.

6   Anticipatory Behavior

While we showed above that the basic behavior in XACS is anticipatory, there are many ways to improve and enhance the possibilities of XACS. First we propose further interaction between the predictive model learner and the state value learner in XACS, next we propose the implementation of further anticipatory mechanisms and the integration of several motivational mechanisms in a motivational coordination module that might be further biased by emotional influences.

6.1   Interactive Learning Mechanisms

Right now the two distinct learning modules XCS and ACS2 do not exchange any information about the importance or relevance of attributes. However, in natural environments insights in the generalization possibilities for the predictive model should also be useful for the state value learner module and vice versa. Thus, the learning mechanisms in XCS as well as ACS2 could be biased based on the insights gained in the other model. For example, the specialization mechanisms in ACS2 could be biased toward specializing task-important attributes or genetic generalization could be biased toward generalizing task-independent attributes. Mutation in XCS could be biased in a similar way. This interaction, however, would rather result in implicit anticipatory mechanisms in the learning component, than in the use of explicit anticipations before action selection.

6.2   Internal Policy Updates

Besides a mutual learning bias, further internal reinforcement learning updates in the state value learner are possible. The currently maintained state list can be compared to the prioritized sweeping approach in Dyna [17]. Beyond single-state updates, it could be useful to extend updates to multiple steps consequently reaching a larger neighborhood. Hereby, the update depth could be made dependent on the surprise level of the stored state represented by the average error value recorded for each attribute. Moreover, it could be useful to apply random updates in the XCS module as previously done in the Dyna implementation [21] or in an application of ACS2 without state values to psychological learning experiments [8].

6.3   Task-Dependent Attention

Another possible future direction is the addition of attentional mechanisms to the learning and behavioral mechanisms in ACS. Psychological evidence suggests that early perception is processed in parallel while late perception undergoes an attention-mediated bottleneck [18]. Similar to this framework, Koch and Ullman proposed a computational approach using a saliency map in which the salience of distinct perceptual properties is combined to an overall importance measure [16]. A winner-takes-all algorithm determines the location in the saliency map that is processed further. Higher cortical influences (possibly anticipatory) on the winner-takes-all algorithm are suggested. Such influences could be implemented for example by the task-dependent predicted salience of attributes resulting in preparatory attention.

In XACS, generalized state values could influence the actual matching and decision making process resulting in similar preparatory attention. Since the specified attributes in the generalized situations essentially denote the task relevant attributes, classifier selection could be restricted to those classifiers that predict perceptual changes in task relevant attributes. Moreover, task irrelevant perceptual attributes could be disregarded from the matching process in ACS2. Predictions could be restricted to the task relevant attributes resulting in further processing speed-ups.

6.4   Multiple Motivations and Emotions

Currently, XACS's learning mechanism is restricted to the only available task in the current environment (i.e. maximization of reinforcement). However, multiple task learning is imaginable in XACS with separate XCS state approximation modules for each of the tasks. For example, XACS could be modeled to continuously find food, water, and shelter in an environment. To accomplish multiple tasks, either the environment would need to provide a reward vector or reinforcement could be generated internally once a particular need was satisfied.

Additionally, behavioral interaction between the different modules could be considered. In the simplest case, the currently strongest motivation—possibly considering respective state values—might win. For example, if the currently strongest motivation was hunger, the animat's policy might be based on the state values in the 'hunger' XCS module.

More effective interactions of the motivations need to be coordinated in an additional module which could be termed a "motivation coordinator". The use of homeostatic variables might be useful in this respect. Each motivation may be triggered by the need to keep the homeostatic variables in equilibrium. Each motivation could influence several homeostatic variables so that the coordination of the motivation is subject to optimization as well. Coordination techniques within an architecture similar to the one outlined above are investigated in [1].

Additional to such a flat motivation coordinator, emotional factors might bias the resulting interactions. Cañamero [10] gives a great overview of how emotions can actually be useful in this respect biasing motivations in implicit anticipation of improved overall behavior. Among others, Cañamero suggests that emotions can be helpful for effective bodily adaptation (for example, rapid response to danger), for action guidance, decision making, and a more varied and flexible behavior (emotions as amplifiers of certain motivations), for learning guidance (highly emotional events are usually remembered easily), and for successful social interaction (signaling relevant events to others). XACS seems to be an ideal framework to investigate the interaction of multiple motivational modules and possible emotional influences in a system that learns environmental structure and the satisfaction of motivations from scratch.

7   Summary and Conclusions

This paper separated the reinforcement learning component in ACS2 introducing an online generalizing state value learner implemented by an adapted XCS [24] version. The resulting system XACS was tested on a blocks-world problem in which it proved to outperform the previous policy learning mechanism in ACS2. It was also shown that non-generalizing state value mechanisms can be outperformed in memory requirements as well as in behavioral performance. The resulting system is an anticipatory behavior system predicting action consequences and biasing action choice on these predictions.

Although the presented results point to a promising direction, we need to note that the current state value generalization mechanism is not very robust as applications in other MDPs suggest. The problem is related to the space sampling problem and the consequent over-generalization problem in the XCS module. Since naïve function approximation mechanisms do not work for the approximation of a state value function in general, it has been suggested that it is necessary to constrain the function approximation further on robustness during dynamic programming updates or explicitly prevent divergence [2]. Research is in progress to further analyze and solve the problem in XCS and XACS.

Nonetheless, the XACS approach has solved large problems, such as the blocks world problem, robustly and effectively. The proposed enhancement towards higher interactivity between state value learner and predictive model learner promises further robustness. Moreover, since XACS is a system that learns online and from scratch, the implementation of an enhanced XACS system with multiple interacting motivational modules possibly influenced by current emotions seems to be a very challenging but also rewarding endeavor. It is for example imaginable, that, dependent on early learning experiences, the emotional patterns of the animat will evolve differently resulting in a, for example, very "shy" or very "bold" animat. Future research will show to what extend the XACS system is suitable for such further investigations.

Acknowledgments

The authors would like to thank Xavier Llora, Martin Pelikan, and Kumara Sastry as well as the whole IlliGAL lab for fruitful discussions and useful comments.

The work was sponsored by the Air Force Office of Scientific Research, Air Force Materiel Command, USAF, under grant F49620-00-0163. Research funding for this work was also provided by a grant from the National Science Foundation under grant DMI-9908252. The US Government is authorized to reproduce and distribute reprints for Government purposes notwithstanding any copyright notation thereon. The work was also funded by the German Research Foundation (DFG) under grant HO1301/4-3.

The views and conclusions contained herein are those of the authors and should not be interpreted as necessarily representing the official policies or endorsements, either expressed or implied, of the Air Force Office of Scientific Research, the National Science Foundation, or the U.S. Government.

References

[1] Avila-García, O., Cañamero, L.D.: A comparison of behavior selection architectures using viability indicators. In: EPSRC/BBSRC International Workshop Biologically-Inspired Robotics: The Legacy of W. Grey Walter, HP Bristol Labs, UK (2002)
[2] Boyan, J.A., Moore, A.W.: Generalization in reinforcement learning: Safely approximating the value function. Advances in Neural Information Processing Systems 7 (1995)
[3] Butz, M.V.: Anticipatory learning classifier systems. Kluwer Academic Publishers, Boston, MA (2002)
[4] Butz, M.V.: Biasing exploration in an anticipatory learning classifier system. In Lanzi, P.L., Stolzmann, W., Wilson, S.W., eds.: Advances in learning classifier systems: Fourth international workshop, IWLCS 2001. Springer-Verlag, Berlin Heidelberg (2002) 3–22
[5] Butz, M.V.: State value learning with an anticipatory learning classifier system in a markov decision process. IlliGAL report 2002018, Illinois Genetic Algorithms Laboratory, University of Illinois at Urbana-Champaign (2002) http://www-illigal.ge.uiuc.edu/.
[6] Butz, M.V., Goldberg, D.E., Stolzmann, W.: The anticipatory classifier system and genetic generalization. Natural Computing 1 (2002) 427–467
[7] Butz, M.V., Goldberg, D.E., Tharakunnel, K.: Analysis and improvement of fitness exploitation in xcs: Bounding models, tournament selection, and bilateral accuracy. Evolutionary Computation (in press, 2003)
[8] Butz, M.V., Hoffmann, J.: Anticipations control behavior: Animal behavior in an anticipatory learning classifier system. Adaptive Behavior (in press, 2003)
[9] Butz, M.V., Wilson, S.W.: An algorithmic description of XCS. In Lanzi, P.L., Stolzmann, W., Wilson, S.W., eds.: Advances in learning classifier systems: Third international workshop, IWLCS 2000. Springer-Verlag, Berlin Heidelberg (2001) 253–272

[10] Cañamero, L.D.: Designing emotions for activity selection in autonomous agents. In Trappl, R., Petta, P., Payr, S., eds.: Emotions in Humans and Artifacts. The MIT Press, Cambridge, MA (in press, 2003)

[11] Gérard, P., Meyer, J.A., Sigaud, O.: Combining latent learning and dynamic programming in MACS. European Journal of Operational Research (submitted, 2002)

[12] Gérard, P., Sigaud, O.: Adding a generalization mechanism to YACS. Proceedings of the Third Genetic and Evolutionary Computation Conference (GECCO-2001) (2001) 951–957

[13] Gérard, P., Sigaud, O.: YACS: Combining dynamic programming with generalization in classifier systems. In Lanzi, P.L., Stolzmann, W., Wilson, S.W., eds.: Advances in learning classifier systems: Third international workshop, IWLCS 2000. Springer-Verlag, Berlin Heidelberg (2001) 52–69

[14] Hoffmann, J.: Vorhersage und Erkenntnis: Die Funktion von Antizipationen in der menschlichen Verhaltenssteuerung und Wahrnehmung. [Anticipation and cognition: The function of anticipations in human behavioral control and perception.]. Hogrefe, Göttingen, Germany (1993)

[15] Kaelbling, L.P., Littman, M.L., Moore, A.W.: Reinforcement learning: A survey. Journal of Artificial Intelligence Research 4 (1996) 237–258

[16] Koch, C., Ullman, S.: Shifts in selective attention: Towards the underlying neural circuitry. Human Neurobiology 4 (1985) 219–227

[17] Moore, A.W., Atkeson, C.: Prioritized sweeping: Reinforcement learning with less data and less real time. Machine Learning 13 (1993) 103–130

[18] Pashler, H.E.: The psychology of attention. MIT Press, Cambridge, MA (1998)

[19] Stolzmann, W.: Anticipatory classifier systems. Genetic Programming 1998: Proceedings of the Third Annual Conference (1998) 658–664

[20] Stolzmann, W.: An introduction to anticipatory classifier systems. In Lanzi, P.L., Stolzmann, W., Wilson, S.W., eds.: Learning classifier systems: From foundations to applications. Springer-Verlag, Berlin Heidelberg (2000) 175–194

[21] Sutton, R.S.: Integrated architectures for learning, planning, and reacting based on approximating dynamic programming. Proceedings of the Seventh International Conference on Machine Learning (1990) 216–224

[22] Sutton, R.S., Barto, A.G.: Reinforcement learning: An introduction. MIT Press, Cambridge, MA (1998)

[23] Watkins, C.J.C.H.: Learning from Delayed Rewards. PhD thesis, King's College, Cambridge, UK (1989)

[24] Wilson, S.W.: Classifier fitness based on accuracy. Evolutionary Computation 3 (1995) 149–175

[25] Wilson, S.W.: Function approximation with a classifier system. Proceedings of the Third Genetic and Evolutionary Computation Conference (GECCO-2001) (2001) 974–981

Autor Index

Lecture Notes in Artificial Intelligence (LNAI)

Lecture Notes in Computer Science